CHEMISTRY RESEARCH AND APPLICATIONS

MERCURY

SOURCES, APPLICATIONS AND HEALTH IMPACTS

CHEMISTRY RESEARCH AND APPLICATIONS

Additional books in this series can be found on Nova's website
under the Series tab.

Additional E-books in this series can be found on Nova's website
under the E-book tab.

ENVIRONMENTAL SCIENCE, ENGINEERING AND TECHNOLOGY

Additional books in this series can be found on Nova's website
under the Series tab.

Additional E-books in this series can be found on Nova's website
under the E-book tab.

CHEMISTRY RESEARCH AND APPLICATIONS

MERCURY

SOURCES, APPLICATIONS AND HEALTH IMPACTS

KI-HYUN KIM
AND
RICHARD J. C. BROWN
EDITORS

nova publishers
New York

For permission to use material from this book please contact us:
Telephone 631-231-7269; Fax 631-231-8175
Web Site: http://www.novapublishers.com

Additional color graphics may be available in the e-book version of this book.

Library of Congress Cataloging-in-Publication Data

Mercury : sources, applications and health impacts / editors, Ki-Hyun Kim and Richard J.C. Brown.
p. cm.
Includes index.
ISBN 978-1-62257-721-7 (hardcover)
1. Mercury--Research. I. Kim, Ki Hyun. II. Brown, Richard J. C.
QD181.H6M 2011
546'.663--dc23
2012030295

Published by Nova Science Publishers, Inc. † New York

CONTENTS

Preface		**i**
Chapter 1	Structural Aspects of Mercury Complexes *M. Melník and P. Mikuš*	**1**
Chapter 2	Organic–Inorganic Mesoporous Monolithic Scaffolds and Their Functionality in the Optical Ion-Sensitive Removal of Mercury Ions *Sherif A. El-Safty and M. A. Shenashen*	**25**
Chapter 3	Isomers in the Chemistry of Mercury Organometallic and Heterometallic Complexes *M. Melnik and P. Mikuš*	**67**
Chapter 4	Generation of Superoxide by the Interaction between Mercury Ions and Endogenous Thiols *Margarita E. Aliaga, Alexis Vásquez-Arce Camilo López-Alarcón, Claudio Olea-Azar Germán Barriga and Hernán Speisky*	**87**
Chapter 5	Effects of Acute Mercury Exposition on the Expression and Function of Organic Anion Transporters in the Kidneys *Adriana M. Torres*	**99**
Chapter 6	Mercury: Environmental Concerns and Management Technologies *I. Padilla, H. Tayibi, O. Rodríguez and A. López-Delgado*	**109**
Chapter 7	Mercury Toxicity Alters The Antioxidant System of Growing Cucumber Seedlings *Denise Cargnelutti, Fernando Teixeira Nicoloso Luciane Almeri Tabaldi, Valderi Luiz Dressler Vera Maria Morsch and Maria Rosa Chitolina Schetinger*	**137**
Chapter 8	Mercury and its Connection with Physicochemical Water Parameters (Case Study of the Rivers of the Northern European Territory of Russia) *Yu A. Fedorov and A. E. Ovsepyan*	**155**

Chapter 9 Mercury Contamination in Reptiles: An Emerging Problem with Consequences for Wild Life and Human Health **173**
Larissa Schneider, William Maher, Aaliyah Green and Richard C. Vogt

Chapter 10 Atmospheric Deposition of Mercury from Waites Island, Horry County, South Carolina, as compared to Regional mercury deposition and South Carolina Fish Consumption Advisories **233**
Jane L. Guentzel

Index **251**

PREFACE

The chemical element mercury (symbol: Hg) is the only metal that is a liquid at standard temperature and pressure. (The only other liquid element under these conditions is bromine.) It is an extremely rare element in the Earth's crust with an average abundance of only 80 parts per billion. However, where mercury ores do occur they can by very concentrated – containing up to 2.5 % mercury by mass. Cinnabar (HgS) is the most commonly found ore of mercury in the earth's crust. The use of these ores, and of mercury itself, by humans has a long history – mercury was found in Egyptian tombs dating from 1500 BC. Early uses were (erroneously) associated with products to improve human health: the ancient Greeks for instance used mercury in ointments. However, by 500 BC mercury was already being used to form amalgams with and extract other metals – one of the earliest examples of human industry leading to anthropogenic emissions of this harmful element.

Pre-industrial wet deposition concentrations of mercury from the atmosphere, associated with the natural background level, were of the order of 4-5 ng/L of ice deposit (although volcanic activity temporarily increased this level many-fold). These concentrations peaked at in the 1980s 20-25 ng/L, as more and more mercury locked away in the Earth's crust was liberated into the air, soil, and water by industrial and manufacturing process. Reductions in mercury usage for many processes and the impact of legislation to decrease usage and emissions have since driven these levels down to about 15 ng/L. However, it is clear from many studies of historical mercury accumulations in peat bogs, lake sediments, ice cores, and biological samples that the global atmospheric mercury burden has increased on average by a factor of 3 ± 1 since the industrial revolution. In some regions localised Hg pollution in soils has increased ten-fold. The impact of this increase on biota is considerable: for example, in animal tissues from the Arctic, mercury has increased by ten times since 1900, although no significant change had been observed previously over the last 700 years.

In 2008, global mercury emissions were estimated at 7500 Mg, with natural sources (including re-emission from natural reservoirs containing mercury deposited from previous anthropogenic activity) amounting to 5200 Mg with oceans the dominant source – contributing 52 % of this total. Fully primary natural emissions from volcanoes and geothermal areas contribute only 2 % of natural source emissions. Of the primary anthropogenic emissions the main contributors are estimated to be: 810 MG was due to fossil-fuel burning, 400 Mg from artisanal small scale gold mining, 310 Mg from non-ferrous metal production, 240 Mg from cement production, 190 Mg from water disposal and 160 Mg from caustic soda production.

A greater understanding of the toxicity of mercury to humans and the environment has led in recent years to a rapid phasing out of many traditional uses of mercury in industry and the introduction of national legislation to limit its trade and transport and to restrict its emissions from stationary sources (such as power stations). Some of the most interesting current uses of mercury are in fundamental research where there is still much work to be done to understand fully the chemistry of mercury and its compounds, and what the use of these compounds might be in areas such as synthetic chemistry, catalysis, and even medicine.

However, the legacy of anthropogenic mercury in the environment will remain for a long time. For instance, even if emissions of mercury are ceased now, it would still take between 100 and 500 years to observe a decline in the global average mercury content in soils and may take thousands of years for the perturbation to be entirely removed. This global burden of mercury from a century of intense industrial activity and the ability of mercury to bio-accumulate in organisms in terrestrial and aquatic biosystems have prompted a much increased global effort to monitor its presence in the environment (in media such as air, soil, and water) in recent years. The main drivers of effort to reduce, monitor, and understand the transport of mercury in the environment are the severe health effects which mercury and its compounds can induce (in humans and other animals): including damage to the brain, kidney, lungs and central nervous system with symptoms typically including sensory impairment, disturbed sensations and a lack of coordination.

Therefore, "Mercury: Sources, Applications and Health Impacts" attempts to cover recent developments and the current state-of-the-art relating to many up-to-date aspects of mercury science. The diverse contributions encompass the monitoring and management of mercury in the environment, the effect of mercury in the environment on animals and plants, and developments in novel chemistry involving mercury and its compounds. As such, we hope that this book provides a relevant and interesting reference text to those with a broad interest in current mercury issues.

Prof Ki-Hyun Kim, Sejong University, Korea
Dr Richard J. C. Brown, National Physical Laboratory, UK
May 2012

In: Mercury: Sources, Applications and Health Impacts
Editors: Ki-Hyun Kim and Richard J. C. Brown
ISBN: 978-1-62257-721-7

Chapter 1

STRUCTURAL ASPECTS OF MERCURY COMPLEXES

M. Melník and P. Mikuš
Department of Pharmaceutical Analysis and Nuclear Pharmacy,
Faculty of Pharmacy, Comenius University in Bratislava, Bratislava, Slovak Republic

ABSTRACT

The structural parameters of one thousand one hundred mercury complexes (of the types: coordination, organometallic and heterometallic) are outlined. The mercury atom in these complexes is found in various oxidation states: 0, +1, +1.33 and +2.Of these the +2 oxidation stateis by far the most common. The coordination arrangements about the mercury atomrange from two to eight ligands. Of these, the diagonal and tetrahedral environments are by far the most common. Five types of isomerism are established: distortion, ligand, coordination number, polymerization and linkage. Of these,distortion is by far the most common (~90%). The mean Hg-Hg bond distance elongatesas theoxidation state of the mercury increases, in the sequence:2.52Å (Hg(I)-Hg(I)) < 2.78Å (Hg(1.33)-Hg(1.33)) < 2.87Å (Hg(II)-Hg(II)); where'-'indicates a bond. The mercury atoms may be bonded to a variety of other metal atoms of both soft and hard character (Ge, Li, Te, Co, Ni, Fe, Ir, Mn, Mo, Os, Pd, Pt, Rh, Ru, Re, W, Nb, Au), illustrating a rich diversity in the chemistry of mercury. Correlations between ligand donor atoms, bond lengths and bond angles are also indicated and discussed.

1. INTRODUCTION

In spite of the fact that Mercury belongs to the 11 B group of elements, it shows such different behavior that it cannot be considered as homologous to zinc and cadmium. These metals are remarkably volatile for heavy metals, and mercury is uniquely so. At ordinary temperatures mercury is a shiny liquid and soluble in both polar and nonpolar liquids. In the biosphere, it is extremely hazardous because bacteria can convert it to the exceedingly toxic soft acid CH_3Hg^+ (which attack the central nervous system, and damages kidney, liver and brain tissue, and cause birth defects, etc.).

Mercury has a wide range of uses: tooth fillings, automatic on - off air-conditioner systems, fluorescent lights, some paints, floor waxes and furniture polishes, as an antibacterial agent, in medicine, as a fungicide on seeds, and to help make plastics, paper, clothing and camera film. Considering these diverse uses, it is very important to be aware of three words: structure-property-activity (catalytic and bio).The interaction of these parameters is important to understand in order to fully develop a full picture of the behavior and effects of the mercury compounds under study. For this reason, we analyzed and classified over one thousand one hundred mercury complexes for which structural parameters are available. There are three groups of mercury complexes: mercury coordination complexes, mercury organometallic complexes and mercury heterometallic complexes.

2. Mercury Coordination Complexes

Mercury coordination complexes have been extensively studied from a structural point of view, with over five hundred and fifty coordination complexes identified in a review [1]. Details of the structures were described in our previous work, therefore this work will be a comprehensive review of the structures of comparable mercury organometallics as well as mercury heterometallic clusters. From the nuclearity point of view, there comprise mono-, di-, tri-, tetra-, hepta-, and polymeric derivatives.

2.1. Monomeric Coordination Complexes

There are almost one hundred monomeric mercury (II) coordination complexes with two "real" identifiable covalent bonds. However, the assignment is often difficult, both because of the impossibility of establishing coordination on the basis of the given data and because the original study lacks the appropriate information. The one hundred monomeric mercury coordination complexes with two short bonds can be divided into three groups:

1. Pure monomers with diagonal coordination around the mercury atom;
2. Diagonal with and additional intramolecular interaction;
3. Diagonal coordination with an additional molecular interaction.

In the series of diagonal coordinated monomeric mercury complexes, the L-Hg-L bond angles range from 170 to 180° with an average of 177.5°.

In the series of compounds with intermolecular interaction, there are examples where in addition to two "real" coordinated bonds to a central mercury atom, there are also two, three or even four additional nearby atoms (Hg-L, ~2.80Å).The most common of these are O-donor ligands. One would expect such an interaction to influence the diagonal L-Hg-L bond angle, which is found to range from 166 to 180° with the average value of 175°.

In the series of compounds with intermolecular interaction, there are examples that have, in addition to two "real" coordination bonds to the central mercury atom one, two, three, four, five and even six mostly O-donor atom ligands (Hg-0~2.83Å). The diagonal bond angle L-Hg-L in these derivatives ranges from 180 to 168° (averaging 174°) with twoO-donors present; 166-158° (av. 163.5°) with three O-donors; 180-173° (averaging 176.5°) with four O-donors; 176.5-170° (averaging 174°) with five O-donors, and 180-177° (averaging 178.5°) with six O-donors. It is noted that the largest deviation from the linearity of the diagonal angle is when three extra atoms interact with a central mercury atom. The overall average value of 173° (with a range 180-158°) found in the series of compounds with intermolecular interactions is smaller than that of 175° (intramolecular interactions) as well as the 177.5° found in the series of "pure" diagonal coordinated monomers.

In general, the diagonal coordination about Hg(II) atoms is built up only by monodentate ligands, with the following chromophores: HgX_2 (X=O, N, C, Cl, S, P, Br, Si, Ge, I or Te), HgNX (X=Cl or Br), HgBP, HgPX (X=Cl, Br or I). Themean Hg-L covalent bond distance elongates with as the covalent radius of coordinating atom/ligand increases in the sequence: 2.05Å (NL, 0.75Å) < 2.20Å (BL, 0.62Å) < 2.31Å (Cl, 0.99Å) < 2.36Å (SL, 1.02Å) < 2.40Å (PL, 1.06Å) < 2.42Å (Br, 1.14Å) < 2.495Å (SiL, 1.17Å) < 2.53Å (GeL, 1.22Å)< 2.67Å (I, 1.33Å).

There are forty examples of three coordinate mercury coordination compounds. As in the previous examples, there are situations in which additional atoms/ligands are close enough to interact with the central mercury atom. In these derivatives, mercury is found only in its +2 oxidation state. From a geometrical perspective, there are examples that contain mercury (II) atoms in a (trigonal-planar) Y-or T-shaped form (the former is the most common by far) with different degrees of distortion. The ligands found are mono- or bidentate. The inner coordination sphere about the Hg(II) atoms are: HgX_3 (X=N, Cl, S, Br, Si, I or Te), PtN_2X (X=O or I), $PtCl_2X$ (X=N, S or P), PtS_2X (X=O, Cl, P or As), HgP_2Cl and HgI_2X (X=N or P). The mean Hg-L bond distance increases in the sequence: 2.17Å(OL 0.73Å) < 2.26Å(NL, 0.75Å) < 2.437Å(SL, 1.02Å) < 2.44Å(PL, 1.06Å) < 2.445Å(Cl, 0.99Å) < 2.515Å(SiL, 1.17Å) <2.545Å(Br, 1.14Å) < 2.555Å(AsL, 1.20Å) < 2.67Å(I, 1.33Å) < 2.695Å(TeL, 1.36Å). As can be seen, except for I and Br, the Hg-L bond distance increases as the covalent radius of coordinated atom/ligand increase. A possible reason for the somewhat longer Hg-I/Br distances could be an additional interaction of these atoms with another mercury atom. In general, the mean Hg-L bond distances found in the three coordinate Hg(II) coordination complexes are longer than that of the two coordinate derivatives.

There are over one hundred and thirty examples of mononuclear tetra-coordinated mercury coordination compounds and in all of these the mercury is in the +2 oxidation state. Tetrahedral geometry is nearly the only configuration found in non-transition compounds where the central atom has no electrons in its valence shell orbitals other than the four pairs forming the σ-bonds to ligands. These vary from regular tetrahedral to those having different degrees of distortion. There are two examples [1] where the Hg(II) atom adopts an almost regular square planar geometry via the tetradentate LS_4 ligand (mean S-Hg-S angle 90.0(±2.4)°), and a tetrahedral geometry is commonly observed, as expected for Hg(II). The square planar geometry found in these cases is enforced by a macrocyclic ligand, and this is unique in the chemistry of mercury. There are examples of mono-, bi- and tridentate ligands, creating various amounts of distortion about the mercury (II) atom. The inner coordination sphere about the mercury (II) atom has the following chromophores: HgX_4 (X=N, Cl, S, P, Br, Se or I); $HgCl_3X$ (X=N or I), HgS_3X (X=Cl or P), HgI_3O, HgO_2X_2 (X=N, Cl, S or P), HgN_2X_2 (X=C, S, Br or I), $HgCl_2X_2$ (X=B, S, P, Se, Br or I), HgP_2X_2 (X=Br or I), $HgBr_2X_2$ (X=Se or As), $HgSe_2I_2$, HgN_2OCl, $HgCl_2ON$ and HgI_2SP. The mean Hg-X distance is the series of halogen derivatives elongates as the covalent radius of X increases in the sequence:

2.49Å(Cl, 0.99Å) < 2.60Å(Br, 1.14Å) < 2.73Å(I, 1.33Å). By contrast, the mean Hg-L distance in the series of monodentate ligands increases in the order: 2.12Å(CN, 0.77Å) < 2.37Å(NL, 0.75Å) < 2.47Å(OL, 0.73Å) ~ 2.47Å(PL, 1.06Å) < 2.57Å(SL, 1.02Å) < 2.68Å(SeL,1.16Å). This does not follow the covalent radius of the donor atom, indicating that most of the ligands play a role that was not found in the less sterically crowded series of di- or trigonal coordinated derivatives. In general, the mean Hg-L distances found for four coordinate derivatives are longer than those in di- and tri- coordinate, as expected. The mean Hg-L distance in the series of bidentate ligands is longer than that of the respective monodentate ligands, increasing in the sequence: 2.42Å(NL) < 2.52Å(PL) < 2.65Å(SL) < 2.75Å(SeL). It is notable that in the series of hetero-bidentate ligands the mean Hg-L distances are somewhat longer when L binds by the "hard" atom rather than binding by the "soft" atom. For example, the difference between the two mean values, Hg-NL(S)(2.505Å) and Hg-SL(N)(2.369Å) increases from 0.136Å for N/S heterobidentate ligands, to 0.224Å (Hg-NL(P)(2.641Å) and Hg-PL(N) (2.417Å) for N/P heterobidentate ligands. When both hetero-atoms of ligand are "hard" or "soft" the mean Hg-L distances show greater differences for smaller atoms that those for large atoms.

In the series of bidentate chelating ligands, the effect of both electronic and steric hindrance of the coordinated atom can be seen in the opening of the L-Hg-L bond angles of the respective metallocycles. For example, in the five-membered metallocyclic rings, the L-Hg-L angles range from 68 to 74° (when L is an N-donor with unsaturation (mostly phenanthroline and 2,2´-bipyridine derivatives), and from 77 to 80° when the N-donor is saturated (mostly ethylenediamine and derivatives), but from 81 to 89° when L is a P-donor. The metallocyclic internal ring angle increases as expected with the number of ring atoms in the order: mean value of 69(2,2)° for four-membered rings, 72(2,2)° for five-membered (unsaturated) and 80(3,8)° for five-membered (saturated), 84(6,8)° for six-membered, and 95(2,2)° for seven-membered. The first number in parenthesis is the mean minus the smallest value and the second is the largest value minus the mean.

In the series of tetrahedrally coordinated HgL_4 derivatives, where L are monodentate ligands, the mean L-Hg-L bond angles depart from the ideal 109.5° bond angle by 4.7°(L=Br) < 7.19(P) < 7.6°(S) < 8.5°(I) < 10.2°(Cl), which indicates that distortion increases in the given order and, excepting I, increases with a decreasing covalent radius of X.

There are six derivatives in which mercury(II) atoms are penta-coordinated. Both of the two principal penta-coordinated geometries, square-pyramidal and trigonal-bipyramidal, have been found in mercury(II) chemistry. There are four examples for the former, with the chromophores HgS_5, HgN_4Cl, HgS_4O and HgN_3O_2, and two examples (HgO_3P_2, HgS_3Cl_2) of the latter geometry. The mean Hg-L distance in this series elongates in the following order: 2.31Å(OL) < 2.36Å(tridentate NL) < 2.37Å(PL) ~ 2.37Å(tetradentate NL) < 2.38Å(Cl) < 2.44Å(bidentate OL) < 2.60Å(tetradentate SL) < 2.605Å(pentadentate SL) < 2.69Å(SL).

There are fourteen derivatives in which six-donor atoms are coordinated to the Hg(II) atoms, with a wide range of geometries: a regular octahedron (HgO_6), tetragonal bipyramidal (HgN_6, HgS_4O_2, HgS_4N_2, HgS_3N_2O), rhombic [HgO_6, HgO_4N_2, HgN_4S_2), trigonal prismatic (HgN_6) and pentagonal pyramidal (HgN_5Cl, HgN_5Br). There are mono-, bi-, tri-, tetra-, penta- and even hexadentate ligands. The L-Hg-L bond angles of the respective metallocycles open in the order: 68(3,3)°(-NC_2N-with unsaturation) < 76(6,6)°(-NC_2S-) < 83(7,7)°(-SC_2S-). The metallocycle intra-ligand ring opens with increasing size of the ring 52.6°(four-) < 76°(five) < 92°(six-membered) as might be expected.

2.2.Dimeric Coordination Compounds

There are twenty Hg(I) and one hundred and twenty-five Hg(II) dimers for which X-ray data are available [1].

2.2.1 DimericHg(I) Coordination Compounds

The Hg-Hg bond distance ranges from 2.467(2) to 2.558(1)Å with a mean value of 2.516Å, indicating a direct bond between two Hg(I) atoms. In most examples (13 compounds) a monodentate N donor ligand completes the diagonal coordination of the Hg(I) atom with a mean Hg-N bond distance of 2.18Å and the mean N-Hg-Hg' bond angle of 170.8°(range 164.5-180°). There is example of the HgOHg' chromophore with an O-Hg-Hg' bond angle of 166.6° and a Hg-O bond distance of 2.14(3)Å. There are two examples, of the HgN_2Hg' chromophore, another two of $HgSe_2Hg'$ and the remaining two with HgO_3Hg'. The mean Hg-Hg bond distances are 2.512, 2.555 and 2.522Å, respectively.

2.2.2 DimericHg(II) Coordination Compounds

There are one hundred and twenty-five dimericmercury(II) coordination compounds for which structural date are available [1]. The structures from the bridge point of view are quite complex. There are several types of bridging, the distorted edge-shared tetrahedral structure being by far the most common. There are various bridging configurations: two oxygen ligands serve as bridges (12 examples), two sulphur ligands (2 examples), a pair of chlorine atoms (26 example), a pair of bromine atoms (12 examples),and a pair of iodine atoms (15 examples). There are correlations between the Hg-Hg distance, the Hg-L-Hg and μL-Hg-μL angles. The Hg-Hg distance elongates as the Hg-L –Hg angle opens and the μL-Hg-μL angle closes, and vice-versa. For example, in the series of di-oxygen bridged compounds, the values are 3.913(4)Å, 105(1) and 74(1)°; 4.297(2)Å, 119.3(3) and 60.7(3)° respectively. For the chlorine bridges the values are 3.773(1)Å, 90.6(1) and 89.4(1)°; 3.896Å, 92.2 and 87.8°; 4.070(5)Å, 95.9 and 84.1(1)° respectively. In the iodine bridged series the values are 3.859(2)Å, 84.8(1) and 95.2(1)°; 4.228(2)Å, 91.7(1) and 88.3(1)° respectively.

There is also correspondence between the bridge angles and the covalent radii of the bridging atoms. If the Hg-Hg distance remains about the same, the Hg-L-Hg angle closes and the μL-Hg-μL angle opens with an increasing radius. Some examples are: 3.913(4)Å, 105(1) and 74(1)° (L=OL, 0.73Å), 3.918(1)Å, 94.1 and 85.9(2)°(Cl, 0.99Å); 3.962, 92.2(2) and 87.8(2)°(Br, 1.14Å), 3.969(2)Å, 83.8 and 96.2(I, 1.33Å). In general the Hg-L-Hg angle opens and the μL-Hg-μL angles closes with decreasing covalent radius of the bridged atom/ligand in the sequence (mean values): 86.8°(I, 1.33Å) < 89.6°(Br, 1.14Å) <92.7°(Cl, 0.99Å) < 109.4°(LO, 0.73Å) for the former and 93.2°<90.4°<87.3°<70.5°, for the latter. The mean Hg-Hg distance of 3.95Å (range from 3.28 to 4.30Å) is considered to preclude the existence of a metal-metal bond.

The mercury atoms are two, three, four (most common), five, six and even seven coordinated. There are mono-, bi-, tri-, tetra-, penta-, hexa- and octa-dentate ligands. In general, the mean Hg-L(bridge) distances are longer than the Hg-L(terminal) values. The Hg-L(bridge) distances for tetrahedrally coordinated derivatives increase in the order: 2.515Å(LO) < 2.382Å(Cl) < 2.778Å(Br) < 2.930Å(I). The Hg-L(terminal) distances increase in the order: 2.350Å(LO) <2.355Å(Cl) < 2.404Å(LP) < 2.510Å(Br) < 2.623Å(LSe) < 2.689Å(I) < 2.728Å(LTe). Both follow the trend of the covalent bond increasing as the radius of the coordinated atom increases. It is noted that the mean Hg-L distances found in the two, three and four coordinated dimericHg(II) compounds are shorter than those found in the corresponding monomeric derivatives. For penta and hexa coordinated derivatives the opposite is true.

2.3.Oligomeric Mercury Coordination Compounds

We previously provided a detailed structural analysis of this group [1], and therefore we will present here only a general overview of this class. There are trimers (x4), tetramers (x9) and even a heptamer (x1). Except for one trimer example, in which a triangular Hg_3^{4+} cluster contains mercury in the unusual formal oxidation state of +1.33, all of the remaining compounds contain mercury in the oxidation state +2. The mean Hg(1.33)-Hg(1.33) bond distance is 2.777Å. Each Hg(1.33) atom has a tetrahedral arrangement (HgP_2Hg_2) with a mean Hg-P bond distance of 2.524Å. Each Hg(II) atom in the remaining trimers are tetrahedrally coordinated (HgS_4) with the mean Hg-S bond distance elongated in the following order: 2.390Å(S) < 2.685Å(μS) < 2.775Å (μ_3S).The Hg...Hg separation of 3.65Å precludes metal-metal bonding. There is one example containing one trimeric and three monomeric units where each Hg(II) atom is tetrahedrally coordinated ($HgCl_4$).

In the series of tetramers, the molecular structure of six examples consist of discrete eight-membered $\{-Hg-X-\}_4$ (X=Cl, S or Se) heterocycles. Each Hg(II) atom is tetrahedrally coordinated, $HgCl_2X_2$ (X=S, P, Se), HgN_2S_2, and HgX_2NCl (X=S, Se). In the remaining tetramers, each Hg(II) atom also hastetrahedrally arrangements ($HgCl_4$, $HgCl_3P$, HgI_3Te and $HgTe_4$). The mean Hg(II)-L bond distance elongates in the following order: 2.37Å(PL) < 2.45Å(NL, μ-SL) < 2.455Å(Cl) < 2.54(μ-SeL) < 2.65Å(μ-Cl) < 2.69Å(I) < 2.77Å(Te) < 2.93Å(μ-I) < 2.95Å(μ-Te). The mean Hg...Hg separation of 3.85Å (range 3.648-4.412μ) precludes metal-metal bonding.

HeptamericHg(II) complexes contain four tetrahedrally coordinated ($HgSe_4$) and three linearly coordinated Hg(II) atoms ($HgSe_2$) held together by one diselenide and eight monoselenide ligands. The mean Hg-Se bond distances (two-vstetracoordinate) are 2.444 vs 2.680Å(μ-Se) and 2.405 vs 2.990Å(μ_3-Se). As in the previous sections, colorless derivatives are the most common by far. However, there is one example each of white, yellow, brown and red compounds. The derivatives crystallized in the monoclinic(x10), triclinic(x4) and orthorhombic(x1) crystal classes.

2.4.Polymeric Mercury Coordination Compounds

There are five Hg(I) and over one hundred polymeric Hg(II) coordination compounds [1]. The series of Hg(I) compounds exist as bent $\{-L-Hg-Hg-L-\}_n$ chains. Each Hg(I) atom is two-coordinated (HgLHg') where L is OL, NL or SL. There is a cooperative effect between the covalent radius of the donor atom, the mean Hg-L bond distance, and the mean L-Hg-Hg bond angle; both the first and second increases, while the angle closes: L=O, 0.73Å, 2.13Å, 173.5° < N, 0.75Å, 2.255Å, 167.6° < S, 1.02Å, 2.508Å, 158.4°. The mean Hg(I)-Hg(I) bond distances are 2.517Å(L=O), 2.499Å(N) and 2.583Å(S). The complexes are colorless (x2), yellow(x2) and pink(x1) and crystallized in the monoclinic(x3), triclinic and orthorhombic each(x1) crystal classes.

The structures of polymeric mercury(II) coordination compounds are very complex, both from the type of bridging as well as the stereochemistry point of view [1]. In terms of the type of bridging atoms/ligands, of which there are several types, almost 1/3 of the examples contain a single atom/ligand: oxygen, chlorine, sulphur, bromine and iodine. In several other examples, two identical bridging atoms serve as bridges: oxygen, chlorine, sulphur, selenium,bromine and a pair of iodine atoms. There are also mixed bridge systems: chlorine plus sulphur, and sulphur plus bromine. In several other examples, mercury(II) atoms are bridged by a single chlorine,bromine and iodine, together with the ligands R_2CS_2, R_2NS_2 or the RCO_2 anion in a syn-syn arrangement. In several other cases, SCN groups serve as the bridge between mercury(II) atoms, and a bidentate ligands serve as diatom or triatom bridges. There are also examples, where a bidentate ligand serves as a four atom link, as a five atom link, as a six atom link, as a seven atom link and even as an eight atom link.

The Hg(II) atoms are two- HgX_2 (X=Cl or S); three- HgX_3 (X=O, S or Te), HgN_2O, $HgCl_2X$ (X=O or S); four- HgX_4 (X=O, N, Cl, S, Br, Seor I), HgO_3P, $HgCl_3X$ (X=N, S or P), HgS_3X (X=N, Cl, Br or I), $HgBr_3N$, HgI_3S, HgC_2P_2, HgO_2Cl_2, HgN_2X_2 (X=Cl or I), $HgCl_2X_2$(X=S or P), HgS_2X_2(X=Br or I), HgP_2I_2, HgO_2NS, HgS_2NX(X=O or P), $HgBr_2ON$; five- HgX_5 (X=Cl or Br), HgO_4N, $HgCl_4X$ (X=S, P or Se), HgO_3S_2, HgN_3S_2, $HgCl_3X_2$ (X=S or P), $HgBr_3S_2$, HgO_2S_2N; six- HgX_6 (X=O or Cl), $HgCl_5N$; HgO_4X_2 (X=N, S or I), HgN_4Cl_2, $HgCl_4X_2$ (X=O or N), HgS_4Cl_2, $HgBr_4N_2$, $HgO_2N_2X_2$ (X=Cl or S); seven- HgO_7, HgO_5N_2, HgN_5Cl_2, $HgCl_5S_2$, and even eight coordinate - HgO_8, of which the tetrahedral environment is by far the most common.

In the series of tetrahedrally coordinate Hg(II) atoms, the mean Hg(II)-L(terminal) distance for a unidentate ligands elongates in the order of 2.18Å(CN) < 2.26Å(NL) < 2.38Å(Cl) < 2.72Å(I). In the series of bidentate ligands, the ordering is 2.365Å(NL) < 2.38Å(OL) < 2.57Å(SL) < 2.575Å(PL). The mean Hg(II)-L(bridge) distance elongates in the order of 2.54Å(OL) < 2.73Å(Cl) < 2.735Å(SL) < 2.78Å(Br) < 2.90Å(I). In the series of five coordinated species, the mean Hg(II)-L(terminal) distance elongates in the following sequence: 2.38Å(PL) < 2.395Å(Cl) < 2.43Å(NL) < 2.44Å(SL) < 2.535Å(SeL) < 2.55Å(Br). The shortestHg-Hg distance of 3.26Å and the mean Hg-Hg distance of 3.86Å found in the series of polymeric mercury(II) coordination compounds would appear to exclude metal-metal bonding.

The complexes crystallized in the following crystal classes: monoclinic(x59) < orthorhombic(x20) < triclinic(x17) < tetragonal(x7) <trigonal(x1). Once again, most of the complexes are colorless except for the following examples: yellow(x11), white(x6), pink(x1) and orange(x1).

3. MERCURY ORGANOMETALLIC COMPLEXES

Over 320 examples of mercury organometallic complexes were classified and analyzed by us [2]. A comprehensive review of their structural parameters is the aim of this chapter.

3.1. Monomeric Organometallic Mercury Complexes

An analysis of almost two hundred and thirty monomeric mercury organometallic complexes showed that the mercury atom is only found in its +2 oxidation state. Coordination about mercury atoms are linear or almost linear, except four examples, in which mercury(II) atoms are three-coordinate (HgN_2C, HgONC(x3)) and in one example the Hg(II) atom is tetrahedrally coordinate (HgP_3C). Monodentate ligands in the series of two-coordinated species created the following inner coordination sphere about mercury(II) atoms: HgC_2(x60), HgOC(x9), HgNC(x68), HgCCl(x47), HgCS(x26), HgCP(x1), HgCBr(x14), HgCSe(1x) and HgCI(x5). The mean C-Hg-X bond angle deviated from the linear 180° in the order of X: 177°(X=Se) < 176.7°(O) < 176.6°(Br) < 175.9°(C) <175.2°(Cl) < 174.4°(N) < 174.1°(S) < 172.6°(I) < 167.5°(P). The most common C donor atoms are the methyl and phenyl groups. The mean Hg-C bond distance elongated in the following sequence:2.035Å(range 2.01-2.05Å)(CN) < 2.06Å(1.96-2.18Å)(Ph) < 2.065Å(1.96-2.14Å)(CF_3) < 2.085Å(2.05-2.11Å)(C_6F_5) <2.09Å(1.96-2.04Å)(CL, excluded mentioned C donor ligands in this sequence) < 2.10Å(1.95-2.15Å)(Me) < 2.135Å(C_6Cl_5). The mean Hg-L bond distance elongated in the order of 2.085Å(CL of all C donor ligands) < 2.09Å(OL) < 2.11Å(NL) < 2.33Å(Cl) < 2.375Å(SL) < 2.43Å(Br) < 2.48Å(SeL) <2.595Å(I). The order follows the covalent radius of the L atom/ligand except for the C donor atom.

These complexes for the most part are colorless with some examples of white (x17), yellow(x7) and even red(x2). A wide variety of crystal classes were found in this series of complexes: monoclinic(x132) > triclinic(x45) > orthorhombic(x43) > tetragonal(x7) > cubic(x2), hexagonal and rhombohedral each(x1).

3.2. Dimeric Organometallic Mercury Complexes

Of fifty examples, only one has the mercury atom in the oxidation state of+1 rather than +2. This derivative has a central Hg_2^{2+} unit (Hg-Hg distance of 2.51Å) which asymmetrically complexes by hexamethylbenzene.

In the series of dimericmercury(II) derivatives there are several types of bridging between the mostly diagonal Hg(II) atoms. In several examples, a single atom of the ligand serves as a bridge: carbon, nitrogen, oxygen and sulphur. There appears to be a relationship between the Hg-Hg distance and the Hg-L-Hg bond angle wherein the former elongates and the latter opens, as for example the following: 3.213(7)Å and 98.9(5)°; 3.441(1)Å and 111.6(3)°; 3.639Å and 126°. In another example, a double bridge using a pair of oxygen donors, chlorine and iodine, perform the bridging, resulting in a geometry of a distorted edge-shared bi-tetrahedra.

By far the most common geometry observed in the diagonal dimericorganomercury(II) structures is one in which both Hg(II) atoms sit on a N-heterocyclic ligand (usually a purine base) and the second coordination site is occupied by a methyl group. In some other examples, two bidentate ligands serve as two atom bridges, or as five or even six atom bridges. In one example, four F_3CCOO^- groups serve as bridges between the Hg(II) atoms in a syn-syn arrangement.

The number of examples of the various geometries decreases in the following order: diagonal (majority, x44) >tetrahedral(x5) >trigonal(x3) >hexa-coordinated (only one). The inner coordination sphere around the mercury(II) atom has the following chromophores: diagonal - HgNC(x25), HgCCl(x8), HgCS(x5), HgC_2(x4) and HgOC(x3); trigonal - HgN_2C, HgS_2C and HgOCCl, each(x1); tetrahedral - HgI_3C(x2), HgO_3C, HgO_2Cl_2 and HgO_2NC,each(x1), and hexa-coordinated - HgO_4C_2.

There are a wide variety of ligands, mono- and non-chelating bidentate, for the former group the methyl group is by far the most prevalent and for the latter the homo-N,N-bidentate (usually a purine base) is the most prevalent. The mean Hg-L(terminal) bond distance for monodentate ligands, elongates in the following sequence: 2.005Å(Ph) < 2.05Å(OL) < 2.07Å(Me) < 2.085Å (CL, excluded Ph and Me) < 2.33Å(Cl) < 2.39Å(SL). The mean Hg-L(bridge) bond distance elongates in the order of2.04Å(μ-OH) < 2.05(μ-NL) < 2.115Å(μ-CL) < 2.365Å(μ-SL). The mean Hg-L bond distances for non-chelating bidentate ligands are N,N´-purine base, 2.10Å; O+N donor ligand 2.07 and 2.14Å and for N+S donor ligands the values are 2.09 and 2.36Å. The mean C-Hg-X bond angle departs from the linear 180º by 3.1º(X=O or S) < 3.3º(Cl) < 4.8º(N) < 5.4º(C). Except in one example, where the Hg-Hg bond distance is 2.791(1)Å, all remaining Hg...Hg separations range from 3.092(1) to 4.109Å (av 3.463Å), excluding one direct bond.

While the majority of the complexes are colorless, there are examples of white(x7) and yellow(x2). The complexes crystallized in the four types of crystal classes: monoclinic(x26), triclinic(x15), orthorhombic(x8) and tetragonal(x1).

3.3. TrimericOrganomercury Complexes

There are almost 30 species of this class of complexes, all of which have mercury in the oxidation state +2 [1]. In several examples, three crystallographic $(HgL)_3$ units are symmetrically linked by a triply bridging atom or ligand, for example $(HgMe)_3$ by O or S atoms, $(HgX)_3$ (X=Cl, NO_3,Br) by carbon ligands {Hg(penicillaminate)}$_3$ by Cl atoms. The resulting geometry ismore or less a flattened trigonalbipyramid. The mean Hg-μ_3L bond distance elongates in this sequence: 2.07Å(C) < 2.08Å(O) < 2.37Å(Cl) < 2.39Å(S). The mean Hg...Hg separation of 3.40Å is too large to indicate bonding, and the Hg-μ_3L-Hg bond angles range from 99 to 120º.

A few examples have nine(-Hg-C-C-)$_3$, twelve-(-Hg-C-C-N-)$_3$ or fifteen- (-HgC-C-C-C)$_3$ membered rings. There are diagonal coordinated mercury(II) atoms with a mean Hg-C bond distance of 2.10Å and a mean value of the C-Hg-C bond angle of 177º. The mean Hg...Hg separation of 3.53Å ruled out a metal-metal bond.

There are eleven examples in which all the Hg(II) atoms are sitting on heterocyclic ligands (mostly purine bases) and the diagonal coordination is completed by a methyl group on each Hg(II) atom. The mean Hg-N and Hg-C(Me) distance of 2.09 and 2.07Å,respectively, together with the mean C-Hg-N angle of 176º, are the same as those found in the dimeric species despite the difference in denticity of the N-heterocyclic ligands. This is quite unique in orgametallic chemistry. As in the other series, most of the complexes are colorless, with just a few white ones. The complexes crystallized in several types of crystal classes: monoclinic(x17), triclinic(x5), orthorhombic(x4), cubic and tetragonal each(x1).

3.4. Oligo- and Polymeric Organomercury(ii) Complexes

We analyzed the structural parameters of this group [2]. There are twelve tetramers, and in four of them a carbon atom is bound to four mercury(II) atoms in a distorted tetrahedron. The tetrahedral angles range from 104.4 to 113.6° and the Hg-Hg distance is 3.36Å. The mean Hg-μ_4C distance is 2.06Å and the mean C-Hg-L angle is 176°. In another two examples, the real center is a chlorine atom with a mean Hg-μ_4Cl distance of 2.995Å. Another two derivatives contain a twisted 22-membered cyclic aggregate. In the remaining four examples, all four Hg(II) atoms are sitting on a purine base with a mean Hg-N distance of 2.09Å. Diagonal coordination about the mercury is completed by methyl groups with a mean Hg-C distance of 2.07Å and mean N-Hg-C angle of 176°.

The pentamer has an Hg_5C_5 core with a ten-membered ring of alternating mercury and carbon atoms. An additional two terminal CF_3 groups complete the tetrahedral coordination of each mercury(II) atom (HgC_4).

In the remaining five examples, diagonal coordinated Hg(II) atoms are linked viaO and C donor ligands toform infinite chain polymers (HgOC) with the mean O-Hg-C bond angle of 174.3°.

4. Heterometallic Mercury Complexes

We analyzed over two hundred heterometallic mercury complexes of known crystal structures [3]. A comprehensive review of the structures found are outlined in this part.

4.1. Hetero – Double Salts

There are twenty-four such species (16 monomers and 8 dimers). Three monomers have mercury(II) with diagonal coordination to two chlorine atoms, with the mean Hg-Cl bond distance of 2.33Å and the mean Cl-Hg-Cl bond angle of 172.2°. In another four examples, mercury(II) atoms have trigonal environments HgL_3 (L=Cl, Si or I) with the mean Hg-L bond distances that elongate as the covalent radius of the coordinated atom increases: 2.415Å(Cl, 0.99Å) < 2.516Å(Si, 1.17Å) < 2.690Å(I, 1.33Å). The most common coordination arrangement about the Hg(II) atoms is tetrahedral, HgL_4(L=Cl, S, P, Se or I) and $HgCl_3I$, with the mean Hg-L distances elongated in the following order: 2.502Å(Cl) < 2.357Å(PL) < 2.590Å(SL) < 2.652Å(SeL) < 2.771Å(I). In general, the Hg-L distances also elongate with an increasing coordination number, as for example: Hg-Cl 2.35Å(bi-) < 2.415Å(tri-) < 2.502Å(tetra-coordinated); and Hg-I 2.690Å(tri-) < 2.771Å(tetra-coordinated).

In this group of double salts of mercury(II), there is an isolated monomeric mercury unit with a heterometal unit in which the coordination ranges from tetrahedral TeC_2NCl, CoN_4, through penta-NaO_5, hexa-RhN_4Cl_2, MgO_6, CrN_6, $TaCl_5O$, KO_6 and CoN_6, hexa-plus hepta-(KO_6 + KO_7) to octa-coordinate CoC_5P_3.

In the dimers, a distorted edge-shared bi-tetrahedral structure aroundmercury(II) is found ($HgCl_4$ (6 species), $HgBr_4$ and HgI_4 each, one specie). The mean Hg-X(terminal) distances are shorter than the Hg-X(bridge) values, and both elongate as the covalent radius of the X atom increases: 2.360 and 2.700Å(Cl, 0.99Å) < 2.518 and 2.724Å(Br, 1.14Å) < 2.696 and 2.936(I, 1.33Å). The mean Hg(II)-Hg(II) distance of 3.90Å rules out a metal-metal bond. The hetero-metal units include, TeC_3, CuN_4, AuN_4, RhN_4Cl_2,CrN_6, CrN_5Cl and KO_6.

4.2. Heterodimeric Mercury Complexes

Forty structures of this class were analyzed and classified by us [3]. The aim of this section is to summarize the structural parameters of heterodimeric mercury complexes. A direct mercury-metal bond is found in most of these examples with the most commonlinkage being aHgX {X=O, C, Cl (most common), S, P, Br or I} moiety mercury-metal bonded to an MX_n (M/X=CoP_4, CoC_7; PtP_2C, PtP_2Br, PtN_3C, PtP_3N, PtN_2C_2O; TeC_5, TeC_3P_2; $IrCl_2P_2C$, IrP_2CClBr, IrC_7; $OsCl_2P_2N$; RuC_3P_2, RuC_{10}; MoC_3P_2, MoC_3N_2Cl, MoC_8, MoC_7P, MoC_7As) moiety. There are also some examples of HgX_2 (X=C, Cl, I, CCl or CS) binding in a similar fashion. The mercury atom is found in the oxidation states 0, +1 and +2.

The Hg(I)-Co(0) distance of 2.481(2)Å is the shortest of this group of derivatives. The mean Hg-M bond distance elongates in the order of 2.569Å(-Pt(II)) < 2.692Å(-Mo(II)),for Hg(0). For Hg(I), the values are 2.490Å(-Co(0)) < 2.499Å(-Pt(II)) < 2.577Å(-Os) < 2.673Å(-Mo). For Hg(II), the values are 2.549Å(Fe(0)) < 2.578Å(Co(0)) ~ 2.578Å(Ir(I)) < 2.627Å (Pt(II)) < 2.660Å(Ru(0)). The mean Hg-M distance thus increases with the covalent radius ofM except for Hg(II)-Ir.Thismay be attributable to the lanthanide contraction.

There are examples in which the Hg(II) and Au(I) centers are bridged by two $RPCS^-$ anions in a syn-syn arrangement, bringing the linearly coordinated metal atoms within a range of2.934(1) to 3.088(1)Å. Structures of some other examples are unique [3],and will be not presented in this paper.

The mercury atoms in the dimeric species are diagonally, trigonally and tetrahedrally coordinated. The mean Hg(0)-L distance increases in the order of 2.09Å(CL) < 2.36Å(Cl) < 2.475Å(SL) < 2.483Å(PL) < 2.720Å(I). The mean Hg(I)-L distances increase in the order of 2.00Å(CL) <2.400Å(Cl) < 2.467Å(Br), and the Hg(II)-L values are 2.10Å(OL) < 2.11Å(CL) < 2.16Å(NL) < 2.335Å(SL) < 2.377Å(Cl). The other metal atom found in these heterobimetallic species is found in coordination sites from two to eight and are sandwiched.

The complexes crystallized in a wide variety of crystal classes: monoclinic(x27), and triclinic (x7), orthorhombic (x3), tetragonal and trigonal each(x1).

4.3. Heterotrimeric Mercury Complexes

There are almost forty species for which structural parameters are available [3]. There are three categories of the complexes: Hg_2M (8 examples), HgM_2 (28 examples) and HgPtW. Eight of the derivatives contain two mercury atoms together with molybdenum, ruthenium, platinum and palladium as a central core with two other "external" mercury atoms. In the former, there is a triangle of Hg-Mo-Hg with Hg-Mo distances from 2.643(8) to 2.690(6)Å and Hg-Mo-Hg bond angles from 70.8 to 76.2(3)°. Also present is the sulphur ligand, the N,N´-diethyldithiocarbamate or ethylsulphide anion, with a Hg-S distance of 2.398(2) and 2.43Å, respectively. Each central Mo atom is also sandwiched between two cyclopentadienyl molecules, with cp-Mo-cp (centroid) angles from 132.7 to 140.5°.

In the $[Hg_2MCl_8]^{2-}$ anion (M=Pt or Pd), central tetrahedrally coordinated M(II) atoms are found edge sharing with Hg(II) atoms. The distorted tetrahedral geometry is completed by two terminal chlorine atoms around each Hg(II) atom.

The most common of the heterotrimeric mercury complexes are those which contain the mercury atom located centrally with the two other usually transition metal atoms. An almost linear array, HgM_2 (M=Ge, Co, Te, Li, Mn, Pt, Mo, W or Ru) is by far the most common arrangement. The mean Hg-M bond distance elongates in the following sequence: 2.483(1)Å (M=Ge) < 2.499(5)Å(Co) < 2.534(2)Å(Te) < 2.573(10)Å(Li) < 2.611(2)Å(Mn) < 2.637(1)Å(Pt) < 2.746(2)Å(Mo) < 2.751(1)Å(W) < 2.794(1)Å(Ru). The M-Hg-M bond angles vary from 170 to 180°.

There are examples in which the M_2Hg array (M=Rh, Pt or Ru) forms a triangle with M-Hg-M angles range from 60 to 95°. The mean Hg-M and M-M bond distance elongates in the following sequence: 2.709 and 2.729Å (M=Pt) < 2.719 and 2.783Å(Rh)< 2.780 and 2.808Å(Ru). The remainders of the HgM_2 complexes are unique structures [3]. An example contains three different metal atoms and is an almost linear array of Pt-Hg-W (171.0(1)°), with Hg-Pt and Hg-W bond distances of2.571(2) and 2.755(1)Å, respectively.

When M is a non-transition element, the mean Hg-M bond distance elongates with increases in the covalent radius of M in the following sequence: 2.483Å (Ge, 1.22Å) < 2.573Å (Li, 1.34Å) < 2.794Å (Te, 1.36Å). When M is a transition element, Hg-M distances do not follow the covalent radii of M, but elongate in the order of 2.500Å (Co, 1.26Å) < 2.534Å (Fe, 1.20Å) < 2.657Å (Pt, 1.28Å) < 2.660Å (Mo, 1.45Å) < 2.714Å (Rh, 1.35Å) < 2.753Å (W, 1.46Å) < 2.768Å (Ru, 1.26Å). There are several possible reasons for this, one of which is the variation of the oxidation state of the Hg atom from 0 to +1 and +2, and the Rh atom from 0 to +1 and +3. Another reason could be the lanthanide contraction, especially with the heavier atoms. In general the mean Hg-M distances in heterotrimeric species are longer than those in heterodimeric species, except for Pt and Ru where the opposite effect is seen. The complexes are found in monoclinic (x20), triclinic(x14), tetragonal(x3) and orthorhombic(x1) crystal classes.

4.4. Heterotetrameric Mercury Complexes

The structural parameters of this group were summarized in [3] and a comprehensive review of the structures is given in this chapter. There are twenty four derivatives that can be subdivided into the five categories: Hg_3M (M=Nb or Cr);Hg_2M_2 (M=Fe (5 examples), Co(4 examples), Ru(2 examples), CuAgMnor Ir each (1 example)); $HgMn_2Sn$ and HgPtGeSn.

The structure of Hg_3Nb consists of a rhomboidal Hg_3Nb cluster. The three N,N´-diethyldithiocarbamate ligands asymmetrically chelate the mercury atoms. The mean Hg-Nb distance of2.790Å is indicative of a metal-metal bond, with the mean Hg-Hg distance being 2.892Å. The Nb atom is sandwiched by cyclopentadienyls.

Tetrameric clusters containing two mercury atoms and a pair of transition metal atoms are by far the most common in this series. In one of Hg_2Co_2 derivatives, the Hg_2^{2+} ion is linearly bridged at each end of two Co{bis(2-diphenylphosphinoethyl)amine} fragments, the Hg-Hg-Co angles having a mean value of 179.1(3)°. The Hg-Hg and Hg-Co bond distances are 2.651(4) and 2.437(9)Å, respectively, are the shortest observed in the tetrameric clusters.

There are examples in which a pair of O atoms (Hg_2Mn_2), a pair of Cl atoms (Hg_2Ir_2), or a pair of Br atoms (Hg_2Ru_2) serves as the bridge between Hg atoms and transitional metal atoms. The mean Hg-M distance elongates with an increasing covalent radius of the atom bridges and simultaneously with decreases in the covalent radius of M: 2.602(2)Å (O, 0.73Å, Mn, 1.46Å) < 2.618(1)Å (Cl, 0.99Å, Ir, 1.37Å) < 2.74Å (Br, 1.94Å, Ru, 1.26Å). In another Hg_2Ru_2 cluster, two non-equivalenttetrahedrally coordinated Hg(II) atoms are bridged by a pair of chlorine atoms, with one of the Hg atoms forming an $HgCl_4$ unit. A pair of Ru atoms is bound to the other Hg atom ($HgCl_2Ru_2$). The mean Hg-Ru and Ru-Ru bond distances are 2.801(2) and 2.827(2)Å. In the Hg_2Co_2 derivative, a pair of iodine atoms serve as bridges between Hg(II) atoms, which have trigonal coordination completed by MeS ligands also acting as bridges between HgI_2 and CoN_5 moieties.

In two Hg_2Te_2 clusters, a pair of S^{2-} anions serve as a triple bridge between a pair of Te(II) atoms and of Hg(I) atoms. The diagonal coordination of each Hg(I) atom is completed by a methyl group with S-Hg-C(Me) bond angle of 177°. The two bridging S atoms bring the Te(II) atoms to within 2.49(2) and 2.771(7)Å with Te-S-Te angles of 67.0 and 75.1°, respectively.

An Hg_2Ag_2 derivative has two pairs of 2-pyrrolidine anions, one pair bridging Ag(I) atoms via O atoms to give a folded Ag_2O_2 ring, and is coordinated to an Hg(I) atom via a terminal N atom. The other pair of bidentate anions form two eight-membered macrocyclic rings (.Ag-O-C-N-Hg-N-C-O.).

Considering the Hg_2Cu_2 derivative, two Cu(II) atoms are bridged by a pair of methoxide groups. Each Cu(II) atom has a square pyramidal geometry formed by five oxygen atoms, two from MeO^- anions, one from NO_3^- and one each from two different $Hg(C_5H_8NO)_2$ units.These last two provide additional bridges between the basal and apical coordination sites of the Cu(II) atoms, thereby forming ten-membered metallocyclic rings.

There are two Hg_2Te_2 derivatives in which two $Te(C_5H_5)(C_5H_3CH_2NMe_2)$ moieties, via the C atom of each penta-ring, form a diagonal coordination around each Hg atom with a mean C-Hg-C bond angle of 178°. Each Te atom is sandwiched (TeC_{10}).

The remaining two derivatives contain a pair of Hg(II) and Co(II) atoms, which are held together by eight thiocyanate groups via the soft S atoms (to Hg) and the harder N atoms (to Co).

There are five examples that contain one mercury atom and three germanium atoms or three cobalt atoms, three platinum atoms, or three ruthenium atoms. In the first case, the mercury atom is coordinated by three $Ge(C_6F_5)$ units, with a mean Hg-Ge bond distance of 2.525Å and a Ge-Hg-Ge angle of 119.4° giving a trigonal planar geometry around the mercury atom.In the second case, a similar trigonal planar array of three $Co(CO)_4^-$ anions is bound around the mercury atom ($HgCo_3$), with Co-Hg bond distances of 2.686Å and Co-Hg-Co angles of 120.0°. In the third case, there is a $Pt_3(\mu_3\text{-}Hg)$ frame with mean Hg-Pt distances of2.917Å and Pt-Hg-Pt angles of 53.6°. The fourth case has a Ru_3 triangle (mean Ru-Ru distance of 2.805Å) and an HgBr fragment spanning a Ru-Ru edge with a mean Hg-Ru distance of 2.740Å. The final example has an ionic structure. The $(GeS)_3Hg$ fragment of the $[\{(C_6F_5)_3GeS\}_3Hg]^-$ anion has symmetry close to C_3v. The mercury has trigonal planar geometry (HgS_3). The $[cp_2Co]^+$cation has an eclipsed sandwich configuration.

Two final examples have four metal atoms linked in a chain, Mn-Hg-Mn-Sn and Ge-Hg-Pt-Sn, in which the M-M distances indicate metal-metal bonding. The values in the former are

Hg-Mn 2.621(3)Å(mean) and Mn-Sn 2.554(3)Å; in the latter the values areHg-Ge, 2.534(3)Å,Hg-Pt 2.617(2)Å and Pt-Sn 2.518(2)Å.

In the series of heterotetrameric mercury complexes, the mean Hg-M distance elongates in the following sequence: 2.530Å<(M=Ge) < 2.603Å(Co) < 2.612Å(Mn) < 2.618Å(Ir) < 2.760Å(Ru) < 2.767Å(Pt) < 2.790Å(Nb). There are several examples in which the Hg...M separation is over 3.00Å and is not considered metal-metal bonding. Homonuclear M-M distances less than 3.00Å are observed in the order of 2.630Å < 2.636Å(Pt) < 2.816Å(Ru) < 2.957Å(Cu).In general, the Hg-M bonding distances are observed to increase with increasing nuclearity of the cluster. The complexes crystallized in three crystal classes: triclinic (14 examples), monoclinic (9 examples) and orthorhombic (1 example).

4.5. Heterooligomeric Mercury Complexes

The structural data for over fifty complexes in this series were summarized in [3]. There are several variations of heteropentameric complexes: Hg_3Rh_2, HgM_4 (M=Te, Ni, Ru), $HgRu_3M$ (M=Te, Mo, Co), $HgOs_3M$ (M=Co, Mo), and $HgGe_2Pt_2$. In Hg_3Rh_2 the five metals form an approximately regular trigonal-bipyramid with mean Hg...Rh distances of 3.970(1)Å and Hg...Hg distances of 3.854(1)Å. The corresponding angles are 58.0° (Hg-Rh-Hg) and 60.9(2)° (Hg-Hg-Rh). In $HgTe_4$ the Lewis acid and ligand (HgMe) caps one of the triangular faces of the Te_4 tetrahedron with Hg-Te distances of 2.606(1), 2.847(1) and 2.960(1)Å. The mean Te-Te distance is 2.627(1)Å.A complex dipositivecation contains a macrocyclic $HgNi_4$ cluster with Hg(II) at the center (mean Hg-Ni, 2.652(2)Å). The four Ni atoms are close and in pairs, with a mean Ni-Ni distance of 2.537(3)Å. Each Ni_2 pair is bridged by one $Ph_2PCH_2PPh_2$ and one MeNC ligand. Each Ni(II) atom is also coordinated by two terminal MeNC ligands.

In five examples, $HgRu_3M$ (M=Fe, Mo, Co) and $HgOs_3M$ (M=Co, Mo), the mercury has a trigonal planar geometry: $HgRu_2M$ and $HgOs_2M$. In $HgRu_2M$ (M=Te, Mo, Co), the mean Hg-Ru and Ru-Ru bond distances are 2.804 and 2.820Å. The Hg-M bond distance elongates as the covalent radius of M increases in the following order: 2.512(3)Å (M=Te, 1.20Å) < 2.571(1)Å (Co, 1.26Å) < 2.743(2)Å (Mo, 1.45Å). In $HgOs_2M$ (M=Co, Mo), the mean Hg-Os and Os-Os bond distances are 2.803 and 2.856Å. The Os-M bond distance also elongates as the covalent radius of M increases: 2.513(4)Å (Co, 1.26Å) < 2.718(3)Å (Mo, 1.45Å). The trigonal planar geometry is distorted such that the mean value of the Ru/Os-Hg-Os/Ru angle is 62°. The corresponding Ru/Os-Hg-M (M=Te, Mo, Co) angles have a mean value of 149°.

A red-brown $HgGe_2Pt_2$ derivative contains a Ge-Pt-Hg-Pt-Ge chain of C_2 symmetry with mean bond distances of 2.648(2)Å(Hg-Pt) and 2.409(2)Å(Pt-Ge), respectively.

Six examples of heterohexameric mercury complexes contain Hg_4M_2 (M=Mo or Te), Hg_2M_4 (M=Te, Co or Pd) and $HgTe_4Mo$, and their structures are complex. For example, the structure of the Hg_4Mo_2 complex consists of centrosymmetric dimers in which tricarbonylmesitylenemolybdenum groups are bonded to mercury (Hg-Mo 2.745(1)Å). The structure of the Hg_4Te_2 derivative is similar to its Mo analogue with a mean Hg-Te bond distance of 2.538(4)Å. The structure of the Hg_2Pd_4 derivative contains a butterfly of $Pd_4(\mu\text{-}CO)_4$. $(PEt_3)_4$ moieties in which the triangular wings are capped by Hg atoms bonded to bromine atoms. The mean values of Hg-Pd distances are 2.711(1) and 2.993(1)Å, and the mean Pd-Pd distance is 2.695(1)Å. The $HgFe_4Mo$ derivative has the Lewis acid ligand

$[HgMo(CO)_3cp]^+$ coordinated via a mercury atom to the tetra-ion butterfly cluster $[Fe_4(CO)_{13}]^{2-}$ (meanHg-Te 2.675(1)Å). The Mo unit retains a distorted four-legged piano-stool configuration with the Hg atom and the three CO groups as the legs.

There are eleven examples of heteroheptameric mercury clusters containing Hg_3Co_4, HgM_6 (M=Ru, Os or Pt), $HgFe_4M_2$ (M=Co, Rh or Sn) and $HgCo_5Ge$. Considering Hg_3Co_4, two{$(Ph_2C_4H_2)(CO)_6Co_2Hg$} moieties bridged metallocyclic units and are symmetrically disposed around a third Hg atom. The mean Hg-Co bond distance of 2.552(3)Å is comparable to the Co-Co bond length at 2.548(3)Å. In the next set of four HgM_6 examples, mercury bridges one side of two triangular M clusters (M=Ru (3 examples) or Os). The Hg atom is surrounded by four M atoms (square planar, distorted tetrahedron) with the dihedral angle between the two cluster planes being 35.8°, 45° and 45.5° (M=Ru), and 65° (M=Os). In another $HgOs_6$ derivative, two μ_3-PPh groups serve as bridges between two {$Os_3H(CO)_{10}$} moieties and a mercury atom, which assumes a near linear geometry (P-Hg-P=171.8(2)°). Each of the bridging groups is bonded to only two of the osmium atoms of the Os_3 triangle. An $HgPt_6$ complex has six Pt atoms at the corners of a trigonal prism whose edges are bridged by isocyanide groups. Each Pt atom also has a terminal isocyanide ligand. The mercury atom occupies the pseudo-center of the prism. The mean Hg-Pt and intra-triangular Pt-Pt distances are 2.943 and 2.643Å, respectively.

In the $HgFe_4Rh_2$ derivative, the Hg atom bridges are one side of two triangular Fe_2Rh clusters. It is thus surrounded by four metal atoms ($HgFe_2Rh_2$) where the dihedral angle between the M_2Hg planes is 71.9°. The mean Hg-Fe and Hg-Rh bond distances are 2.576(3) and 2.737(2)Å,respectively. The Fe-Fe bond distance is 2.609(2)Å. The structure of the $HgFe_4Sn_2$ cluster contains an unsupported FeSnFeHgFeSnFe skeleton in the $[\{(OC)_4FeSnCl_2Fe\ (CO)_4\}_2Hg]^{2-}$ anion. The mean Hg-Fe and Fe-Sn bond distances are 2.571(1) and 2.545Å, respectively.

The $HgCo_5Ge$ cluster contains a common Ge atom bridging the Co-Co bonds, both of a $Co_2(CO)_7$ and a $Co_2(CO)_6$ unit, which is also bridged by an $HgCo(CO)_4$ group. The mean M-M bond distance elongates in the order of 2.368(2)Å(Co-Ge) < 2.605(2)Å(Hg-Co) < 2.644(3)Å (Co-Co).

Seven heterooctameric mercury clusters can be arranged into the following subgroups: Hg_4M_4 (M=Mn, Re or Ir); Hg_2M_6 (M=Ru or Pt) and Hg_2Au_5Pt. The Hg_4M_4 (M=Mn or Re), contain a Hg_4M_4 eight-membered ring with a four-fold alternating molecular axis. The four Hg atoms have an almost coplanar arrangement with mean Hg-Hg distances of 2.888(2)Å (M=Mn) and 2.92Å (M=Re) and Hg-M-Hg angles of 66.2(2)° (M=Mn) and 67.6° (M=Re),respectively. The non-linearity of the M-Hg-M groups with markedly small angles at the Hg atom (157.2° Mn; 158.8° Re), together with short M-Hg distances (2.640 and 2.74Å, respectively), indicate metal-metal bonding. An Hg_4Ir_4 cluster has Hg-Se-Hg linking the four {$(dmpe)_2Ir(\mu\text{-}Se)_2Hg$} units to make up an eight membered$(HgSe)_4$ ring. In addition, there are four other nearly flat four-membered $Ir(\mu\text{-}Se)_2Hg$ rings. The whole assembly has the appearance of a windmill.

In the Hg_2Ru_6 derivative, two tetrameric clusters (Ru_3Hg) are in a butterfly arrangement and joined by two bromine atoms, which bridge the two Hg atoms asymmetrically to give a dimer. The mean Hg-Ru and Ru-Ru bond distances are 2.736(2) and 2.840(3)Å,respectively. There are two Hg_2Pt_6 clusters, dark-green and dark-violet in color. The former contains a cage-type cluster where a Hg-Hg (2.872(7)Å) unit is incorporated into a trigonalantiprismatic platinum core (mean Hg-Pt = 2.950Å). Two Pt_3 triangles (mean Pt-Pt

= 2.664(4)Å) are supported by three 1,6-bis(diphenylphosphino)hexane ligands. The latter consists of two triangular $\{Pt_3(\mu\text{-}CO)_3(PPhPr^i_2)_3\}$ units, each capped by a mercury atom. The two units are joined via the mercury atoms. The mean Hg-Pt and Pt-Pt bond distances are 2.994(1) and 2.658(1)Å, respectively.

In the dark orange Hg_2Au_5Pt cluster, the central Pt atom is bound to peripheral $AuPPh_3$ units, two peripheral $HgNO_3$ units and one PPh_3 ligand to give an eight coordinate Pt derivative. The geometry of the cluster core can be described as a fragment of a Pt-centered icosahedron. The mean M-M distance elongates in the order: 2.659(3)Å (Hg-Pt) < 2.680(3)Å (Pt-Au) < 2.906(3)Å (Au-Au) < 2.951(4)Å (Hg-Au).

There are four heterononameric derivatives, with Hg_3Co_6, Hg_2Ru_7, $HgNi_8$ and $Hg_3Ge_4Ni_2$ frameworks. A red Hg_3Co_6 derivative has two $\{Co_2(CO)_4(PhC_2H)_2.\ [HgCo(CO)_4]_2\}$ bridged metallocyclic units disposed about a third Hg(I)atom, with Co(I)-Hg(I)-Co(I) angle of 180°. The mean Hg-Co bond distance is 2.563(4)Å and Co-Co distance is 2.553(6)Å.

The structure of the orange Hg_2Ru_7 derivative indicates a slightly distorted octahedron with four CO groups and two cis-$HgRu_3(CO)_9(\mu_3\text{-}C_2Bu^t)$ fragments bound to a central ruthenium atom,with mean Hg-Ru bond distances of 2.657(1) and 2.820(1)Å,respectively. A black $HgNi_8$ complex contains two distorted cubic Ni_8 clusters. All the faces of the cube are occupied by μ_4-PPh ligands. In one of the cubics, a nickel atom is bonded to an external mercury atom (Hg-Ni = 2.650(3)Å). The mean Ni-Ni distance is 2.525(2)Å. The fourth example ($Hg_3Ge_4Ni_2$) has a chain structure GeHgNi(Ge)HgNi(Ge)HgGe with mean intermetallic distances of 2.454(3)Å(Hg-Ni), 2.516(3)Å(Hg-Ge),and 2.329(4)Å(Ni-Ge).

There is one example of a deca-metal cluster Hg_6Rh_4, which has an octahedron of Hg_6 with four tetrahedrally related faces capped by $Rh(PMe_3)_2$ units. The Hg-Hg and Hg-Rh distances range from 3.131(3) to 3.149(3)Å, and 2.690(4) to 2.724(4)Å, respectively.

Three undeca-metal clusters exist with Hg_3Fe_8 Hg_2Au_8Pt and $HgOs_{10}$ fragments. In the dark-brown Hg_3Fe_8 example, the central Hg(I) atom lies on a crystallographic inversion center, so that the Fe(1)-Hg(1)-Fe(1) angle is exactly 180°. The mean Hg-Fe and Fe-Fe bond distances are 2.624(3) and 2.680(4)Å, respectively. In a red Hg_2Au_8Pt derivative, the metal frame has approximately D_{2d} symmetry in a centered square antiprism capped on the squares. The Pt atom is in the center and is surrounded by eight $AuPPh_3$ groups and two Hg atoms. The Au atoms are on the vertices of the squares and the Hg is on the caps. The mean M-M distances are 3.004(1)Å (Hg-Au), 2.987(1)Å (Hg-Pt) and 2.632(1) (Au-Pt). The mean Au-Au distances are 2.917(1) and 3.353(1)Å,respectively.

Two dodeca-metal clusters are found: the yellow Hg_4Mo_8 and dark red Hg_2Ru_{10} frameworks. The former has a slightly distorted cubane structure and was the first example of a bimetallic cubane structure. The Hg-Mo distances are 2.626(3) and 3.177(3)Å,respectively. In the latter cluster two distorted bridged-butterfly $Ru_5C(CO)_{14}(\mu\text{-}Cl)$ units are linked by an $Hg_2(\mu\text{-}Cl)_2$ unit. The mean Hg-Ru bond distances are 2.617(3) and 2.866(3)Å,and Ru-Ru 2.877(4)Å.

The bimetallic red trideca-metal $HgRu_{12}$ cluster consists of two distorted $Ru_6C(CO)_{16}$octahedra linked by two edge-related ruthenium atoms to a central mercury atom. Each of these two octahedra contains an interstitial carbon atom. Of the sixteen CO groups, around each Ru_6 unit, twelve are terminally bonded and four form asymmetric bridges. The mean Hg-Ru and Ru-Ru distances are 2.839(2) and 2.891(2)Å, respectively. In the structure of another red bimetallic pentadeca-metal $Hg_9Co_6(CO)_{18}$ cluster, the metal framework is a rectangular trigonalprism with theCo atoms at each corner and an Hg atom at the midpoint of each edge. The mean Hg-Co bond distance is 2.562Å.

There are two isostructural anionic Hg_2Os_{18} clusters, having a metal core consisting of a Hg_2 fragment joining two $Os_9C(CO)_{21}$units, differing only in their degree of distortion. In the $[Hg_2Os_{18}C_2(CO)_{42}]^{2-}$ anion, the Hg-Hg bond distance is 2.744(5)Å and in the $[Hg_2Os_{18}C_2(CO)_{42}]^{4-}$ anion the value is 2.820(2)Å. The mean Hg-Os and Os-Os bond distances, anion 2- vs anion 4- are 2.777(5) and 2.845(5)Åvs 2.819(2) and 2.854(2)Å.

Finally, there are three examples with 21 metal atoms; Hg_3M_{18}(M=Ru or Os) and $HgOs_{20}$. The molecular structure of the $[Hg_3M_{18}C_2(CO)_{42}]^{2-}$ anion consists of two tricapped octahedral nona-ruthenium or nona-osmium subclusters fused by a bi-face capping $(Hg_3)^{2+}$ unit. The mean Hg-Ru, Hg-Hg and Ru-Ru distances are 2.689(6), 2.925(5) and 2.833(6)Å respectively. For the osmium analogous the mean values are 2.711(5)Å (Hg-Os), 2.927(6)Å (Hg-Hg) and 2.856(3)Å (Os-Os).

Examination of the data for the heterooligomeric mercury clusters shows the mean Hg-M (bonding less than 3.00Å) distance elongating in the following sequence: 2.516Å(M=Ge) < 2.533Å(Ni) < 2.562Å(Co) < 2.650Å(Mn) < 2.651Å(Fe) < 2.711Å(Pd) < 2.719Å(Mo) < 2.722Å(Rh) < 2.740Å(Re) < 2.776Å(Ru) < 2.794Å(Os) < 2.864Å(Pt) < 2.951Å(Au). The homo-bimetallic distance elongates in the order of 2.531Å(Ni) < 2.641Å(Co) < 2.652Å(Fe) < 2.655Å(Pt) < 2.695Å(Pd) < 2.833Å(Ru) < 2.853Å(Os) < 2.863Å(Hg) < 2.912Å(Au). There are also several other hetero-metallic distances that are shorter than 3.00Å: 2.368Å(Co-Ge); 2.409Å and 2.680Å(Pt-Ge, Au); 2.498, 2.545 and 2.570Å(Fe-Co, Sn, Rh). These illustrate the ability of mercury to bond to both transition metal and non-transition metal atoms to create a wide range of cluster compounds.

These clusters crystallized in the following crystal classes: triclinic (24 examples), monoclinic (19 examples), tetragonal (4 examples), orthorhombic (3 examples) and cubic (1 example).

4.6. Heteropolymeric Mercury Complexes

Structural parameters were summarized in [3]. A comprehensive review of this group of complexes is given in this chapter. Almost forty complexes belong to this category. The most common heterometal atom is cobalt, and the most common ligand is thiocyanate - used as a bridging ligand. There are several colorless examples in which the CN groups bridge the mercury atom to a non-transition metal such as Li(I), Na(I), K(I) and Ba(II). The CN attaches to Hg(II) via the soft end giving a mean Hg-C bond distance of 2.196Å. The C-Hg-C bond angles range from 104 to 115°. The nitrogen end of the ligand binds the harder non-transition metals. The coordination around these metal atoms are tetrahedral (Li and Na), seven coordinate (K) and eight coordinate (Ba). In another colorless HgSr derivative, selenocyanate bridges Hg(II) and Sr(II) atoms, via the soft end (Hg-Se=2.641Å) and the hard end (Sr-N=2.667Å), respectively. Each Hg(II) atom is tetrahedrally coordinated. The Sr(II) atoms have Archimedean antiprismatic coordination using the N atoms of four SeCN groups, and four pyridine ligands.

There are several colored bimetallic mercury derivatives: purple Cu(II),yellow Cu(I) and Fe(II), and pink-red Co(II). Here the thiocyanate groups serve as bridges between Hg(II) and the transition metal, using the soft end and the hard end, respectively, giving zig-zag chains.

Each Hg(II) atom is tetrahedrally coordinated with a mean Hg-S distance of 2.528Å and the maximum deviation of the S-Hg-S angle from the ideal value (109.5°) is 19.4°. Except for Cu(I) andCo(II), which are four coordinated, the heterometal atoms are six coordinated with different degrees of distortion depending on the nature of the additional two ligands.

In several other examples, there are two Hg(II) atoms plus additional metal atoms. These have hetero-bis donor ligands with N plus O donor sites. The ligands bind to Hg(II) via the N atom,with the O atom binding the heterometal atom. The Hg(II) is linearly coordinated with a mean Hg-N bond distance of 2.03Å and N-Hg-N angle of 176.4°. The remaining ligands, mostly nitrate or perchlorate groups, complete the coordination around the other metal atom. These range from six-Cu(II),eight-Er(III), nine-Eu(III) and Ge(III), and ten-coordinate Th(IV). Heterobidentate C_4H_6NO ligands are found in a series of Hg_3M_2 bound in a manner identical to the previously described cases. The mean Hg-N distance is 2.03Å and the mean N-Hg-N angle is 175°. Nitrate groups complete five-fold coordination about Zn(II),six-about Co(II), and nine-about Tb(III)and Er(III).

In general, the mean Hg-L bond distances for the tetrahedrally coordinated sites in the series of heterometallic derivatives are longer than those found in the momomeric derivatives [1,2], 2.196Å versus 2.127Å (L=CN) and 2.528Å versus 2.466Å (SCN).On the other hand, for the two coordinated derivatives, the mean Hg-N distance of 2.03Å is shorter than that found for either the mononuclear (2.05Å) [1] or organometallic (2.11Å) [2] mercury derivatives.

Heteropolymeric mercury complexes crystallized in a wide variety of crystal classes: monoclinic (20 examples), tetragonal (5 examples), triclinic (4 examples), orthorhombic (3 examples) and cubic (2 examples).

5. Isomers in the Chemistry of Mercury Complexes

Systematic studies in the field of stereoselectivity of complexes have accumulated increasing interest over the last 50 years. Stereoselectivity is very often related to important stereospecificty of biological systems, catalysis and stereochemical effects in technical processes. Isomers can be broadly classed into two major categories, structural and stereoisomers. The former can be divided into ionization, hydrate, coordination, linkage and polymerization sub-categories, and the latter can be divided into geometric (cis-trans, fac-mer), optical, distortion and ligand isomerism.

5.1. Mercury Coordination Complexes – Isomerism

In the chemistry of mercury coordination complexes were found over sixty-five examples, which exist as isomers. Previously, we have surveyed a detailed analysis of the isomers [4],therefore, only a brief comprehensive view will be presented in the paper. The isomers included are distortion (73%), polymerization (20.6%); coordination number (3.2%) and ligand (3.2%).

5.1.1. Distortion Isomerism

The coexistence of two or more species, even within the same crystal, differ mostly by degree of distortion and is typical of a general class of distortion isomerism [5]. There are 46 complexes that are classical examples of such isomerism. The mercury oxidation states in these isomers are +1 in only two examples and +2 in all the remaining examples. Each Hg(I) atom has a trigonal planar arrangement ($HgSe_2Hg'$) with a direct Hg-Hg bond (2.553(1)Å(α-) and 2.5579(8)Å(β-)). The sum of Hg-Se(x2) plus Hg-Hg′ bond distances are 8.06Å(α) and 8.125Å(β). This indicates that the α-isomer is somewhat more crowded than the β-isomer.

The coordination number around the Hg(II) atoms ranges from two to six. In general the sum of the Hg(II)-Ln (n=2-6) bond distances elongate as the covalent radii of coordinated atoms increases. In the series of $HgX._2$ (isomer 1 vs isomer 2, chromophore, covalent radius) the order is 4.09 vs 4.10Å ($X=N_2$, 1.50Å) < 4.58 vs 4.70Å (Cl_2, 1.98Å) < 4.66 vs 4.70Å (S_2, 2.04Å) < 4.66 vs 4.77Å (ClP, 2.05Å) < 5.03 vs 5.09Å (Ge_2, 2.44Å). For HgX_3 the order is: 7.20 vs 7.23Å (Cl_3, 2.97Å) < 7.47 vs 7.48Å (S_2P, 3.10Å) < 7.59 vs 7.60Å (Br_3, 3.42Å). For HgX_4 the order is 9.16 vs 9.18Å (O_2N_2, 2.96Å) < 9.12 vs 9.38Å (N_2C_2, 3.04Å) < 9.53 vs 9.72Å (N_2OCl, 3.26Å) < 9.83 vs 9.87Å (N_2S_2, 3.54Å) < 9.89 vs 9.92Å (Cl_4, 3.96Å) < 9.88 vs 10.09Å (Cl_2P_2, 4.10Å) < 10.24 vs 10.28Å (S_4, 4.08Å) in monomeric derivatives; and 9.91 vs 9.97Å (O_2Cl_2, 3.44Å) < 10.41 vs 10.47Å (S_4, 4.08Å) < 11.23 vs 11.29Å (I_4, 5.32Å) in di- to polymeric derivatives.

Both electronic and steric factors are observed in the chelated derivatives as monitored by variation in the L-Hg-L internal angles of the respective metallocyclic rings. The mean chelating L-Hg-L bond angles open (less squeezed isomer vs more squeezed isomer)in the order of 46.6 vs 52.8° (-OCO-) < 66.8 vs 70.7° (-SCS-) < 72 vs 73° ($-OC_2N-$) < 79.1 vs 79.5° ($-NC_2S-$) < 89.3 vs 91.2° ($-SC_3S-$).

5.1.2. Polymerization Isomerism

There are thirteen examples of this class of isomerism [4]. Eight of them contain monomers and dimers within the same crystal; one contains three monomeric units and a trimeric unit; two derivatives contain monomeric and polymeric units; and one contains a dimeric and polymeric units. The mercury atoms are only found in oxidation state +2. In monomeric units, the Hg(II) atoms are found with the chromophores: HgX_2 (X=Cl, P or I),HgX_4 (X=Cl or I), HgI_3O and HgN_5X (X=Cl or Br). In the dimers, each Hg(II) is tetrahedrally coordinated: HgX_4 (X=Cl, Br or I) and HgO_2Cl_2.In the trimer ($HgCl_4$) and in the polymers, $HgCl_3S$, $HgCl_5$ and HgN_4Cl_2, the Hg...Hg separations of ~3.9Å ruled out a metal-metal bond.

5.1.3. Coordination Number Isomerism

There are two derivatives that are examples of this class of isomerism [4]. One of the derivatives contains three monomeric units within the same crystal with the inner coordination spheres around Hg(II) atoms of HgI_3, HgN_2I and HgN_2I_2. For the other one, which contains two monomers, the chromophoresare $HgCl_4$ and HgN_4Cl.

5.1.4. Ligand Isomerism

There are two yellow moieties, orthorhombic $Hg(Pr_2OCS_2)_2$ and monoclinic $Hg(Pr_2^iOCS_2)_2$, which are classical examples of the ligand isomerism [4].

The isomers are for mostly part colorless(x38) and yellow(x20) with some examples of white(x2), red and yellow each(x1). These isomers crystallized in the several crystal classes: monoclinic(x38), triclinic(x21), orthorhombic(x4), tetragonal and trigonal each(x1).

5.2. Organometallic Mercury Complexes – Isomerism

In the chemistry of organometallic mercury complexes over sixty examples that exist as isomers were found. We have previously surveyed a detailed analysis of the isomers [6],therefore, only a brief review will be outlined in this section. The isomers included are distortion (97%) and ligand (3%).

5.2.1. Distortion Isomers

There are ten monomeric derivatives that exist in two isomeric forms. In two colorless $Hg(CF_3)_2$, cubic and orthorhombic, each Hg(II) atom has a diagonal coordination with the mean Hg-C bond distances of2.109 and 2.139Å,respectively. In another six pairs of isomers, two monodentate ligands, one with a C donor site and the other one with an N donor site created a diagonal arrangement around each Hg(II) atom with mean Hg-N and Hg-C bond distances (α- vs β-isomer): 2.075 and 2.04Å vs 2.10 and 2.085Å. This indicated that the inner coordination spheres in α-isomers are somewhat tighter than in β-isomers. The N-Hg-C bond angles range from 171.8(7) to 178.3(6)° (averaging 176°) in α- and from 173(1) to 178.5(7)° (averaging176°) in β-isomers. In another pair of isomers, C monodentate ligands with chlorine atom forma diagonal coordination around each Hg(II) atom with Hg-C, and Hg-Cl bond distances(α vs β) 2.34 and 2.53Å vs 2.15 and 2.50Å. The C-Hg-Cl bond angles are 178 and 180°, respectively.

In a much distorted trigonal plane around the Hg(II) atom a pair of Hg{(η^2-ethylenedioxyl)-o-tolyl}(Ph) was found (HgONC). The α-isomer contains two crystallographically independent molecules within the same crystal. The sum of Hg-L(x3) bond distances are 6.66Å (α-, molecule 1), 6.69Å (α-, molecule 2) and 6.68Å (β-isomer).

Almost thirty monomeric derivatives contain crystallographic independent molecules; one contains five, another four, two derivatives contain three and the remainder is two independent molecules. In all of them each Hg(II) atom is diagonally coordinated by two monodentate ligands with the chromophores: HgC_2, HgNC, HgCCl, HgCS, HgCBr and HgCl. The mean value of the sum of Hg-L(x2) bond distance (more crowded moleculevs less crowded molecule) increases as the covalent radii of the coordinate atom increases in the sequence: 4.10 vs 4.20Å (HgNC,1.52Å) < 4.13 vs 4.19Å (HgC_2, 1.54Å) < 4.38 vs 4.44Å (HgCCl, 1.76Å) < 4.45 vs 4.45Å (HgCS, 1.79Å) < 4.69 vs 4.70Å (HgCI, 2.10Å). The mean Hg-L bond distance (more crowded molecule vs less crowded molecule) elongates as the covalent radius of coordinated atom increases, except the C atom, in the order: 2.07 vs 2.13Å (N, 0.75Å) < 2.06 vs 2.09Å (C, 0.77Å) < 2.35vs 2.37Å (Cl, 0.99Å) < 2.36 vs 2.38Å (S, 1.02Å) < 2.36 vs 2.43Å (Br, 1.14Å) < 2.59 vs 2.61Å (I, 1.33Å).

There are ten derivatives (two di-, five tri-, two tetra- and one polymeric) that contain two crystallographically independent molecules within the same crystal. In two colorless monoclinic dimers (each contains two independent molecules), both Hg(II) atoms sit on an N-heterocyclic ligand and the second coordination site is occupied by a methyl group (HgNC). Unfortunately, in one example, only Hg...Hg separations are available with the values of

3.388(1) (molecule 1) and 3.392(1)Å (molecule 2). For the other dimer, only average values are available, which include both molecules.

Structures of trimers that contain two crystallographically independent molecules are different. In one, three $\{HgMe\}_3$ units are symmetrically linked by a triply bridging oxygen atom, with the mean Hg-O bond distance of2.08Å and Hg-C bond distance of 2.15Å. The Hg-Hg separation is 3.44 and Hg-O-Hg angle of 115°. In another trimer, three $\{Hg(ONO_2)\}_3$ units are linked by the C atom of the CCOOH group. Each Hg(II) atom is diagonally coordinated (HgOC). The mean values of the sum of Hg-O plus Hg-C bond distances are 4.16Å (molecule 1) and 4.21Å (molecule 2), with mean Hg...Hg separations of 3.359 and 3.351Å, respectively. The structure of the third trimer contains a fifteen-membered ring where three mercury(II) atoms are bridged by 2,2´-diphenylene groups. Each Hg(II) atom has a diagonal coordination (HgC_2), with mean Hg-C bond distances of 2.116 and 2.124Å, respectively. The mean Hg...Hg separations are 3.419 and 3.422Å, respectively.

The structures of the remaining two trimer derivatives are similar. In each of them, three $\{HgMe\}_3$ units are sitting on N-heterocyclic ligands. Each Hg(II) atom has a diagonal coordination (HgNC).

There are two tetramers, triclinic and monoclinic. In the triclinic tetramer, the real center is a chlorine atom with mean Hg-μ_4Cl bond distances of 3.019 and 3.076Å, respectively. Each Hg(II) atom has a trigonal arrangement (T shape).($HgCl_2C$). In the monoclinic tetramer, four $\{HgMe\}_4$ units are sitting on a purine base with the mean Hg-N bond distance of 2.07Å. Each Hg(II) atom has a diagonal coordination (HgNC). The mean Hg-C bond distance is 2.06Å,and the mean N-Hg-C bond angle is 175°.

In orthorhombic $\{Hg_3(\mu\text{-O})(\mu\text{-CCHO})\}_n$ diagonal coordinated Hg(II) atoms are linked via O and C donor ligands to form an infinite chain polymer. The mean Hg-O and Hg-C bond distances (molecule 1 vs molecule 2) are: 2.17 and 2.08Å vs 2.15 and 2.11Å, respectively.

5.2.2. Ligand Isomers

There are two colorless monoclinic examples $Hg(o\text{-tolyl})_2$ and $Hg(p\text{-tolyl})_2$ that are isostructural. Each Hg(II) atom is diagonally coordinated (HgC_2) with the mean Hg-C bond distances of 2.09 and 2.08Å, respectively. The former isomer is almost linear, 178.0(4)°, and the latter is exactly linear 180°.

5.3. Mercury Heterometallic Complexes -Isomerism

We analyzed the isomers of this group of complexes [6], and a brief view on the isomers will be the aim of this section.

5.3.1. Distortion Isomers

There are nine examples of this type of isomerism. The Hg_2Mo cluster exists in two isomeric forms, orange and red. In both, the molybdenum is a central core with two other "external" mercury atoms (Hg-Mo-Hg) with mean Hg-Mo bond distances of2.676Å (orange) and 2.677Å (red). The Hg-Mo-Hg bond angles are 73.1 and 70.8°, respectively. Each of the SEt group is terminally coordinated to the mercury atom and completes diagonal coordination of each Hg(II) atom (HgSHg´). Each molybdenum atom is sandwiched by two cyclopentadienyl rings with cp-Mo-cp (centroid) angles of 140.5°. The S-Hg-Mo bond angles are 171.2° (orange) and 167.6° (red).

There are six derivatives that contain two crystallographically independent molecules within the same crystal: HgMo, $HgPt_2$, $HgCo_2$, $HgCr_2$, $HgFe_2$ and $HgCo_3$. In the triclinic yellow HgMo dimer, two units $(Ph_3P)Hg$ and $Mo(CO)_3$(cyclopentadienyl) are held together by direct Hg-Mo bond with the distance 2.685(1)Å (molecule 1) and 2.687(1)Å (molecule 2). The mercury atoms are diagonally coordinated (HgPMo) with Mg-P bond distances and P-Hg-Mo bond angles (molecule 1 vs molecule 2) of 2.490(1)Å and 169.7(1)° vs 2.476(2)Å and 160.8(1)°.

In the triclinic orange $HgPt_2$trimer, the metal atoms form a triangle with Pt-Hg-Pt angles of 59.96(2) and 60.78(2)°. Two terminally coordinated chlorine atoms completed a highly distorted tetrahedron of each mercury atom ($HgCl_2Pt_2$). The mean Hg-Cl and Hg-Pt bond distances (molecule 1 vs molecule 2) are 2.429 and 2.714Å vs 2.447 and 2.704Å. Each platinum is five coordinated ($PtP_2ClHgPt'$) with Pt-Pt bond distances of 2.712(1) and 2.736(1)Å,respectively. The structure of $HgCo_2$trimer, consists of a linear Co...Hg...Co arrangement bridged by diethylphosphanate groups with Co...Hg separations of 4.065(1)Å (molecule 1) and 4.078(2)Å (molecule 2). The six O atoms of the Hg atom form a tetragonal distortion with Hg-O bond distances of 2.270 and 2.292Å, respectively. Each Co atom has a CoC_5P_3 inner coordination sphere. The sum of Hg-O(x6) and (Co-C(x5) plus Co-P(x3)) bond distances (molecule 1 vs molecule 2) are 13.62 and 16.90Å vs 13.74 and 16.92Å, indicating that molecule 1 is somewhat tighter than molecule 2.

In another trimer an almost linear Cr...Hg...Cr arrangement is bridged by phenyl groups. The Hg atom is linearly coordinated C-Hg-C, 180°, with the mean Hg-C bond distances of 2.066Å (molecule 1) and 2.074Å (molecule 2). Three CO groups around each Cr atom completed a four-coordination about each Cr atom (CrC_4). The sum of Hg-C(x2) and Cr-C(x4) bond distances (molecule 1 vs molecule 2) are: 4.13 and 7.77Å vs 4.15 and 7.78Å.

In the triclinic orange Hg_2Fe_2 tetramer, two substituted cyclopentadienyls serve as bridges. Each mercury atom has an almost linear coordination (HgC_2). The Hg...Hg separations are 3.323(2) and 3.362(2)Å,respectively. Each iron atom is sandwiched (FeC_{10}). In the monoclinic $HgCo_3$ tetramer the mercury atom is coordinated by three $Co(CO)_4$ units with a mean Hg-Co bond distance of2.686Å and Co-Hg-Co angles of 120.0°.

5.3.2. Linkage Isomers

A monoclinic HgAu dimer exists in two isomeric forms, a light yellow and yellow. In the former one, the$Ph_2(S)PCH_2$ anion is bonded to the Hg atom via C and the other anion via the S atom (HgCS) and correspondingly to the Au atom (AuSC). In the yellow form, twoPh_2 $(S)PCH_2$ anions are bonded to the Hg atom via C atoms (HgC_2) and to the Au atomvia S atoms (AuS_2).

CONCLUSION

A brief survey of the crystallographic and structural parameters of mercury complexes (coordination, organometallic and heterometallic) is outlined. There are nearly one thousand one hundred examples, for which the parameters are available. Almost 12% of the examples exist in the isomeric forms. The mercury atom is found in the oxidation state 0, +1, +1.33 and +2.Of these +2 is by far the most common. Nuclearities range from monomer to polymers.

Coordination numbers about mercury range from two to eight, four and two being the most common, with ligands ranging from monodentate to octadentate with a wide variety of donor atoms, from which N,C,S and P donor sites and chlorine are the most common. The complexes crystallized in the several crystal classes, monoclinic, triclinic, orthorhombic, tetragonal, trigonal and cubic, from which monoclinic and triclinic are the most common. Most mercury complexes are colorless or yellow, with some example of pink, white and red. The heterometallic complexes are colorless when the other metal atom is a non-transition element (Li, Na, K, Sr, Ba, Te, Cd, Zn, Th, Eu, Gd, Tb). When the second metal is a transition element, there are colorless derivatives (Ag, Au, Ir, Ru), but the majorit are yellow to red (Fe, Ni, Co, Os, Pt, Pd, Rh, Mo, W, Mn, Re). There are also a few examples of green, blue, violet, brown and even black.

Five types of isomerism are established: distortion, ligand, coordination number, polymerization and linkage, from which the first one is the most common (~90%).

The mean Hg-Hg bond distance elongates when an increase of the oxidation state of the mercury is in the order of 2.52Å (range 2.48-2.56Å)(Hg(1)-Hg(I)) < 2.78Å (range 2.76-2.80Å)(Hg(1.33)-Hg(1.33)) < 2.87Å(Hg(II)–Hg(II)), indicating a metal-metal bonding interaction. In most Hg(II) derivatives, the mean Hg(II)-Hg(II) distanceis 3.90Å, excluding a metal-metal bond.

A summary of the mercury-metal distances below 3.00Å is given in Table 1. Although mercury is classified as a soft, or class B, metal atom, it can be seen that it is found bonded to a variety of other metal atoms of both the soft and hard (class A) variety, illustrating a rich diversity in the chemistry of mercury amongst its 1100 or so structurally determined complexes.

Table 1. Summary of Hg-M bond distances [Å][a]

Hg-M	Shortest	Longest	Mean
- Ge	2.48	2.53	2.51
- Co	2.43	2.69	2.54
- Ni	2.44	2.68	2.55
- Li	2.56	2.58	2.57
- Fe	2.51	2.86	2.58
- Ir	2.58	2.62	2.60
- Mn	2.58	2.66	2.62
- Mo	2.56	2.77	2.68
- Os	2.58	2.89	2.685
- Pd	2.71		2.71
- Pt	2.50	2.98	2.72
- Rh	2.59	2.80	2.72
- Ru	2.60	2.95	2.74
- Re	2.74		2.74
- W	2.75	2.76	2.755
- Te	2.78	2.80	2.79
- Nb	2.79		2.79
- Au	2.93	2.96	2.95

a. Only values below 3.00Å .

References

[1] C.E. Holloway and M. Melník, *Main Group Met. Chem.,* 17 (1994) 799; and ref's therein.

[2] C.E. Holloway and M. Melník,*J. Organomet. Chem.,* 495 (1995) 1; and ref's therein.

[3] C.E. Holloway and M. Melník, *Main Group Met. Chem., 17* (1994) 741; and ref's therein.

[4] M. Melník, *Centr. Eur. J. Chem.,* 8 (2010) 469; and ref's therein.

[5] M. Melník, *Coord. Chem. Rev.,* 47 (1982) 239.

[6] M. Meník and P. Mikuš, *Centr. Eur. J. Chem.,* in press.

In: Mercury: Sources, Applications and Health Impacts
Editors: Ki-Hyun Kim and Richard J. C. Brown
ISBN: 978-1-62257-721-7

Chapter 2

ORGANIC–INORGANIC MESOPOROUS MONOLITHIC SCAFFOLDS AND THEIR FUNCTIONALITY IN THE OPTICAL ION-SENSITIVE REMOVAL OF MERCURY IONS

Sherif A. El-Safty**[*1,2,3] and M. A. Shenashen[1]

[1]Materials Research Laboratory for Environmental and Energy, National Institute for Materials Science (NIMS), Ibaraki, Japan
[2]Graduate School for Advanced Science and Engineering, Waseda University, Tokyo, Japan
[3]Department of Chemistry, Faculty of Science, Tanta University, Tanta, Egypt

AIM OF THE CHAPTER

The toxic effects of mercury (Hg) to humans are well known, such as damage to the nervous system. Mercury is an element and therefore not degradable into non-toxic compounds or elements. It is released mainly from mining, industries, fossil fuel combustion, and volcanoes and may, via anthropogenic and natural activities, be transformed into ionic forms and bioaccumulated in biota and bíomagnified through the food chain, notably in aquatic environments.

The engineered control of 3D mesoporous cage monoliths has led to widespread advances in the adsorption of toxic metal ions, especially Hg(II) ions. With the aim of controling the design functionality of mesoporous cage monoliths as a platform for fabricating a Hg(II) ion optical sensor, this chapter first focuses on the successful up-to-date developments in different techniques, designs, and studies which are potentially useful in enhancing the effectiveness of Hg(II) removal methods. The approach proposed in this chapter highlights a technique that depends on the use of 3D mesoporous cage monoliths as optical sensors for Hg(II) metal ions. The key to designing optical nanosensors is the ability to immobilize chromophore probes with different functional characteristics, such as densities,

[*] Corresponding author: E-mail: sherif.elsafty@nims.go.jp & sherif@aoni.waseda.jp.

accessibility, and intrinsic mobility, onto these 3D mesoporous cage monolith pore surfaces, yielding a facile and reliable synthesis design. This practical design of an optical sensor based on mesoporous silica shows evidence of controlled assessment processes, which involve evaluation of intrinsic properties of adsorption (e.g. color change, long-term stability, adsorption efficiency, and reusability).

MERCURY

Mercury in the Environment

Heavy metal ions play an important role in fundamental physiological processes in organisms ranging from bacteria to mammals. Certain metals, the so called toxic 'heavy metals', have no biological function and are toxic in small amounts to humans and animals such as birds, fish and mammals. "Mercury is one of the most dangerous and ubiquitous pollutants in the global environment" [1a]. It poses a very serious problem to the global environment because of Hg's toxicity, non-degradability, and persistent threat to human health and the environment. Mercury and its compounds, even at low concentrations, are very dangerous neurotoxins for living organisms. [1].

Global Mercury Cycle in the Environment

Mercury sources are both natural and anthropogenic. Natural occurrences of volatized mercury include volcanoes and associated geothermal systems, fires, and mercury evaporated from seawater. Since the industrial revolution, the atmospheric levels of mercury have increased far above natural levels due primarily to emissions from the anthropogenic sources. Anthropogenic sources include gold mining, solid waste incineration, fossil fuel combustion, and numerous industrial processes. Tracking the transport and cycling of Hg in the environment is needed to minimize potential human and ecosystem health effects. Although anthropogenic Hg emissions have been reduced by half in the last few decades, ongoing contamination is still a worldwide problem [2].

Mercury circulates around the environment primarily in three different forms: elemental mercury (Hg^0 - as a vapor), monovalent mercury (Hg^{+1}), and divalent mercury (Hg^{+2}). All forms of mercury are toxic, can create health problems in humans, and have a variety of different effects based on the chemical species introduced into the body. Mercury cycles through land, air, and water, undergoing a number of complex chemical and physical transformations. The oxidation of Hg^0 in the atmosphere is an important mechanism resulting in the deposition of mercury onto land and water. Hg^0 vapor is released into the atmosphere in gaseous form from rocks, soils, water, volcanoes, and human activity (natural occurrences and anthropogenic sources). The mercury circulation, started by the release of mercury vapor into the atmosphere, is transported in the atmosphere for a year or more by the effect of wind, and then is re-deposited onto soil and water, after which further cycling may occur. Hg^0 vapor is photochemically converted into inorganic Hg^{+2}, which can combine with water vapor and travel back to the Earth's surface as rain and be deposited in soils and bodies of water. Once in the soil, mercury accumulates until a physical event causes it to be released again. Whilst it circulates in the environment and changes its chemical form, mercury is an element and therefore is persistent and not degradable into less toxic forms [3]. Inorganic mercury can be converted into barely soluble mercury sulfide which, in turn, can be converted by bacteria that process sulfate into methylmercury. In the environment, sulfate-reducing bacteria take up mercury in its inorganic form and through metabolic processes convert it to methylmercury. Sulfate-reducing bacteria are found in anaerobic conditions, typical of the well-buried muddy sediments of rivers, lakes, and oceans where methylmercury concentrations tend to be highest. So, while sulfate-reducing bacteria are respiring and producing sulfide that will sequester available metal cations (including inorganic mercury) into a nontoxic precipitate, they may also be methylating a portion of this same mercury pool into a more toxic, bioavailable form. This process may play an important role in the Methylation of mercury. Methylmercury is much more toxic than inorganic mercury. Organisms require a long time to eliminate methylmercury, leading to more bioaccumulation than for other forms of mercury. The path from the release of elemental mercury (Hg^0) to the formation of ionic mercury may involve photochemical processes, deposition, and the formation of methylmercury at the sediment–water interface as shown in Fig. 1.

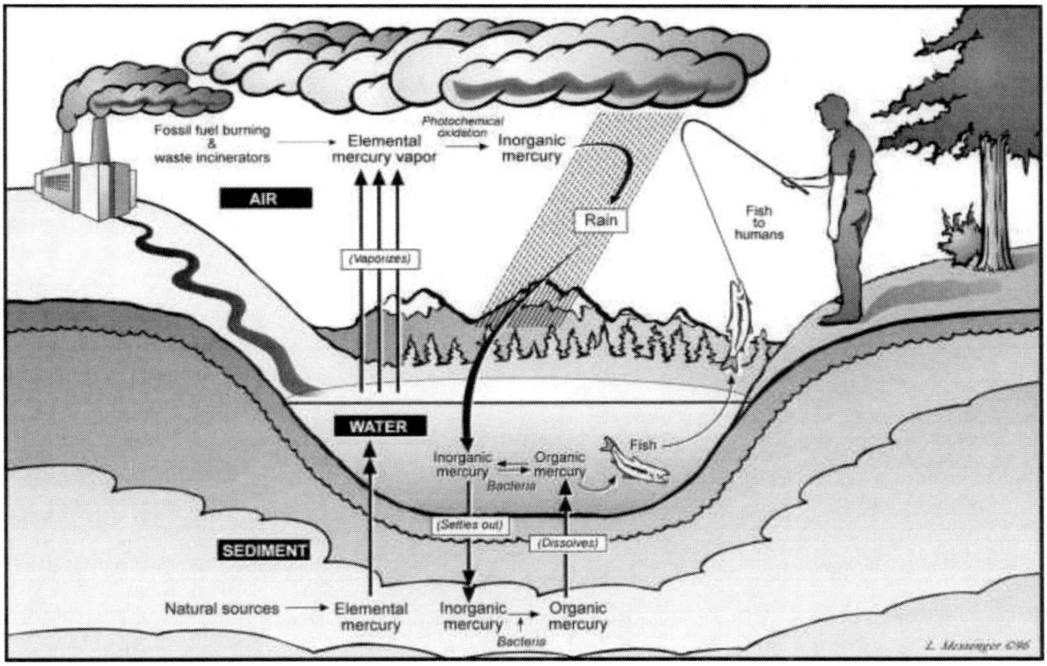

Figure 1. Mercury cycles in the environment as a result of natural and human (anthropogenic) activities. Image adapted from: http://people.uwec.edu/piercech/Hg/mercury_water/cycling.htm.

Toxicity of Mercury

Considering heavy metals are non-biodegradable, the accumulation of these metals at various trophic levels through the food chain leads to human health problems. Mercury has attracted growing interest, especially its presence in samples of human food products, because of the various levels of toxicity of its different chemical forms. Organic mercury compounds, of which methylmercury is the most common, are of particular concern because of their enhanced toxicity. Mercury contamination of freshwater ecosystems remains a major problem in industrialized nations, with deposition from natural and anthropogenic sources and their subsequent methylation, leading to high mercury concentrations in fish [4]. Human and wildlife health concerns surrounding the consumption of mercury-contaminated fishe require considerable research and monitoring efforts by government agencies. Losses caused by mercury-contaminated products (mainly fish and marine mammals) are estimated at billions of dollars globally [5]. Methylmercury accumulates in muscle tissues where it binds with the sulfhydryl groups of muscle proteins [6].

As reported, human exposure to high levels of environmental concentrations of elemental mercury and methylmercury results in adverse effects to the central nervous system (CNS), such as hallucinations and delirium (caused by elemental mercury), blindness, deafness, and impaired levels of consciousness (caused by methylmercury) [7]. Effects on the gastrointestinal tract and respiratory system from acute inhalation exposure to elemental mercury have also been noted in humans. The methylated forms of mercury are especially mobile in humans because of complexation reactions with cysteine, which increases the transport of mercury throughout the body. Children may suffer from mental retardation, cerebral palsy, and convulsions. Mercury also causes genetic defects, chromosome breakage, and it interferes with cell division, resulting in abnormal chromosome distribution. Mercury impairs pulmonary and kidney functions, as well as causing chest pain and dyspnea [7]. Human exposure to methylmercury is mainly through dietary intake, especially via fish and shellfish that bio-accumulate this compound. Mercury exposure can also lead to personality changes, tremors, loss of motor control, and possibly death [8]. It also affects reproduction and survival rates in both birds and fish [9].

Mercury Removal Technologies

Several technologies have been used to remove heavy metals from mercury-contaminated aqueous streams. Among the most commonly used physicochemical techniques are chemical precipitation, chemical oxidation and reduction, ion exchange, filtration, electrochemical treatment, reverse osmosis (membrane technologies), evaporative recovery, and solvent extraction [10]. The detection or "sensing" of toxic pollutants at the nanoscale level is a most intriguing development. Chemical surface functionalization of ordered nanoscale structures can be used to produce materials that have desired surface properties for applications such as sensing, catalysis, separation, and for use in nanoreactors [11]. In the past few years, many analytical methods, such as atomic absorption spectrometry [12], inductively coupled plasma-mass spectrometry [13], continuous flow cold vapor atomic fluorescence spectroscopy [14], inductively coupled plasma atomic emission spectrometry [15], electrochemical methods [16], X-ray fluorescence spectrometry [17], anodic stripping voltammetry [18] and UV-Vis spectrometry [19] have been applied to detect low-level Hg(II) concentrations.

Adsorption

Adsorption processes have the potential to achieve high mercury removal efficiencies at low cost and at low effluent mercury levels. According to Pearson's principle of hard and soft acids and bases, sulfur has a high tendency to combine with mercury. Thus, huge efforts have been focused on the incorporation of thio groups into mesostructured silica materials. Many adsorbents have been studied for removing Hg(II) from aqueous solutions including: activated carbon, silicates, polymers, and biomass, such as bacteria and chitosan. In addition, other materials including clay, organic matter, iron oxide, ironpyrite, suspended solids/sediments in natural water, and natural minerals, have also been extensively studied because of their adsorptive properties for mercury [20].

Activated Carbon

Activated carbon is historically the most common material used for removing mercury from aqueous solutions. Activated carbon comprises a group of materials that possess highly developed internal surface areas and porosity. Activated carbon is made from a variety of carbonaceous materials such as wood, peat, lignite, and coal, and from other cheaper materials [21]. Activated carbon can be impregnated with various elements orcompounds to increase its mercury removal efficiency [22]. They can remove mercury from aqueous solutions via four main chemical mechanisms, namely: ion exchange, reduction/precipitation, complexation (using sulfur-impregnated activated carbon), and pore capture [23].

Considering the relatively high cost of activated carbon, research is continually being conducted to find cheaper sorbent alternatives. This often includes modification of material structures and surfaces with functional groups to enhance adsorption efficiency. Manohar et al. modified natural clay and tested it for mercury removal; the modified clay adsorbed 80% more mercury over a limited pH range of 4–6 [24].

Bioadsorbents

The uptake of metals by biomass can take place actively through a metabolism-dependent process (bioaccumulation) or a passive and usually rapid metabolism-independent process (biosorption – taking several minutes) [25]. The use bioadsorbents has been investigated for removing heavy metals from solutions. The application of chitosan, crab shell, chitin, anaerobic granular biomass, rice husk, maize cobs, rose waste biomass, tea factory waste biomass, and sewage sludge for heavy metal removal has been explored [22, 26]. However, these materials are not easily available in all places and at all times. Various microorganisms that have been tested for biosorption include fungi, algae, yeast and bacteria. They have been investigated for removing heavy metal ions, especially mercury, from aqueous [27] and other solutions. These biosorbents possess metal-sequestering properties and decrease the concentration of heavy metal ions in solution from the ppm to the ppb level. Therefore, biosorption using microorganisms is an ideal technique for the treatment of high volumes of complex wastewater with low metal concentrations.

Other Adsorption Processes

Various adsorption methods, apart from activated carbon and bioadsorbents, were used and reported as mercury adsorbents, such as the following [28]:

- Bicarbonate-treated peanut hull carbon (BPHC).
- Modified Hardwickia.

Ion Exchange

The majority of ion exchange–based removal processes involve forming negatively charged $HgCl_3$–complexes and the subsequent removal of this species through an anionic resin. These removal processes are therefore more suited to treatment plants using wastewater that contains high background chloride levels. For wastewater with low background chloride levels, cation exchange resins are effective in removing mercuric ions [29]. Monteagudo mentioned that the Hg(II) is removed from water through the ion exchange method using crandallite-type compounds as the ion exchanger because the surface area and pore diameter of crandallitics have no effect. The data indicated that Hg(II) replaced the Ca(II) and Sr(II) ions in the crandallite [30]. Ion exchange membrane bioreactor (IEMB) can be used for mercury removal. The charge on the exchange membrane must be used for pollutant transport, along with the oppositely charged counter-ion [31].

Chemical Reduction

Mercury ions may be reduced to elemental mercury using other metals higher in the electromotive series. The advantage of sedimentation processes are their ability to separate the resultant mercury metal directly, via gravity or filtration. Metals such as zinc, aluminum, and tin have been used to study mercury reduction [32]. Monomethylmercury is removed from sediments by reduction into volatile elemental mercury and by disproportionation into volatile dimethyl mercury using H_2S, although the significance of the latter reaction has yet to be demonstrated under in situ conditions [33].

Precipitation

Precipitation with various chemicals or chelating agents is often used to remove heavy metals such as mercury from industrial wastewater. In this process, the addition of metal sulfide to the waste stream converts soluble mercury into the relatively insoluble mercury sulfide form. Precipitation is often used in conjunction with subsequent removal technologies, such as filtration, gravity settling, or dissolved air flotation. pH adjustments affect the precipitation process. The wastewater is stirred with additional amounts of metal sulfide in a reaction vessel, where the mercury is precipitated as mercury sulfide. The precipitated solids can then be removed by gravity settling in a clarifier. Flocculation, with or without a chemical coagulant or a settling aid, can be used to enhance the removal of precipitated solids. Patterson [34] reported that mercury removal via the addition of magnesium sulfide produces discharge concentrations of 10–100 μg/L from a feed concentration of 10 mg/L without any additional treatment. Given their strong affinity for mercury, most chelating or complexing compounds contain sulfur functional groups. Matlock et al. investigated the removal of mercury from aqueous solution using a sulfur compound; he reported that this compound is capable of reducing a 50 mg/L mercury solution, at pH 6, to 0.02 mg/L [35].

Coagulation

The coagulants used for mercury removal include aluminum sulfate (alum), iron salts, and lime. For alum and iron, the dominant mercury removal mechanism is most likely through adsorptive co-precipitation [36]. Conventional techniques for the removal of heavy metals from wastewater, such as chemical precipitation, ion exchange, activated carbon adsorption, and separation processes, have limitations because they become inefficient and expensive, especially at heavy metal concentrations less than 100 ppm [37].

Membrane Separation

Several membrane processes have been used for water and wastewater treatment to remove mercury. These include ultrafiltration, charged filtration, crossflow microfiltration, magnetic filtration, and reverse osmosis. Negatively charged filtration membranes result from the dissociation of subgroups within the membrane structure. The negative charge of the filtration membrane (for ultrafiltration, microfiltration, and reverse osmosis) minimizes fouling of the membrane. Crossflow microfiltration uses a moving membrane. In both ultrafiltration and crossflow microfiltration, pretreatment of the mercury-containing feed stream is required because these processes cannot remove the mercury alone through size exclusion. Modified silica membrane disks containing triazene ligands are used to determine the presence of preconcentrated Hg(II) ions in the actual water samples [38].

Sensor

A sensor is a device that measures or detects a physical or chemical parameter and converts it into a signal, which is subsequently read by an observer or by an instrument. Sensors have high sensitivity and selectivity; they also provide on-line and real-time analysis. This has revolutionized the field of chemical analysis, particularly analysis of blood and serum samples in critical care applications [39]. Sensors with faster response times and higher selectivity are desired to monitor ultratrace levels of toxic metal ions. For these and many other reasons, sensors are being developed based on several energy-transduction principles (such as radiant, electrical, mechanical, and thermal energies) [40].

Advances in fabrication methods have involved many processes, such as the evaluation of intrinsic properties (such as sensitivity, linearity, hysteresis, selectivity, and shelf life) and manufacturability (costs and tolerances).New research methods are being applied to reduce response times. However, the complex nature of the interactions between function and composition, preparation and method, and the end-use conditions for the sensor materials, often make their rational design for "real world" applications difficult [41].

Optical Chemical Sensors

The field of optical chemical sensors has been a growing research area over the last three decades. Experts in the field, who have highlighted the advantages of optical sensing over other transduction methods, have published a wide range of books and review articles. Optical chemical sensors employ optical transduction techniques to generate analyte information. Optical sensors are most effective because of their accurate detection of pollutants at low concentrations even without close control of experimental environments or the use of sophisticated equipment [42]. The most widely used techniques employed in optical chemical sensors are those involving optical absorption and luminescence, but sensors based on other spectroscopic techniques, as well as on optical parameters, such as refractive index and reflectivity, have also been developed [43]. Fabrication of solid-state colorimetric sensors for visual detection is much less advanced, even though colorimetric sensors allow on-site, real-time qualitative/semi-quantitative detection without complicated analytical instruments [44].

The key components in this sensing method are optical sensors that can efficiently detect toxic ions with good sensitivity and selectivity and in real-time [45]. In addition, solid sensors are often provide remarkably enhanced sensitivity and selectivity with respect to comparable indicators or probe molecules. However, such developments in the form of compact instrument-free ion-sensors are still at the research level and are currently being investigated using different techniques, such as modified sol–gel membranes, molecular imprinted polymers, and nanothin film techniques [46]. These novel features are currently required to analyze ultratrace levels of environmental pollutants [47]. The development of sensing processes entails invention, optimization, and commercialization, and remains a significant challenge in materials science [48].

Optical Chemical Sensors Based on Nanomaterials

Given their regularly arranged three-dimensional (3D) pores and unique features (large surface areas, highly ordered pore structures, and very narrow pore-size distributions), nanomaterials, especially mesoporous materials, are an interesting class of materials for technological applications. These attractive and interesting features make these materials broadly applicable in heterogeneous catalysis, separation, and sensing, as well as in biomedicine for separation applications [49]. Porosity, the fraction of pore volume, is the major requirement for a solid to be used in catalysis, storage, and separation, because it correlates to a high surface area on a given material. Porous materials are categorized into three different types according to pore size: micropore refers to a region less than 2 nm, mesopore refers to a region from 2 nm to 50 nm, and macropore is a region higher than 50 nm [49, 50]. Since the fabrication of the classic MS41 materials with pore sizes ranging from 2 to

10 nm using cationic surfactants, significant effort has been expended to create new levels of hierarchical design in materials with utmost control over the mesophase geometry and pore morphology by adjusting surfactant templates and synthesis conditions [51]. The flexibility in controlling 3D geometrical structures with uniformly shaped cylindrical and cage-like pores is of interest to advanced adsorption and catalytic applications. This is because 3D pore functionality and connectivity should efficiently transport guest species to network sites [52].

New technologies can offer completely new approaches to chemical analysis. Pushing the limits of these approaches can transform research technologies into practical methods and it often leads to newer technologies. Porous monolithic architectures provide a very robust, open and tunable periodic scaffold at the nanometer scale [49, 50]. Nanometer-sized materials with engineered features, including size, shape, composition, and function play lead roles in emerging applications in a number of areas [52, 53]. The flexibility in controlling 3D geometrical structures with uniformly shaped cylindrical and cage-like pores are of interest as potential catalysts and sorbents. This is because the 3D morphologies and cage functionalities should efficiently transport guest species by diffusion much more directly and easily to the network sites [43, 54]. Among all 3D mesostructured materials, ordered cubic mesostructures with cage-like structures are desirable in terms of their potential applications. Considering the well-defined pore sizes of spherical cavities and the connectivity among pores [55], the 3D cage structures are promising in many practical applications, such as the separation of proteins, electronics, catalytic surfaces and supports, sensing, and inclusion chemistry [55]. In addition, the mesostructured features have led to extensive work in creating composites with active components embedded inside the pores [43, 54]. These innovations have led to new or vastly improved methods by enabling new levels of instrument control and data processing, improving sensitivity, and adding versatility and selectivity to binding assays. These nanoscale materials are expected to improve chemical analysis substantially with unique optical, mass transport properties, specificity, and amplification of chemical processes through microfabrication [56]. Our recent efforts were geared towards fabricating optical chemical nanosensors that incorporate simplicity in terms of fabrication time and low-cost materials, but still retain the optical devices for specific ''real world'' sensing of toxic metal ions [45a, 57].

The color changes when chromophores react with metal ions have been widely used as signaling events, because they require inexpensive or no equipment with color changes being observed using direct visualization [50d, 58]. Manipulation of matter at the nanoscale level has enabled the development of nanosensors that exhibit high strength, as well as recognition and signaling of a broad range of chemical species. [45, 59, 60]. The advanced functionality obtained by manipulating chromophore probes into nanomaterials as sensing receptors has received attention in the design of chemical sensor arrays for responsive recognition of several species, such as metal cations, [54,56] and charged and neutral organic molecules [59].

Optical Chemical Sensors Based on Fluorescent Chemosensors

Fluorescent chemosensors consist of a receptor part and a fluorophore part. The receptor is mainly responsible for recognizing the analyte, whereas the fluorophore part converts the recognition events into optical signals. Fluorescent chemical sensors for Hg^{2+} have been synthesized based on various fluorophores, such as dansyl, pyrene, rhodamine, fluorescein, anthracene, naphthyl, and nile blue [57b, 61]. Rhodamine-based dyes are known for their excellent spectroscopic properties, with large molar extinction coefficients and high fluorescence quantum yields. The rhodamine framework is an ideal model for constructing ON–OFF fluorescent and chromophoric chemosensors because of its unique structural properties [62]. Rhodamine B and its derivatives are well known for many applications, such as fluorescent chemosensors for Hg(II).

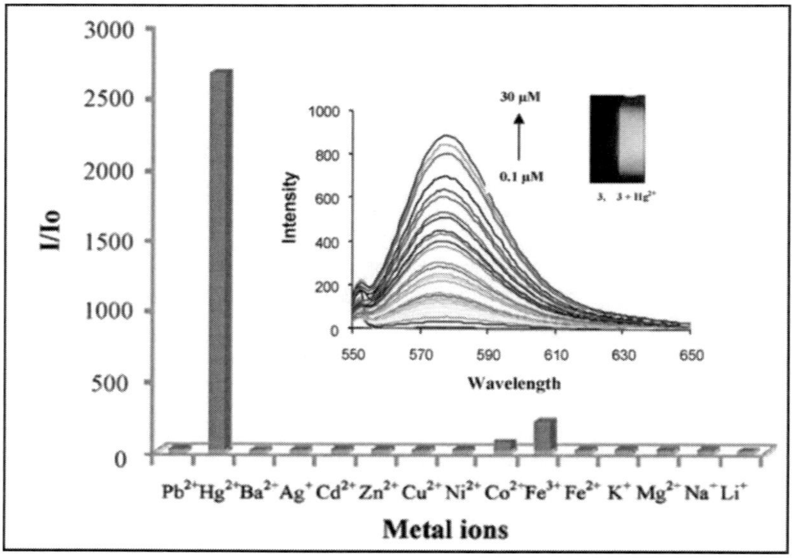

Figure 2. Receptor fluorescence at 576 nm with different metal ions Adapted from [64], (Copyright: © 2010 Elsevier B.V.).

A rhodamine–terphenyl derivative has been synthesized and used as a selective and sensitive chemosensor for Hg^{2+} ions under direct observation, as well as UV-Vis and fluorescent responses by Bhalla [64] *Fig. (2)*. This material shows fluorescence selectivity and colorimetric change upon contact with Hg(II) ions, which indicates that rhodamine–terphenyl is a selective and a sensitive "on–off" type fluorescent marker for Hg(II) ions. To understand the distinct toxicities to organisms of Hg(II), rhodamine B selenolactone has been synthesized as a fluorescent probe for imaging Hg(II) in live cells *Fig. (3)* [65]. Depending on the deselenation process, the fluorescent probe produces a fluorescence-on reaction with Hg (II) and it is constructed through the insertion of Se atoms inside the rhodamine structure. The applicability of this method was confirmed with Hg(II) imaging in HeLa cells. Thiooxorhodamine B has used as a fluorescent sensor and it has demonstrated high selectivity for sensing Hg(II) [66].

The functionalization of rhodamine-6G with pyrene and its application as a selective colorimetric and fluorometric sensor for Hg(II) was illustrated by Ahamed and Ghosh [67] *Fig. (4)*. Although the functionalization of rhodamine-6G with pyrene shows selective sensing of Hg(II) in water, the generation of acid in the solution is responsible for the color change, as well as fluorescence changes in water. The fluorescent urea-based modified rhodamine probe was synthesized for the detection of Hg(II) ions. The addition of Hg(II) ions into the urea-based modified rhodamine showed a highly selective fluorescence enhancement *Fig. (5)* [68].

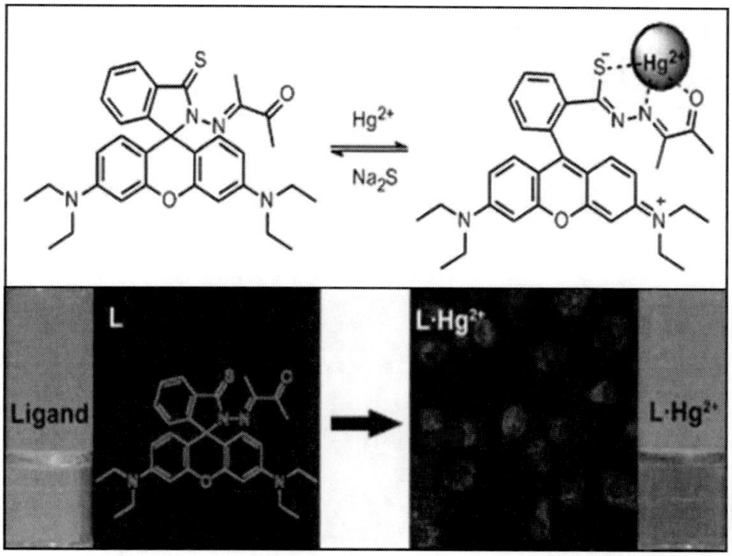

Figure 3. Confocal fluorescence images of intracellular Hg(II) in A549 cells. (a) Bright-field transmission micrograph; (b) Confocal fluorescence micrographs stained with receptor; (c) Subsequently exposed to Hg(II) for 40 min. (d) Fluorescence emission spectra. Adapted from [66], (Copyright: 2011 Elsevier Ltd.).)

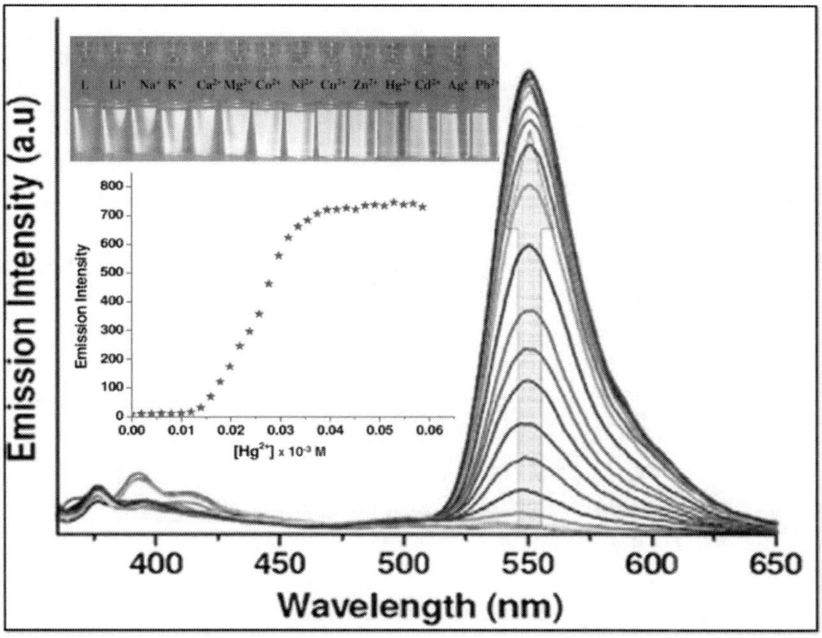

Figure 4. Pyrene and rhodamine-6G as an integrated system for selective colorimetric and fluorometric sensing of Hg(II). Adapted from [67], (Copyright:2011 Elsevier B.V).

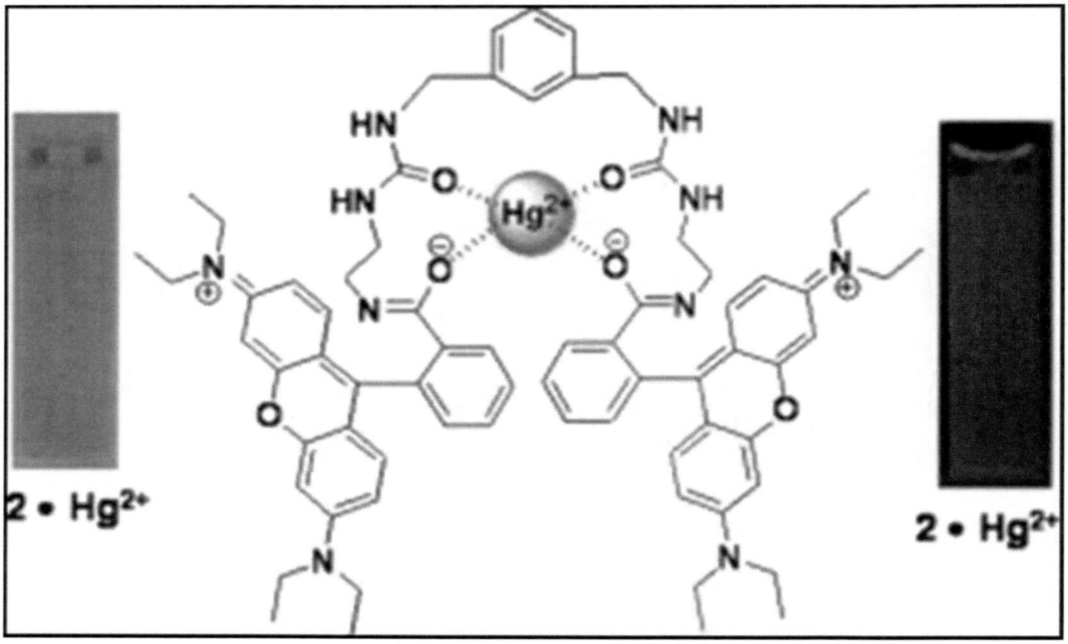

Figure 5. Rhodamine derivatives bearing urea group as a Hg(II) sensor. Adapted from [68], (Copyright: 2007 Elsevier Ltd.).

Given the strong, selective, and coordinate interaction captured between Hg(II) and bis-thymine, the metal-mediated T Hg T that forms stable DNA duplexes has attracted significant interest. A T-rich Hg-specific oligonucleotide (MSO) probe (5′-TTCTTTCTTCCCCTTGTTTGTT-3′), which includes two T-rich domains spaced by a "mute" oligonucleotide sequence spacer, was designed to bind with Hg(II) selectively [69]. Moreover, the high selectivity for Hg(II) of polythiophene-containing thymine moieties was studied [70]. Hg(II) ions reportedly specifically link with otherwise impossible T–T pairs and form T–Hg(II)–T complexes [71]. Zhu and Li [72] demonstrated thymine (T)-rich, Hg-specific oligonucleotide (MSO) probe, and Au nanoparticles (NPs) as sensitive electrochemical sensors for the detection of Hg(II) in aqueous solutions. The MSO probe contains seven thymine and forms a hairpin structure in the presence of Hg(II) through the Hg(II)-mediated coordination of T-Hg(II)-T base pairs. With the introduction of Au NPs, the efficiency of this sensor increases and can electrochemically detect Hg(II) as low as 0.5 nM.

Gavina and Palomares [73] demonstrated that Hg(II) ions strongly affect benzothiazolium hemicyanine dye because of the presence of sulfur, a mercury-coordinating element. The selective, colorimetric, and fluorometric mercury detection using benzothiazolium hemicyanine dye in aqueous solution has a response limit of 100 ppb under fluorescence spectroscopy *Fig. (6).* The fluorescence imaging of *Arabidopsis thaliana* cultured under different Hg(II) ion concentrations using rhodamine B thiolactone as a highly selective and sensitive probe for Hg(II) ions has been demonstrated [74]. The results indicate that the fine distribution of Hg(II) ions in *A. thaliana* and tat Hg(II) ions accumulate mainly in the root cap and the meristematic zone, and only a small amount reaches the maturation zone *Fig. (7).*

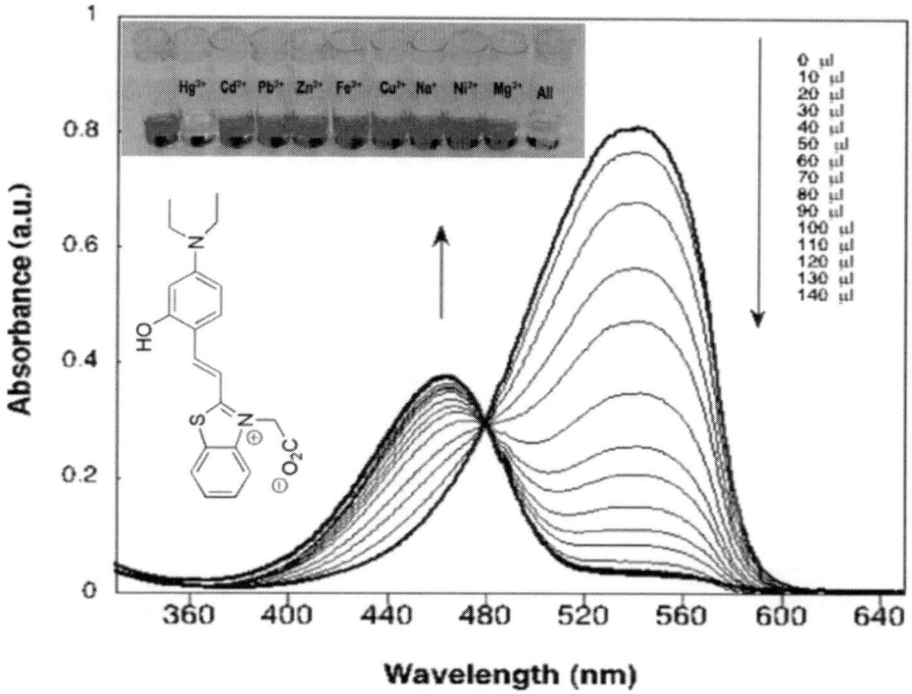

Figure 6. Effect of Hg(II) concentrations on UV-vis spectra of organic molecular probe in aqueous solution. Adapted from [73], (Copyright: © 2006 American Chemical Society).

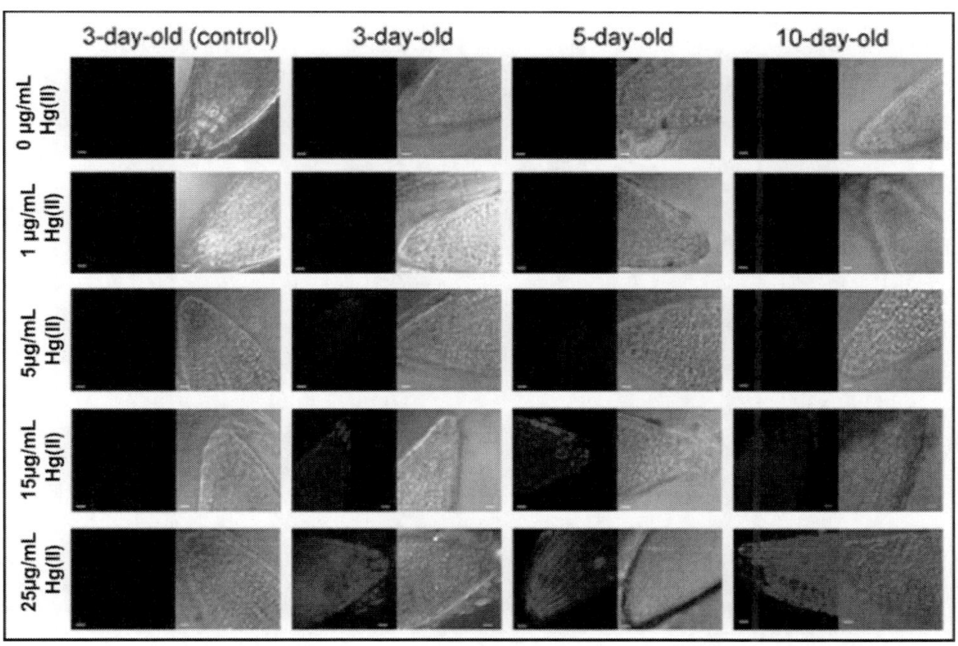

Figure 7. Fluorescence images (the left of each column) of the root cap and meristematic zone of A. thaliana cultured in varying HgCl2 concentrations. Adapted from [74], (Copyright: © 2010 Elsevier B.V.)

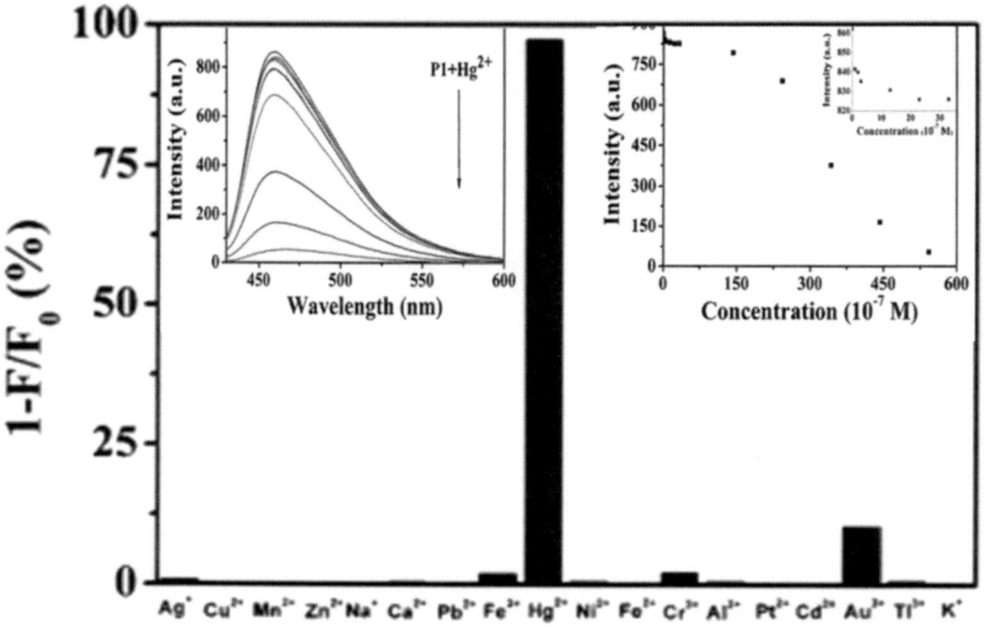

Figure 8. Fluorescence quenching efficiencies of polymer (1-F/F0) in the presence of various main group and transition metal ions. Adapted from [76], (Copyright: ©2011 Elsevier Ltd.).

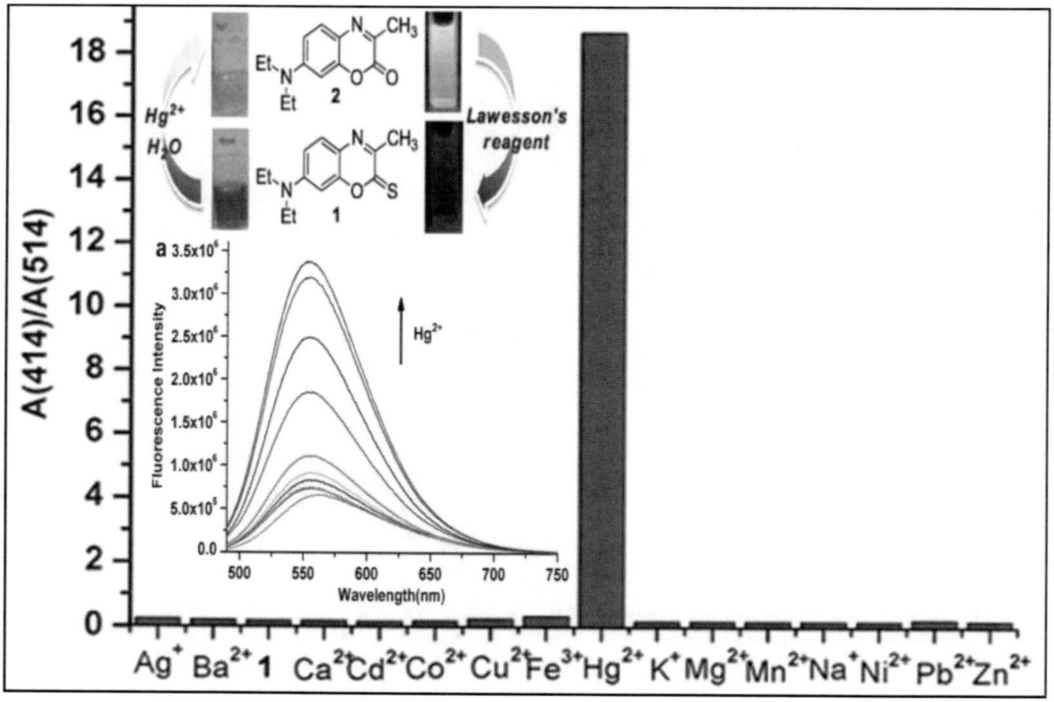

Figure 9. The ratio of A(414)/A(514), color change, emission spectra, and UV-Vis absorption spectra of sensorin in the presence of Hg(II) ions and other competing ions. Adapted from [77], (Copyright: © 2010 Elsevier Ltd.).

Conducting polymers, especially π-conjugated polymers, have attracted great attention because of their simple preparation, low cost, and wide range of applications. Many of the water-soluble π-conjugated polymers have been prepared as detectors for Hg(II) under neutral pH, such as polymers with carboxylic acid functional groups and carboxylic-functionalized thiophene that contain oligophenylenevinylene [75]. Polyaniline that has been synthesized via a Pd catalyst displays high sensitivity and selectivity toward Hg(II) [76]. Aniline, in the polyaniline backbone, shows high selectivity toward Hg(II) in the presence of competitive ions *Fig. (8)*. Oxazine-thione has been synthesized as a highly selective and sensitive probe for the detection of Hg(II) ions in water-compatible solvents with different pH values. Using UV-Vis and fluorescent responses, the detection of Hg(II) ions was possible by the naked eye [77]. Azobenzene was studied as an optical chemosensor for detecting Hg(II). This material has high sensitivity and selectivity for Hg(II) over several competitive metal ions, such as common alkali, alkaline earth, and transition metal ions *Fig. (9)* [78]. Fan and Wang [79] reported that a cationic conjugated polymer shows a "turn-on" sensor for the detection of Hg(II) in aqueous solution with high sensitivity and selectivity. This sensor has several important features: it rapidly detects micromolar Hg(II) ions making them visible even by direct observation. Nanomolar sensitivity was achieved using a fluorometric method and does not require fluorescent labels on the oligonucleotide probe, which should significantly reduce its costs. This is based on a target-induced conformational change of oligonucleotide and phenylethanolamine-N-methyl transferase (PMNT).

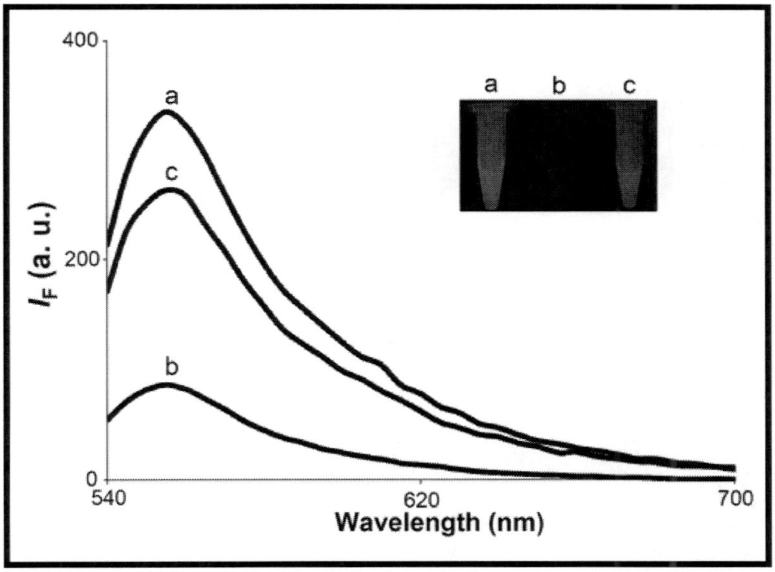

Figure 10. Fluorescence spectra of solutions of (a) R6G and (b, c) BSA@R6G/MPA-AuNPs in the (b) absence and (c) presence of Hg(II). Adapted from [82], (Copyright: 2011 American Chemical Society).

To study the application of nanoparticles (NPs) on the detection of Hg(II), Huang et al. used mercaptopropionic acid (MPA)-modified gold NPs with 2,6-pyridinedicarboxylic acid (PDCA) as a colorimetric method for detecting mercury (II) in aqueous solutions [80]. Lee et al. studied DNA-modified Au NPs for detecting Hg(II) and found that the material is highly selectivity and sensitive, with color changes at a given temperature or melting temperature

(Tm) in aqueous media [81]. Modified Au NPs have been developed by introducing bovine serum albumin (BSA) into a solution of rhodamine 6G and a 3-mercaptopropionic acid-based fluorescence sensor for the detection of Hg(II) and its organic compounds in aqueous solutions [82]. BSA is used to protect the sensor from serious aggregation in high salt solutions *Fig. (10)*. The uptake of mercury ions on the surface of Au NPs induces the release of R6G molecules into the solution; thus restoring the fluorescence of R6G. The presence of masking agents increases the efficiency of the sensor to detect specific levels of organic mercury species. Su and Wang [83] reported BSA-functionalized fluorescent Au NPs as a sensor for on-site, real-time Hg(II) detection *Fig. (11)*. Through fluorescence quenching, the probe shows that aqueous Hg(II) can be detected at 0.1 nM, and the results demonstrated low interference and high sensitivity.

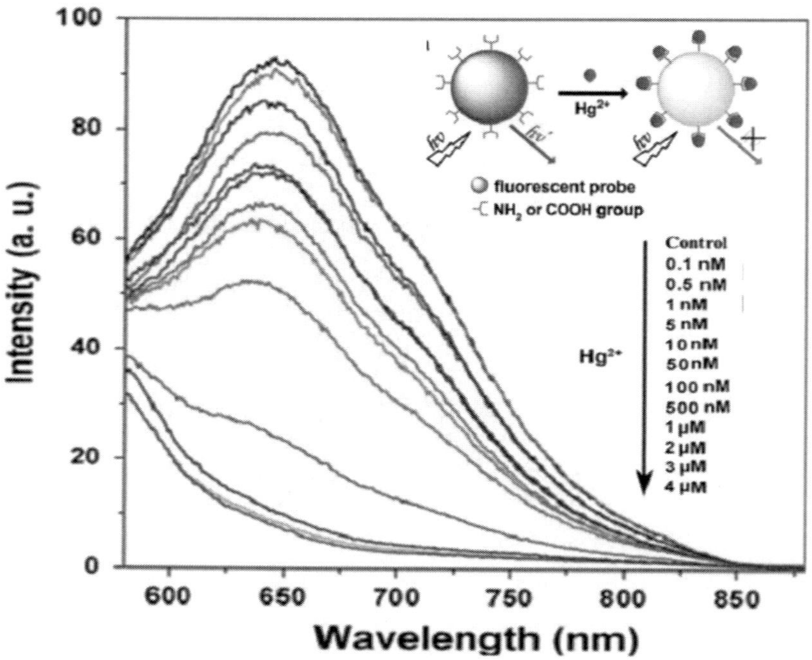

Figure 11. Fluorescence response of BSA-GNPs upon the addition of Hg(II). Adapted from [83], (Copyright: The Author(s) 2010. This article is published with open access at Springerlink.com).

Darbha and Ray [84] demonstrated that the nonlinear optical (NLO) properties of mercaptopropionic acid/homocysteine/2,6-pyridinedicarboxylic acid (MPA-HCys-PDCA)-modified gold nanoparticles could be used for the rapid, easy, and reliable screening of Hg(II) ions in aqueous solution, with high sensitivity (5 ppb) in aqueous solution *Fig. (12)*. The modulation of fluorescence quenching efficiency of rhodamine B–capped thioglycolic acid–functionalized gold nanoparticles by Hg(II) ions was studied. This was shown to be a sensitive method for the detection of Hg(II) ions in aqueous solution was studied [85]. The method shows that the surface of the Au NPs was completely quenched, with the fluorescence of negatively charged rhodamine B (RB) resulting from fluorescence resonance energy transition (FRET) and collision *Fig. (13)*.

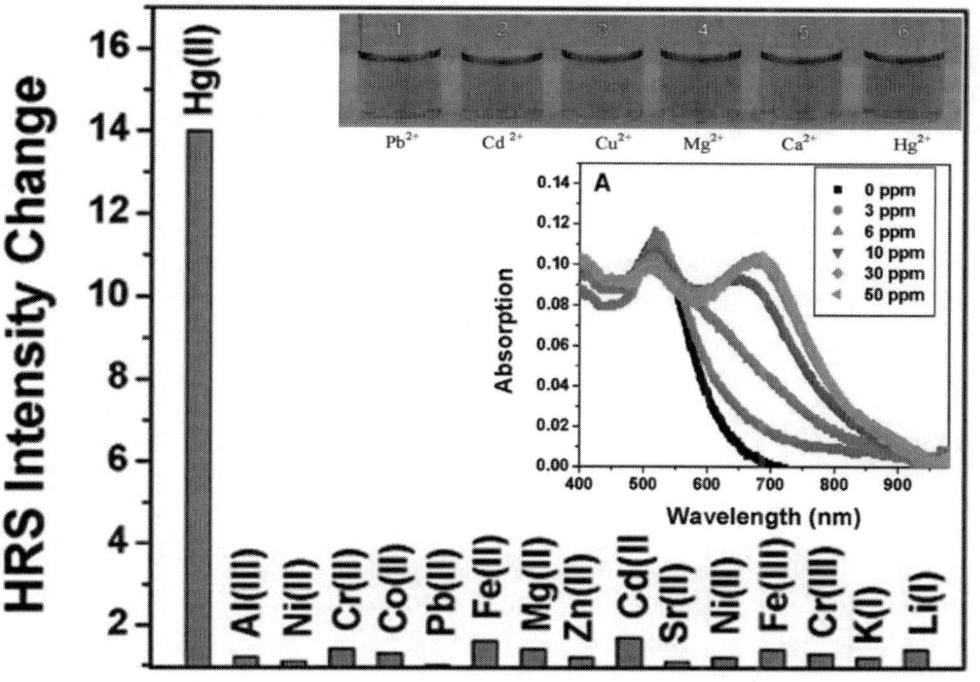

Figure 12. Hyper Rayleigh Scattering intensity, absorption, and photographic image changes upon the addition of 40 ppm different metal ions on the gold nanoparticle-MPA-HCys-PDCA solution. Adapted from [84], (Copyright: 2008 American Chemical Society).

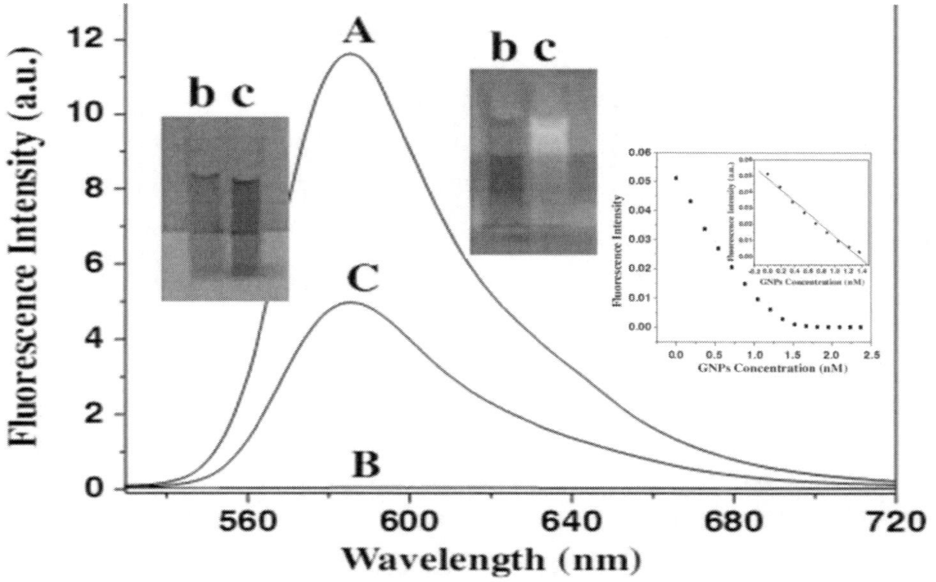

Figure 13. Fluorescence spectra of solutions of RB (A) and RB–GNPs (B, C) in the absence (B) and presence (C) of Hg(II) (100 μM). Adapted from [85], (Copyright: Springer-Verlag 2008).

Optical Sensors Based on Colorimetric Probes

Recently, several colorimetric probes, so-called "naked eye chemosensors," have been developed for detecting small quantities of mercury [86]. Thiospirolactone rhodamine derivatives have been reported as ideal chromophores to construct reversible sensors that are able to control the spiro-ring opening processes because of the coordination between Hg(II) and S atoms in acidic conditions [87]. Moreover, the combination of chromophores and 'soft' chelator units provides high affinity for Hg(II) under physiological conditions [88]. Several groups have shown that molecular sensing based on ruthenium complexes is an attractive approach, because of the sensitivity and mercury-selectivity of such complexes [89]. However, the use of such complexes might be seen as a necessary step towards the development of cheap sensing technology. The development of selective and sensitive methods for detecting mercury ions has received much attention because the mercury ion is one of the most toxic metal ions widely distributed in the environment. However, most fluorescent sensors for Hg^{2+} ions have limitations in terms of aqueous solubility, selectivity, sensitivity, interference from other heavy metal ions, and response type [87a, 90]. Fluorescent chemosensors consist of a receptor part and a fluorophore part. The receptor is mainly responsible for recognizing the analyte and the fluorophore converts the recognition events into optical signals. Fluorescent chemical sensors for Hg(II) have been synthesized based on various fluorophores, such as dansyl, pyrene, rhodamine, fluorescein, anthracene, naphthyl, and nile blue [62].

In the development of such optical sensors, El-Safty et al. recently introduced chromophore probes onto large cage mesostructures that had had their surfaces modified with functionalizing agents [59a]. Although these optical cage sensors exhibit improved detection of trace amounts of specific metal ions down to subnanomolar concentrations [59a], the field of optical sensor technology still demands a simple yet general approach to the design of reliable optical chemical sensor–based nanostructure materials for the fast, sensitive, selective, and inexpensive recognition of a broad range of toxic metal ions [59a, 91].

Design of Optical Sensor Probes

The most common method used for designing optical chemical sensor arrays is the grafting technique [60, 49a, 92]. The remaining challenges in these grafting techniques include sophisticated immobilization procedures involving receptor probes and the retention of specific activity and flexibility of the electron acceptor/donor strength of the immobilized probes. This mobility can lead to easy generation and transduction of optical color signals in response to probe–metal binding events. The key to designing optical nanosensors is the ability to immobilize chromophore probes with different functional characteristics (i.e., hydrophobic and hydrophilic), easy accessibility and intrinsic mobility onto these 3D highly ordered mesoporous (HOM) pore surfaces. These qualities indicate a facile and reliable synthesis (*Scheme: 1*). For all mesoporous silica structures, the surfactant and the incorporated hydrocarbons were removed by calcination at 450 °C (1 h under a nitrogen atmosphere and 6 h in oxygen). Ordered and disordered structures, shape- and size-controlled cages, and cylindrical pore geometries are all possible by controlling the type of surfactant phases. In general, the successful fabrication of HOM monoliths is proof of the effective use of the direct-templating method for the synthesis of various mesoscopically ordered silicas.

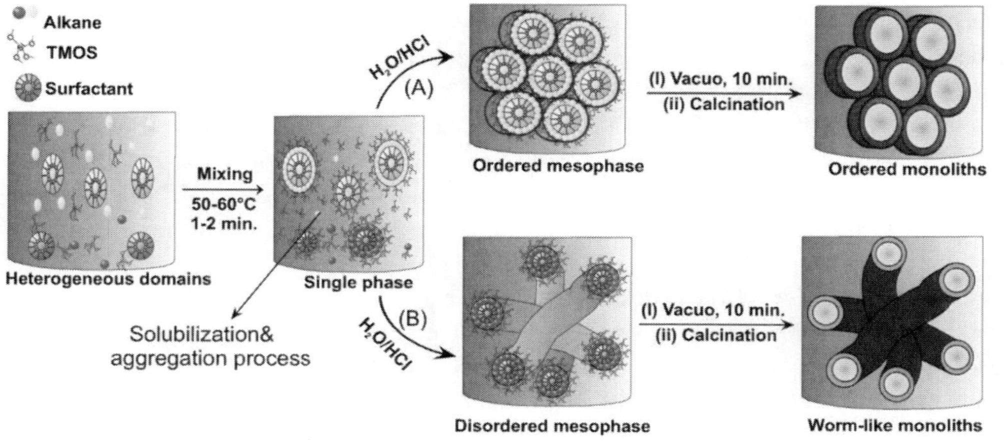

Scheme 1. A simple and facile fabrication of transparent HOM monoliths through instant direct-templating of micro-emulsion liquid crystalline phases formed by addition of alkanes to the copolymer L64/ TMOS composition mixture. Adapted from [50d], (Copyright: © 2008 American Chemical Society).

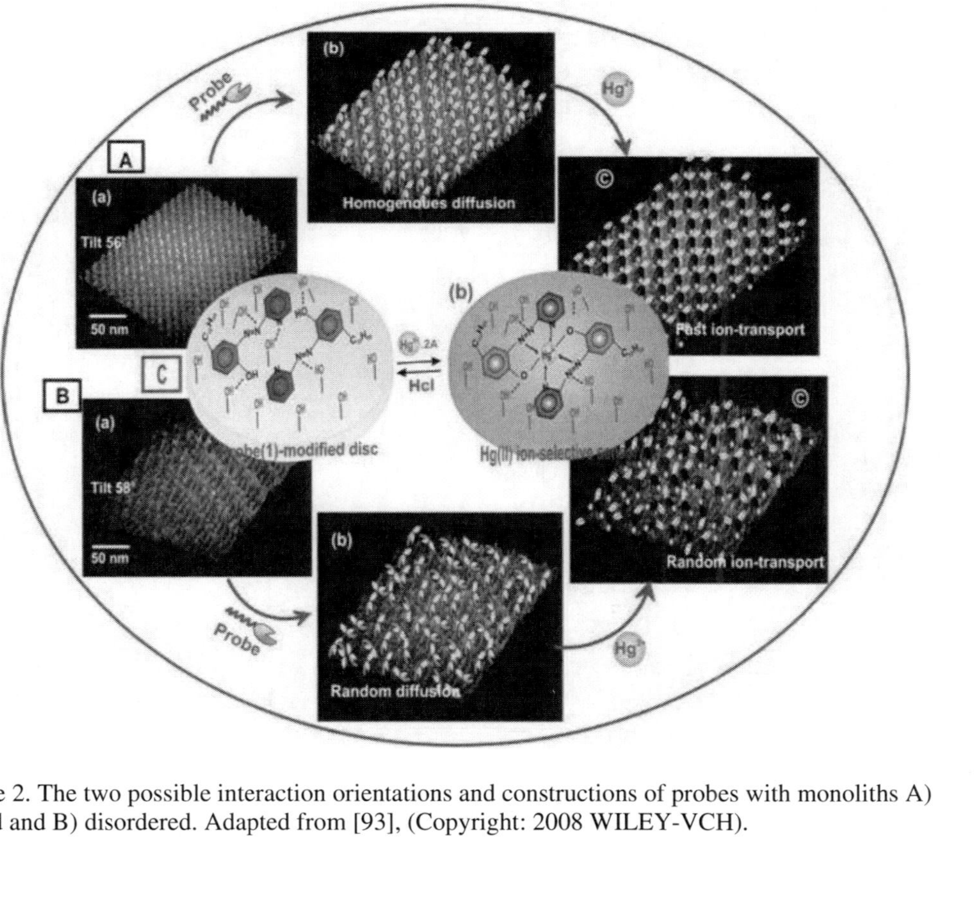

Scheme 2. The two possible interaction orientations and constructions of probes with monoliths A) ordered and B) disordered. Adapted from [93], (Copyright: 2008 WILEY-VCH).

The design of colorimetric nanosensor membrane strips was based on the simple physical adsorption of chemically responsive azo dyes, followed by strong dye-metal ion analyte interactions (Scheme 1). In such a design, the effect of the physical properties and structural ordering of 3D bicontinuous cubic Fd3m geometry on optical sensor fabrication was evident [Scheme 1(B-C)]. In addition, the high dimensional stability, rigidity, and diffusivity of membrane disc platforms were substantially affected by ion-sensing strips during the visual detection of ultratrace Hg(II) ions (Schemes 2 and 3). A controlled design of ordered and wormhole strips occurred through sufficiently physisorbed interactions (Scheme 2C) between the silanol groups of the silica membrane discs and the heteroatoms of the azo-receptor molecules, leading to high loading capacity and accessibility of probes, as previously reported with silica nanocarriers [59].

El-Safty et al. [59f] have designed optical nanosensors based on the building-blocks approach, which neither changed the 3D pore geometries nor led to significant collapse in the mesoscopically oriented ordered frameworks (Scheme 4). Scheme 4 shows the general, simple synthesis of optical nanosensors based on the building-blocks approach. For instance, the polarity of silica surface matrices of 3D HOM monoliths was first fine-tuned by the dense dispersion of cationic surfactants, such as dilauryldimethyl ammonium bromide (DDAB) (Scheme 4A). However, the bonded physisorption phases of DDAB on the silica surfaces mainly occur via short-range electrostatic interactions (i.e., van der Waals and H-bonding interactions), leading to the formation of positively charged silica surfaces that appear as an "island" (Scheme 4B). The successful dispersion of large amounts of DDAB led to increased polarity of the surface carriers, thereby enhancing their ability to attract and adsorb probe molecules compared with surfaces that were grafted using an expensive silane- or thiol-coupling agent (Scheme 4C).[59a, 91]

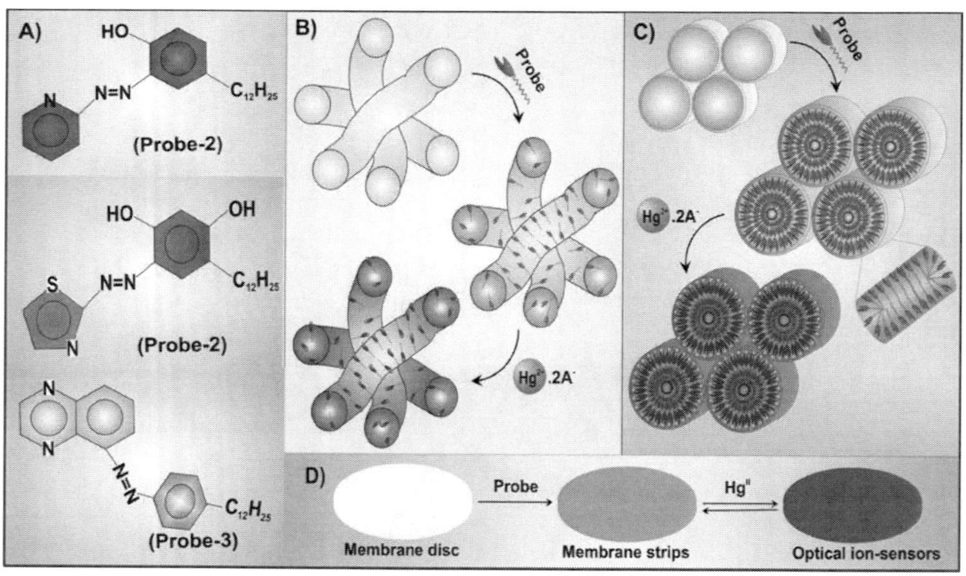

Scheme 3. A) Structure of chemically-synthesized azo chromophore diffusion of hydrophobic probes 1–3 onto B) worm-like and C) ordered pore structures during the direct immobilization of the probe onto membrane-disc arrays and D) systematic designs of Hg(II) ion-sensor membrane-disc strips with their possible distribution. Adapted from [93], (Copyright: 2008 WILEY-VCH).

El-Safty et al. [58b] demonstrated another design for optical nanosensors based on disc-like monoliths by using two-step inclusion procedures. Firstly usinggraft-controlled surface modification of the cage disc-like monoliths with N-Trimethoxysilylpropyl-N,N,N-trimethylammonium chloride (TMAC), and then by immobilization of the chromophore, such as tetraphenylporphine tetrasulfonic acid (TPPS), into disc-like monolithic HOM. This ledto the creation of optical chemical nanosensors that have efficient sensing functionalities (Scheme 5) in terms of sensitivity, selectivity, and response time for Hg(II) ions. Considering the potential leaching of hydrophilic (TPPS) chromophores through the washing cycle, the TPPS probe could not be directly embedded into the silica surface matrices without tuning the surface polarity. The introduction of TPPS probe molecules, successfully resulting in strong ionic interactions between the TPPS chromophore (containing sulfonate acid groups) and the carriers (charged by the functionalized ammonium ion (Scheme 5)) leading to a stable sensor during the sensing assays of Hg(II) ions.

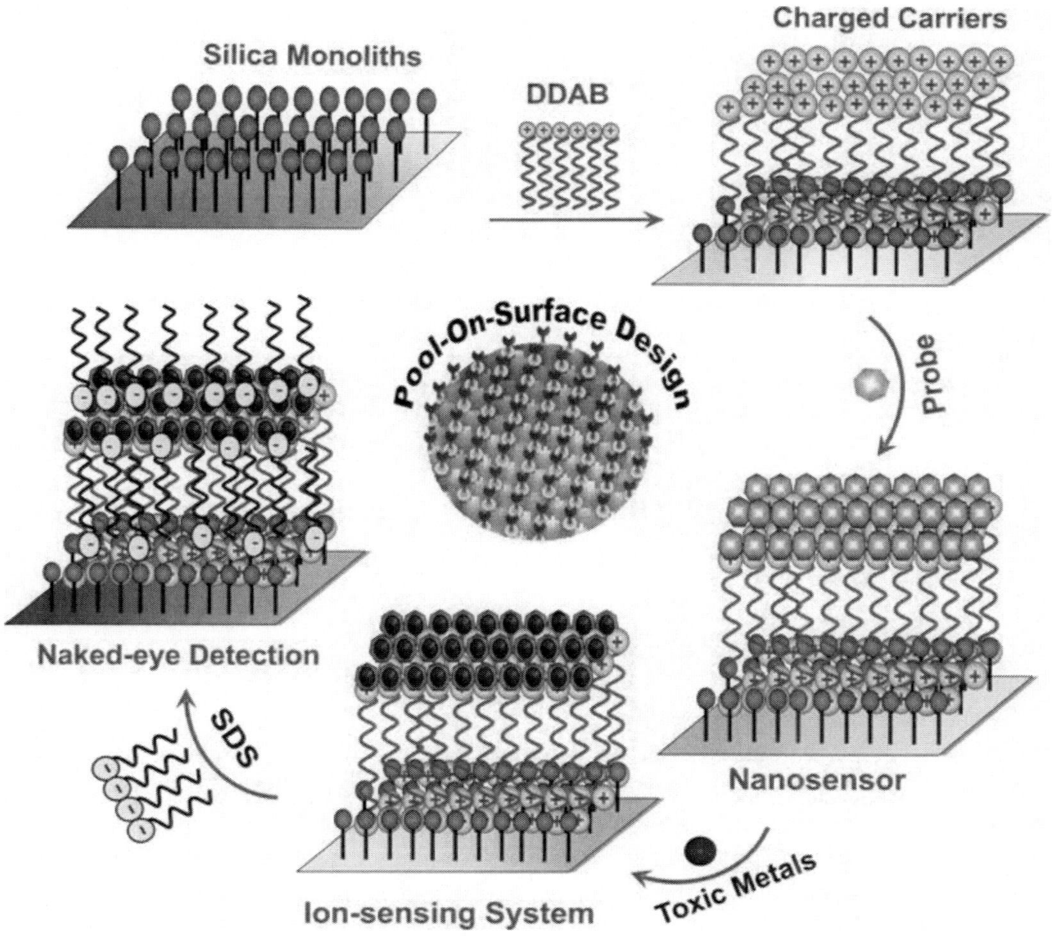

Scheme 4. Representative scheme of the sequential construction of optical nanosensors by doping hydrophobic (DPC and DZ) and hydrophilic (PR and TPPS) chromophores that have different molecular sizes and characteristics onto 3D HOM pore surface "carriers" through the building-blocks approach (A-C). Model design (D) represents the pool-like sensing recognition of Cr(VI) and Pb(II) ions . Adapted from [59e], (Copyright: 2008 WILEY-VCH).

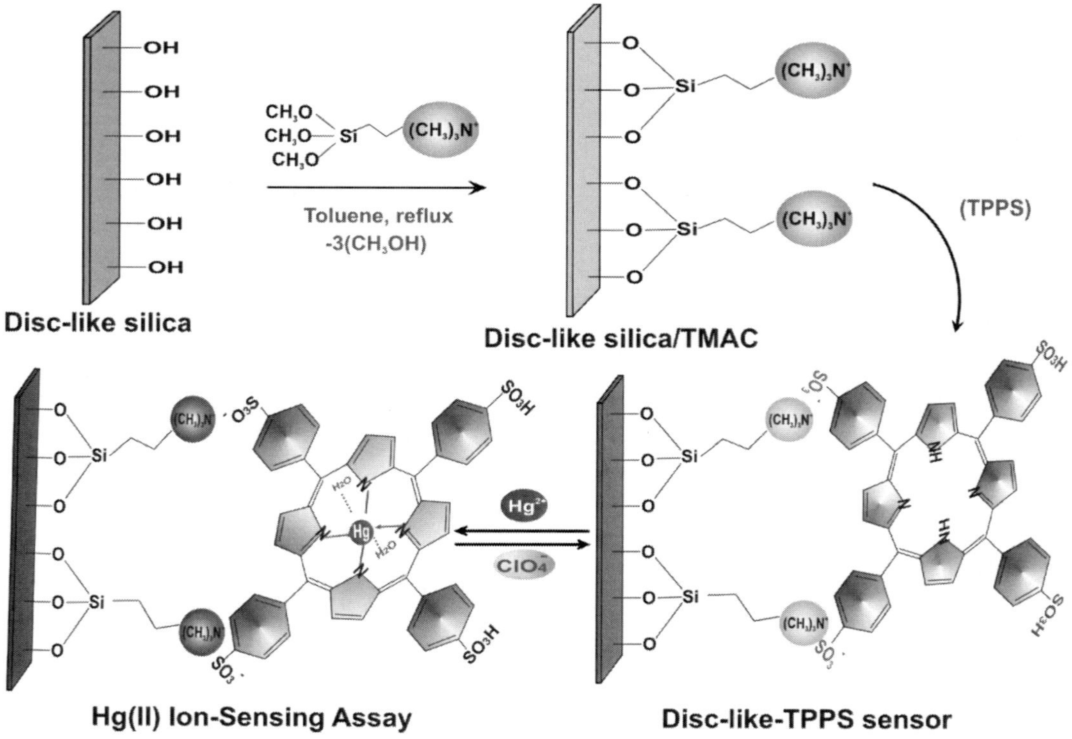

Scheme 5. Disk-like sensor for Hg(II) ion-sensing assay. The grafting technique controls the sensor design by the immobilization of two organic moieties, TMAC and TPPS, into the mesopore disc-like monoliths. The reversibility of the [Hg–TPPS]n+ complex occurred by repeated use of 0.01 M ClO4- as stripping agent without any leaching of the TPPS probe from the HOM pore surface. Adapted from [58b], (Copyright: Springer Science+Business Media, LLC 2009).

ADVANTAGES OF THE HG(II) ION-OPTICAL SENSORS

El-Safty et al. [59a] reported that the synthesized cubic Fm3m nanostructured cage-monolith was used as an optical sensor for the highly sensitive and selective detection of toxic metal ions, including Hg(II), at subnanomolar concentrations and on the timescale of a few minutes.

To the best of our knowledge, this is the first visual detection method for detecting toxic ions at low-level concentrations based on the color change of probe sensors (Scheme 6). Such cage monoliths with spherical nanosized cavities show evidence of a high adsorption capacity for probe molecules, efficient transport of toxic species through a much more direct and easier diffusion to the network sites. They also showa high degree of detection functionality for these toxic ions without loss of cage character. The potential use of these cage materials as colorimetric sensors for the naked-eye detection of metal ions has generated considerable evidence for the accessibility and the ability to retain probe molecules in the 3D cage cavity (Scheme 6).

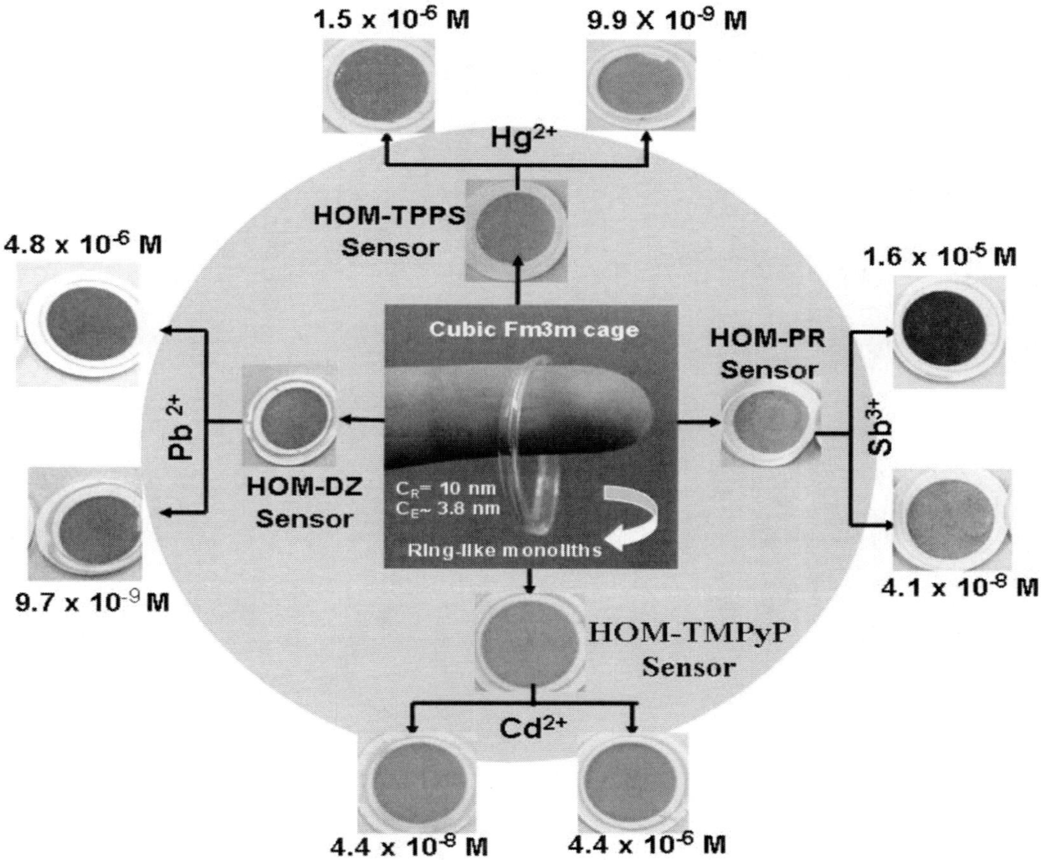

Scheme 6. A schematic representation of the colorimetric detection systems of optically monolithic cage sensors, and their respective color changes over a tunable and wide detection range of toxic analyte ions. Adapted from [59a], (Copyright: 2006 Wiley-VCH).

1. Simple Workability for Ultratrace Detection

El-Safty et al. [93] synthesized different azo chromophores with long hydrophobic tails based on thiazolylazo-, phenylazo-, or pyridylazo compounds, such as 4-n-dodecyl-6-(2-pyridylazo)-phenol (1), 4-n-dodecyl-6-(2-thiazolylazo)-resorcinol (2), and 8-(4-n-dodecyl-phenylazo)- quinoxaline (3) molecules, which eliminated these problems, thus providing an eco-friendly sensing method for the visual detection of Hg (II) ion in the aqueous phase *Fig. (14)*. The amphiphilic azo probes 1, 2, and 3 transform the HgII ion-sensing system into a smart, stable assembly without introducing organics. El-Safty et al. [93] mentioned that optically visible and spectrally intense transitions were observed when the aqueous medium of the Hg(II) analyte and probes 1, 2, and 3 were maintained at pH 7, 8, and 6.2, respectively, *Fig. (14A-C)*. The change in the probe spectra with the formation of $[Hg\text{-}(probe)]^{n+}$ complexes indicates a perturbation of the π-electron distribution, which leads to decreased energy in the π–π^* transition.

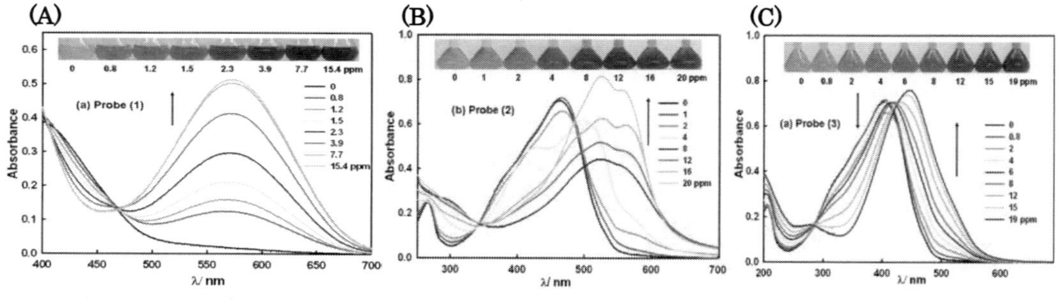

Figure 14. Changes in UV-Vis absorption spectra and colorimetric response profile (insert) of 15 mM probe solutions (1–3) in ethanol upon titration with standardized concentrations of Hg(II) ions ,at 25 °C. A constant reaction time of 15 min was maintained for stable color development in all solutions, using a 15 ml volume. Adapted from [93], (Copyright: 2008 WILEY-VCH).

2. pH-Dependence for Sensitive Signaling Response

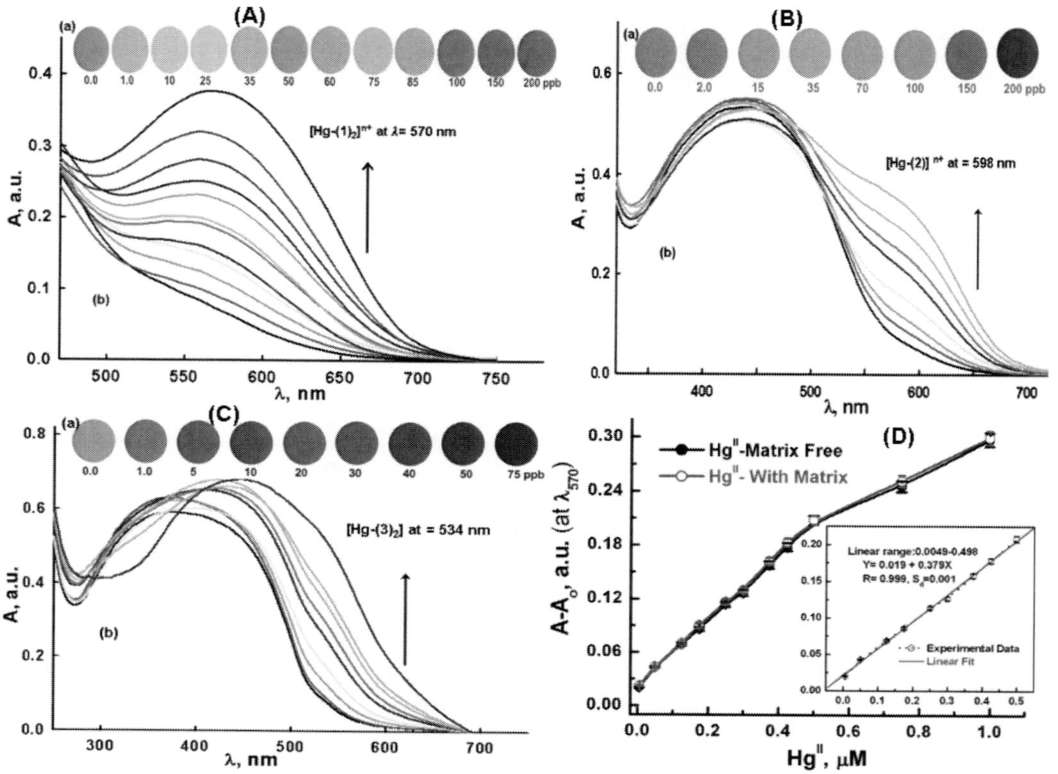

Figure 15. Concentration-proportionate color-transition map and (a) reflectance spectra (b) observed for Hg(II) ion-sensor membrane strips A) 1, B) 2, and C) 3. D) Calibration plot for sensor strip 1 of the spectral absorbance measured at λ=570 nm with different Hg(II) concentrations. Adapted from [93], (Copyright: 2008 WILEY-VCH).

The Hg(II) ion-sensing system with sensor strips 1, 2, and 3 was studied over a wide range of solution pH. The results indicate that membrane strips 1, 2, and 3 have high functionality in terms of selectivity and sensitivity for mercury ions at pH 7, 8, and 6.2, respectively. The inverse correlation between signal response and the thickness of the sensor membrane strips was studied using probe assemblies 1, 2, and 3. The notable ion spectral responses of these sensor strips ensure excellent optical isolation for ultratrace concentrations of Hg(II) analytes, with bright color transitions at the same frequency as the human eye *Fig. (15)*. The nanoscale membrane-disc strips showed phenomenal behavior in terms of selectivity and sensitivity by forming stable chelate complexes with Hg(II) ions. The quantification of nanosensor strips 1–3 for sensing Hg(II) ions over a wide range of concentrations was monitored through UV-Vis spectroscopy. The reflectance spectra of nanosensor membranes 1, 2, and 3 exhibited a bathochromic shift from λ max = 445, 405, and 370 nm to low-energy sidebands at 570, 598, and 534 nm, respectively, with addition of a wide range of Hg(II) ion concentrations (4.0×10^{-9} to 1.0×10^{-6} M).

3. Ion-Selectivity System for Stripping of Hg (II) Ion among Competitive Matrices

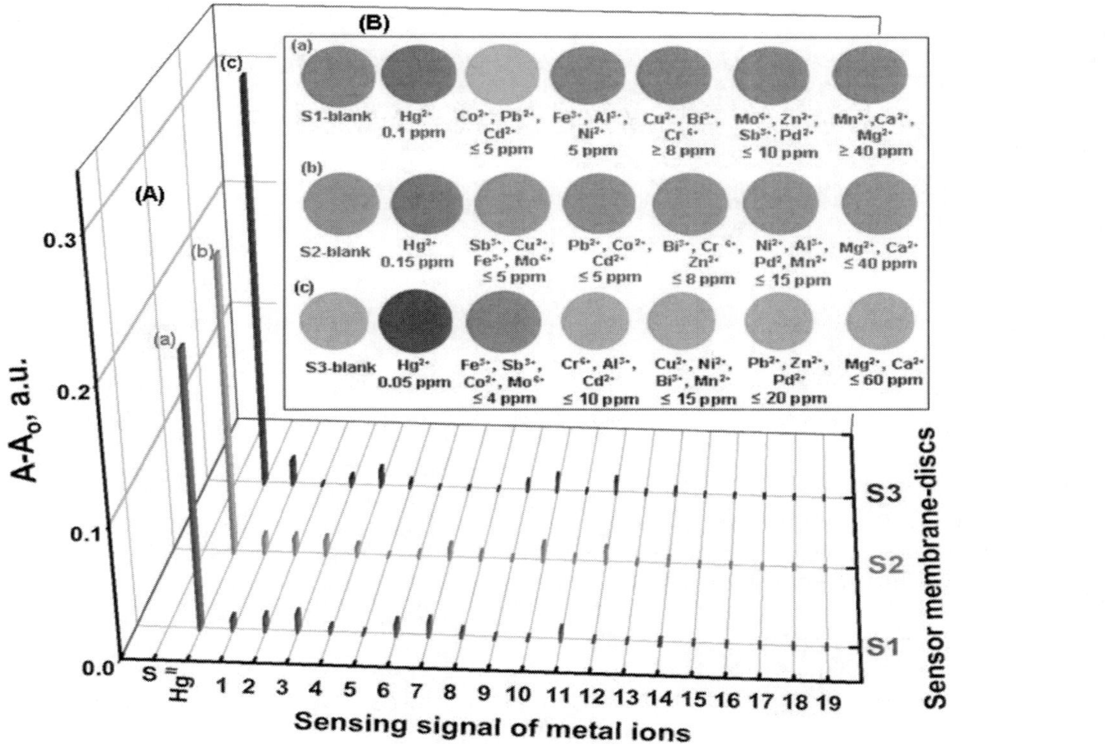

Figure 16. (A) Sensing signal responses and (B) (a-c) color profiles for sensor membrane strips 1, 2, and 3 in the presence of Hg(II) at 0.1, 0.15, and 0.05 ppm concentrations, respectively; and in the presence of other metal ions. Graph (A) represents the difference in reflectance spectra of sensor discs 1–3 and blank (S) at (a) λ=570, (b) λ=598, and (c) λ=534 nm, respectively. The selectivity profile of Hg(II) ions was obtained by ordered sensor membrane-disc strips 1, 2 and 3 at pH values of 7, 8, and 6.2 after 15 min of contact time. Adapted from [93], (Copyright: 2008 WILEY-VCH).

One of the key issues in membrane sensor workability is the Hg(II) ion-selectivity system. Initially, when a series of transition metals and Group I and Group II metal ions and anionic interferants were added to membrane sensors 1–3 at high concentrations (range: 5–7000 ppm) and at varying pH values (7, 8, and 6.2, respectively), no significant changes in colorimetric spectra and visible color patterns of the strips were observed *Figs. (16A and 16B)*. However, after adding the Hg(II) analyte, the membrane sensor strips showed significant changes in visual color and signal intensity, indicating specificity of the strips for Hg(II) ions at ultratrace concentrations.

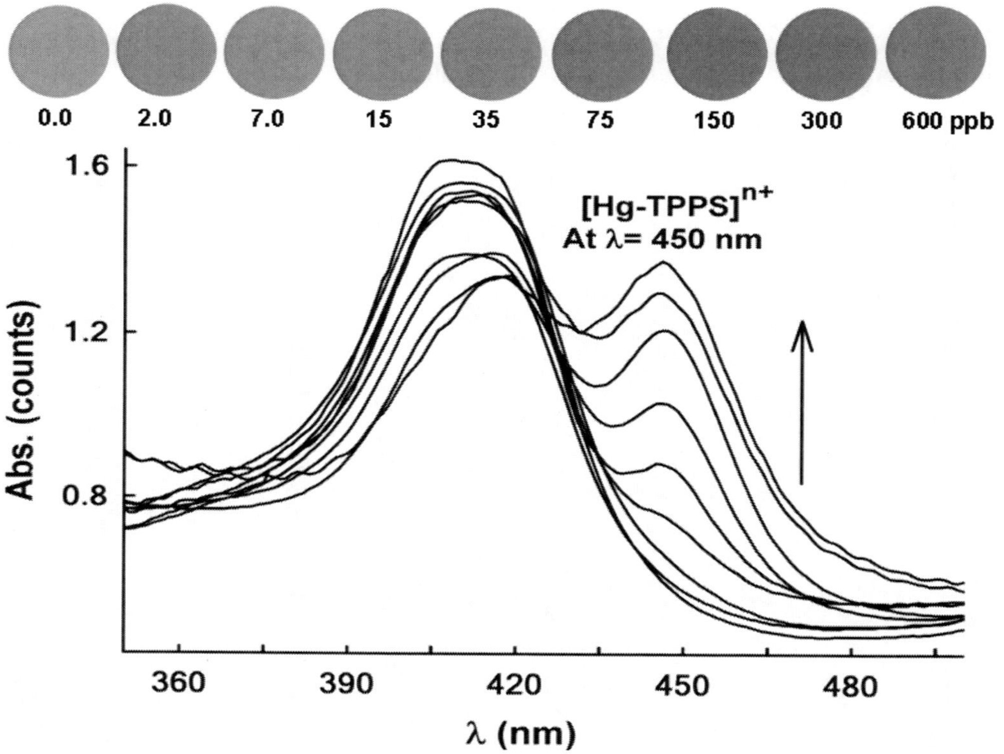

Figure 17. Representative concentration-dependent changes in disc-TPPS sensors in the color map and in UV-Vis reflection spectra of [Hg–TPPS]n+ complex formation at λ = 450 nm, with the addition of various concentrations of Hg(II) ions. Adapted from [58b], (Copyright. Springer Science+Business Media, LLC 2009).

One of the practical uses of disc-like sensors is their ability to show high selectivity for Hg(II) ions when combined with different ions (especially interfering components) under experimentally controlled conditions [58b]. To investigate the effect of extraneous ions in the simultaneous selective detection of Hg(II) ions, particularly at low concentrations, controlled sensing experiments were conducted in which competing cations and anions were added at high concentrations (range: 2–6000 ppm) to silica-disc-TPPS sensors *Fig. (17)*. In *Fig. (18)*, the results show that no significant changes in reflectance spectra at λ= 597 nm and no visible color patterns from the sensors were observed. In fact, the selectivity for Hg(II) ions of disc-like sensors, over active multi-component ions and species, indicated high levels of thermodynamic binding of Hg(II) ion to N-chelate TPPS ligands and fast Hg-to-TPPS binding kinetics at an optimized pH of 9 *Fig. (18)*.

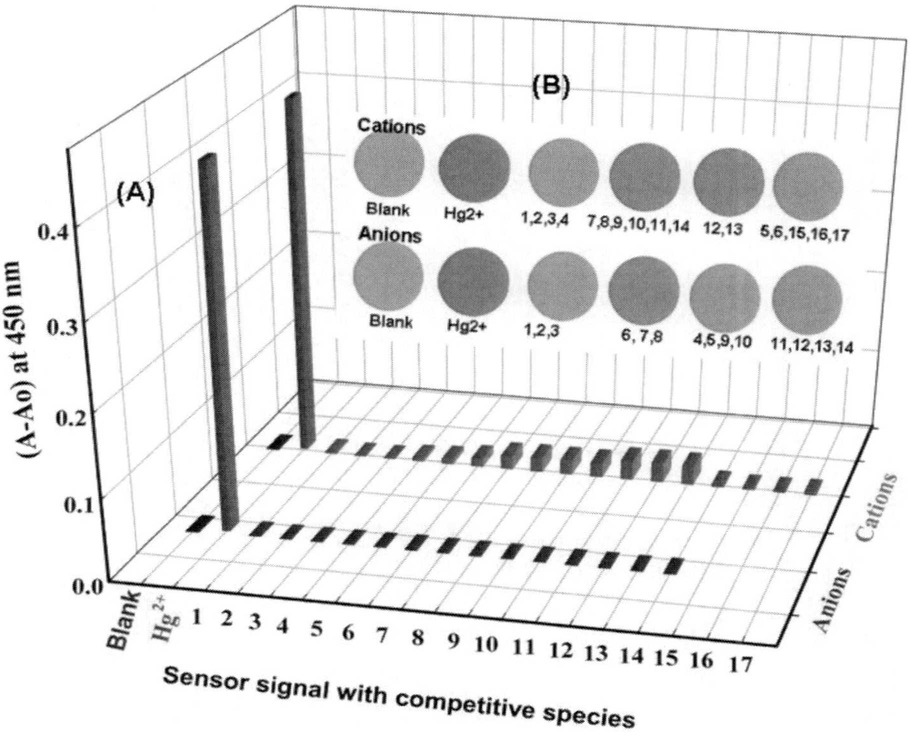

Figure 18. Representative selectivity profiles of disc-sensors for 0.075 ppm Hg(II) ions with different cations, surfactants, and anions. Adapted from [58b], (Copyright: Springer Science+Business Media, LLC 2009).

4. Reducibility and Reversibility of Optical Sensors

A major advantage of monoliths is their reducibility and reversibility after multiple regeneration and reuse cycles with adsorbents. In this waymonoliths can extend the control of adsorption and removal of toxic metal ions from aqueous solution even after several cycles. Monoliths that retain high Hg(II) adsorption efficiency have been introduced after a number of reuse cycles, i.e., “dead-end adsorption” [as shown in ***Fig. (19)***] [45b, 94]. Both sets of parameters were extensively studied by repeated highly sensitive-detection regeneration cycles. In these treatment procedures, the use of specific concentrations of stripping agents, such as EDTA or Cl^-, SCN^-, and CN^- anions, made the removal of metal ions (decomplexation) possible. Our experiments show that more than six cycles of Hg(II) analyte removal results in a 90% reduction in observed sensitivity ***Fig. (19)***. The decrease in sensitvity is due to the influential effects of stripping agents on sensing functions over several cycles. [59a, 95]

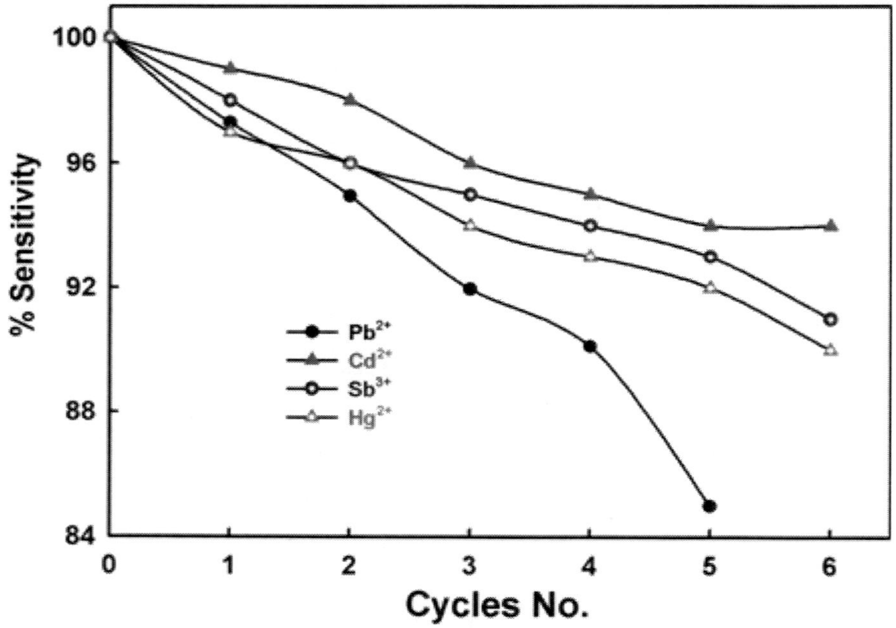

Figure 19. Sensitivity of cage sensors 1 (a), 2 (b), 3 (c), and 4 (d) after complexation-regeneration cycles of analyte ions at 0.3 ppm. Cycle no. = number of cycles. Adapted from [59a], (Copyright: 2006 Wiley-VCH).

Conclusion and Outlook

Mercury has been recognized to have serious effects on human health and the environment. This recognition has led to numerous studies that deal with the properties of various forms of mercury, as well as the development of methods for Hg detection and removal. Conventional technologies for the treatment of aqueous mercury include precipitation, coagulation, reduction, membrane separation, ion exchange, and adsorption. Among these technologies, adsorption has been widely studied because it is easy to operate and cost-effective. The field of optical chemical sensors has been a growing research area over the last three decades. The key component in this sensing method are optical sensors that efficiently detect Hg(II) ions in real-time with high sensitivity and selectivity. Nanosensor design techniques are identified as the best current process for detecting trace amounts of Hg(II). The key to designing optical nanosensors is the ability to immobilize chromophore probes with different functional characteristics, such as density, accessibility, and intrinsic mobility, onto the pore surfaces of these 3D mesoporous cage monoliths. This yields an easy and reliable synthesis process. A critical assessment of the potential nanosensor designs in the areas of environmental detection of Hg(II) analytes has been delivered in this chapter. The practical design of an optical sensor based on mesoporous silica shows evidence of controlled assessment processes and involves the evaluation of the intrinsic properties of adsorption (e.g., color change, long-term stability, adsorption efficiency, and reusability).

The development of sensing processes entails invention, optimization, and commercialization. This remains a significant challenge in materials science and nanotechnology because, the combination of these properties is difficult to achieve with chemically designed sensors, even though we have made progress with such an endeavor. The desire for simple and smart detection, as well as eco-friendly solid sensors is still limited by few technological challenges, which in recent years have been found to surround the production of relevant solid materials. Thus, research efforts are currently focused on making nanosensors with probes that can ultimately be used in basic laboratory assays and in portable devices for use inhousehold and commercial devices. Even though commercial returns from environmental optical sensors are substantially less than from household applications, current research effort is targeted at the application of optical sensors for detecting Hg(II) and other environmental hazards in relevant environmental media.

REFERENCES

[1] (a) Yang, H. Zhou, Z. Huang, K. Yu, M. Li, F. Yi, T. Huang, C. (2007) "Multisignaling Optical-Electrochemical Sensor for Hg2+ Based on a Rhodamine Derivative with a Ferrocene Unit" *Org. Lett.* 9(23): 4729-4732 b) Von Burg, R. (1995) "Inorganic mercury" *Journal of Applied Toxicology,* 15: 483-493. DOI: 10.1002/jat.2550150610. c) Moreno, F.N.; Anderson, C.W.N.; Stewart, R.B.; Robinson, B.H. (2005). "Mercury volatilisation and phytoextraction from base-metal mine tailings" *Environmental Pollution* 136: 341–352. Doi:10.1016/j.envpol.2004.11.020. f) Zenga, X.; Donga, L.; Wua, C.; Mua, L.; Xuea, S-F.; Tao, Z. (2009). "Highly sensitive chemosensor for Cu(II) and Hg(II) based on the tripodal rhodamine receptor" *Sensors and Actuators B* 141(2): 506–510. Doi:10.1016/j.snb.2009.07.013. e) Wang, C.; Tao, S.; Wei, W.; Meng, C.; Liu, F.; Han, M. (2010). "Multifunctional mesoporous material for detection, adsorption and removal of Hg2+ in aqueous solution" *Journal of Materials Chemistry,* 20: 4635–4641. DOI**:** 10.1039/C000315H. f) Nascimento, A.M.A.; Chartone-Souza, E. (2003). "Operon mer: Bacterial resistance to mercury and potential for bioremediation of contaminated environments" *Genetic and Molecular Research.* 2(1): 92-101. g) Shokrollahi, A.; Ghaedi, M.; Shamsipur, M. (2009). "Highly selective transport of mercury(II) ion through a bulk liquid membrane" *Química Nova* 32(1):153-157 ISSN 0100-4042

[2] Pacyna, E.G.; Pacyna, J.M.; Pirrone, M. (2001). "European emissions of atmospheric mercury from anthropogenic sources in 1995. *Atmospheric. Environment* 35: 2987–2996.

[3] (a) Schroeder, W.H.; Munthe, J. (1998). "Atmospheric Mercury-An Overview" *Atmospheric Environment*. 32(5): 809–22. (b) Mason, R.P.; Sheu, G-R.(2002). "Role of the ocean in the global mercury cycle" *Global Biogeochemical Cycles 16: 41093* (c) Goodsite, M.E.; Plane, J.M.C.; Skov, H. (2004). "A theoretical study of the oxidation of Hg^0 to $HgBr_2$ in the troposphere" *Environmental Science and Technology*. 38(6): 1772–1776.

[4] Boening, D.W. (2000). "Ecological effects, transport, and fate of mercury: a general review" *Chemosphere* 40(12):1335-1351.

[5] Hylander, L.D.; Goodsite, M.E. (2006). "Environmental costs of mercury pollution" *Science of the Total Environment* 368 352–370. doi:10.1016/j.scitotenv.2005.11.029.

[6] (a) Benounis, M.; Jaffrezic-Renault, N.; Halouani, H.; Lamartine, R.; Dumazet-Bonnamour, I. (2006). "Detection of heavy metals by an optical fiber sensor with a sensitive cladding including a new chromogenic calix[4]arene molecule" *Materials Science and Engineering, C: Biomimetic and Supramolecular Systems* 26(2-3): 364-368. (b) Harris, H.H.; Pickering, I.; George, G.N. "The Chemical Form of Mercury in Fish" *Science* 301(5637):1203. DOI: 10.1126/science.1085941 (c) Tag, K.; Riedel, K.; Bauer, H.J.; Hanke, G.; Baronian, K.H.R.; Kunze, G. (2007). "Amperometric detection of Cu^{2+} by yeast biosensors using flow injection analysis (FIA)" *Sensors and Actuators B* 122(2):403–409. doi:10.1016/j.snb.2006.06.007.

[7] (a) Agency for Toxic Substances and Disease Registry (ATSDR). (March 1999). *"Toxicological Profile for Mercury"* U.S. Department of Health and Human Services Public Health Service. (b) WHO (1990) *Environmental Health Criteria* 101, Methyl Mercury, Geneva, World Health Organization 68.

[8] Clarkson, T.W.; Magos, L. (2006). "The toxicology of mercury and its chemical compounds" *Critical Reviews in Toxicology* 36(8):609-662. 10.1080/ 10408440 60084561.

[9] Scheuhammer, A.M.; Meyer, M.W.; Sanheinrich, M.B.; Murray, M.W. (2007). "Effects of environmental methylmercury on the health of wild birds, mammals, and fish" AMBIO: *A Journal of the Human Environment* 36(1):12-19. doi: *http://dx.doi.org/10.1579/0044-7447(2007)36[12:EOEMOT]2.0.CO;2.*

[10] (a) Liyuan, C. (2002). "Study of gelatitinous supports for immobilizing inactivated cells of Rhizopus oligosporus to prepare biosorbent for lead ion" *International journal of Environmental Studies.* 5: 1-6. b) Ahalya, N.; Ramachandra, T.V.; Kanamadi, R.D. (2003). "Biosorption of Heavy Metals". *Journal of Chemical and Environmental Research.* 7:71-78. c) Aguado, J. Arsuaga, J.M. Arencibia, A. (2005)."Adsorption of Aqueous Mercury (II) on Propylthiol-Funcionalized Mesoporous Silica Obtained by Co-condensation" *Industrial and Engineering Chemistry Research*, 44: 3665-3671. DOI: .

[11] (a) Nam, J.-M.; Thaxton, C.S.; Mirkin, C.A. (2003). "Nanoparticle-based bio-bar codes for the ultrasensitive detection of proteins" *Science*, 301:1884. DOI: 10.1126/science.1088755. b) Elghanian, R.; Storhoff, J.J.; Mucic, R.C.; Letsinger, R.L.; Mirkin, C.A. (1997). "Selective Colorimetric Detection of Polynucleotides Based on the Distance-Dependent Optical Properties of Gold Nanoparticles" *Science*, 277:1078-1081. DOI: 10.1126/science.277.5329.1078. c) Wolfbeis, O.S. (2005). "Materials for fluorescence-based optical chemical sensors" *Journal of Materials Chemistry* 15: 2657-2669. DOI: 10.1039/B501536G. d) El-Safty, S.A.; Ismail, A.A.; Matsunaga, H.; Mizukami, F. (2007). "Optical nanosensor Design with Uniform Pore Geometry and Large Particle Morphology" *Chemistry - A European Journal* 13, 9245-9255.

[12] Tseng, C.M.; Diego, A.D.; Martin, F.M.; Amouroux, D.; Donard, O.F.X. (1997). "Rapid determination of inorganic mercury and methylmercury in biological reference materials by hydride generation, cryofocusing, atomic absorption spectrometry after open focused microwave-assisted alkaline digestion" *Journal of Analytical Atomic Spectrometry* 12:743-750. DOI: 10.1039/A700956I.

[13] (a) Li, Y.F.; Chen, C.Y.; Li, B.; Sun, J.; Wang, J.X.; Gao, Y.X.; Zhao, Y.L.; Chai, Z.F. (2007). "Elimination efficiency of different reagents for the memory effect of mercury using ICP-MS" *Journal of Analytical Atomic Spectrometry* 21 (2006) 94–96. DOI: 10.1039/B511367A. (b) Vallant, B.; Kadnar, R.; Goessler, W. (2007). "Development of a new HPLC method for the determination of inorganic and methylmercury in biological samples with ICP MS detection" *J. Anal. At. Spectrom.* 22: 322-325. DOI: 10.1039/B615463H .

[14] Chen, Y.W.; Tong, J.; D'Ulivo, A.; Belzile, N. (2002). "Determination of mercury by continuous flow cold vapor atomic fluorescence spectrometry using micromolar concentration of sodium tetrahydroborate as reductant solution" *Analyst* 127:1541-1546. DOI: 10.1039/B207896A.

[15] Anthemidis, A.N.; Zachariadis, G.A.; Michos, C.E.; Stratis, J.A. (2004). "Time-based on-line preconcentration cold vapour generation procedure for ultra-trace mercury determination with inductively coupled plasma atomic emission spectrometry" *Analytical and Bioanalytical Chemistry*" 379(5-6): 764-769. DOI: 10.1007/s00216-004-2593-2.

[16] (a) Pérez-Marín, L.; Otazo-Sánchez, E.; Macedo-Miranda, G.; Avila-Pérez, P.; Alonso Chamaro, J.; López-Valdivia, H. (2000). "Mercury(II) ion-selective electrode. Study of 1,3-diphenylthiourea as ionophore" *Analyst* 125:1787-1790. DOI: 10.1039/b003502p. (b) Yantasee, W.; Lin, Y.H.; Zemanian, T.S.; Fryxell, G.E. (2003) "Voltammetric detection of lead(II) and mercury(II) using a carbon paste electrode modified with thiol self-assembled monolayer on mesoporous silica (SAMMS)" *Analyst* 128:467-472. DOI: 10.1039/B300467H. (c) Caballero, A.; Lloveras, V.; Curiel, D.; Tárrage, A.; Espinosa, A.; Garcia, R.; Vidal-Gancedo, J.; Rovira, C.; Wurst, K.; Molina, P.; Veciana, J. (2007). "Electroactive Thiazole Derivatives Capped with Ferrocenyl Units Showing Charge-Transfer Transition and Selective Ion-Sensing Properties: A Combined Experimental and Theoretical Study" *Inorganic Chemistry*. 46 (3):825–838. DOI: 10.1021/ic061803b.

[17] Bennun, L.; Gomez, J. (1997). "Determination of mercury by total-reflection X-ray fluorescence using amalgamation with gold" *Spectrochimica Acta - Part B Atomic Spectroscopy* 52(8):1195-1200.

[18] Ugo, P.; Moretto, L.M.; Bertoncello, P.; Wang, J. (1998)."Determination of Trace Mercury in Saltwaters at Screen-Printed Electrodes Modified with Sumichelate Q10R" *Electroanalysis* 10(15):1017–1021. DOI: 10.1002/(SICI)1521-4109(199810)10:15<1017::AID-ELAN1017> 3.0.CO;2-D.

[19] Tan, J.; Yan, X.P. (2008). "2,1,3-Benzoxadiazole-based selective chromogenic chemosensor for rapid naked-eye detection of Hg^{2+} and Cu^{2+}" *Talanta* 76(1):9-14 DOI: 10.1016/j.talanta.2008.01.056.

[20] (a), Namasivayam, C.; Kadirvelu, K. (1999) "Uptake of mercury (II) from wastewater by activated carbon from an unwanted agricultural solid by-product: coirpith" *Carbon* 37(1): 79-84. DOI: 10.1016/S0008-6223(98)00189-4. (b) Guerra, D.L.; Airoldi, C.; Viana, R.R. (2008). "Performance of modified montmorillonite clay in mercury adsorption process and thermodynamic studies" *Inorganic Chemistry Communications* 11: 20–23. doi:10.1016/j.inoche.2007.09.029. (c) Das, S.K.; Das, A.R.; Guha, A.K. (2007) "A Study on the Adsorption Mechanism of Mercury on Aspergillus versicolor Biomass" *Environmental Science and Technology* 41:8281–8287. (d) Vieira, R.S.;

Beppu, M.M. (2006) "Interaction of natural and crosslinked chitosan membranes with Hg(II) ions" *Colloids and Surfaces A: Physicochem. Eng. Aspects* 279: 196–207. (e) Yin, Y.J.; Allen, H.E.; Li, Y.; Huang, C.P.; Sanders, P.F. (1996) "Adsorption of mercury(II) by soil: Effects of pH, chloride, and organic matter" *Journal of Environmental Quality* 25(4): 837-844. ISSN: 00472425. (f) Bonnissel-Gissinger, P.; Alnot, M.; Lickes, J.P.; Ehrhardt, J.J.; Behra, P. (1999). "Modeling the Adsorption of Mercury(Ii) on (Hydr)Oxides Ii: Alpha-Feooh (Goethite) and Amorphous Silica" *Journal of Colloid and Interface Science* 215(2):313-322. (g) Tiffreau, C.; Lutzenkirchen, J.; Behra, P. (1995). "Modeling the Adsorption of Mercury(II) on (Hydr)Oxides .1. Amorphous Iron-Oxide and Alpha-Quartz. *Journal of Colloid and Interface Science* 172(1):82-93.

[21] (a) Budinova, T.; Savova, D.; Petrov, N.; Razvigorova, M.; Minkova, V.; Ciliz, N.; Apak, E.; Ekinci, E. (2003). "Mercury Adsorption by Different Modifications of Furfural Adsorbent" *Industrial and Engineering Chemistry Research* 42 (10):2223–2229. DOI: 10.1021/ie020707p (b) Cox, M.; El-Shafey, E.I.; Pichugin, A.A.; Appleton, Q. (1999). "Preparation and characterisation of a carbon adsorbent from flax shive by dehydration with sulfuric acid" *Journal of Chemical Technology and Biotechnology* 74(11):1019–1029. DOI: 10.1002/(SICI)1097-4660(199911)74:11<1019::AID-JCTB152>3.0.CO;2-N (c) Mohan, D.; Gupta, V.K.; Srivastava S.K.; Chander, S. (2000). "Kinetics of mercury adsorption from wastewater using activated carbon derived from fertilizer waste" *Colloids and Surfaces A: Physicochemical and Engineering Aspects* 177:(169- 181).

[22] Anoop Krishnan K.; Anirudhan, T.S. (2002). "Removal of mercury(II) from aqueous solutions and chlor-alkali industry effluent by steam activated and sulphurised activated carbons prepared from bagasse pith: kinetics and equilibrium studies" *Journal of Hazardous Materials*, 92(2): 161-183. doi:10.1016/S0304-3894(02)00014-6

[23] Cox, M.; El-Shafey, E.; Pichugin, A.; Appleton, Q. (2000). "Removal of mercury(II) from aqueous solution on a carbonaceous sorbent prepared from flax shive" *Journal of Chemical Technology and Biotechnology* 75(6):427-435. DOI: 10.1002/1097-4660(200006)75:6<427

[24] Manohar, D.; Anoop Krishnan, K.; Anirudhan, T. (2002). "Removal of mercury(II) from aqueous solutions and chlor-alkali industry wastewater using 2-mercaptobenzimidazole-clay" *Water Research* 36: 1609–1619. PII: S0043-1354(01)00362–1.

[25] (a)Godlewska-Zykiewicz, B.; Kozowska, M. (2005). "Solid phase extraction using immobilized yeast Saccharomyces cerevisiae for determination of palladium in road dust" *Analytical Chemical Acta.* 539(1):61-67. b) Fourest, E.; Roux, JC. (1992). "Heavy metal biosorption by fungal mycelial by-products: mechanisms and influence of pH" *Applied Microbiology and Biotechnology* 3:399–403.

[26] (a)Miretzky, P.; Fernandez Cirelli, A. (2009). "Hg(II) Removal from Water by Chitosan and Chitosan Derivatives: A Review" *Journal of Hazardous Materials* 167:10-23 (b) Campos Gavilana, K.; Pestovc, A.V.; Garciab, H.M.; Yatluk, Y. Roussya, J.; Guibala, E. (2009). "Mercury sorption on a thiocarbamoyl derivative of chitosan" *Journal of Hazardous Materials*165(1-3): 415-426. (c) Malkoc, E.; Nuhoglu, Y. (2005). "Investigations of Ni(II) removal from aqueous solutions using tea factory waste" *Journal of Hazardous Materials* 127120-128. (d) Iftikhar, A.R.; Bhatti, H.N.; Hanif,

M.A.; Nadeem, R. (2009). "Kinetic and thermodynamic aspects of Cu(II) and Cr(III) removal from aqueous solutions using rose waste biomass" *Journal of Hazardous Materials* 161:941-947. (e) Lopez, A.; Lazaro, N.; Priego, J.M.; Marques A.M. (2000). "Effect of pH on the biosorption of nickel and other heavy metals by Pseudomonas fluorecens 4F39" *Journal of Industrial Microbiology and Biotechnology* 24: 146-151.

[27] (a) Carro, L.; Anagnostopoulos, V.; Lodeiro, P.; Barriada, J.L.; Herrero, R.; De Vicente, E.S. (2010). "A dynamic proof of mercury elimination from solution through a combined sorption-reduction process" *Bioresource Technology* 101: 8969-8974. (b) Rezaee, A.; Ramavandi, B.; Ganati, F.; Ansari, M.; Solimanian, A. (2006). "Biosorption of mercury by biomass of filamentous algae Spirogyraspecies" *Journal of Biological Sciences* 6: 695-700. (c) Tuzuna, I.; Bayramoglub, G.; Yalcınb, E.; Basarana, G.; Celikb, G.; Arıca, M.Y. (2005). Equilibrium and kinetic studies on biosorption of Hg(II), Cd(II) and Pb(II) ions onto microalgae Chlamydomonas reinhardtii" *Journal of Environmental Management* 77:85-92 (d) Khambhaty, Y.; Mody, K.; Basha, S.; Jha, B. (2008). "Hg(II) removal from aqueous solution by dead fungal biomass of marine Aspergillus niger: Kinetic studies" *Separation Science and Technology* 43(5):1221-1238.

[28] (a)Namasivayam, C.; Periasamy, K. (1993). "Bicarbonatetreated peanut hull carbon for mercury (II) removal from aqueous solution" *Wafer Research* 27(11): 1663-l668 b) Sen, A.K.; De, A.K. (1987). "Adsorption of mercury (II) by coal fly ash" *Wafer Research* 21(8):885-888.

[29] (a)Noh, Y.D.; Komarneni, S. 45(2011). "Mercury(II) exchange by highly charged swelling micas, sodium Engelhard titanosilicate-4, and sodium titanosilicate" *Environmental Science and Technology* 45(16):6954-6960. DOI: 10.1021/es200712r (b) Lopes, C.B.; Otero, M.; Coimbra, J.; Pereira, E.; Rocha, J.; Lin, Z.; Duarte, A. (2007). "Removal of low-concentration Hg^{2+} from natural waters by microporous and layered titanosilicates"; *Microporous and Mesoporous Materials* 103:325-32. (c) Monteagudo, J. Ortiz, M. (2000). "Removal of inorganic mercury from mine waste water by ion exchange" *Journal of Chemical Technology and Biotechnology* 75(9):767–772. DOI: 10.1002/1097-4660(200009)75:9<767

[30] Monteagudo, J.; Duran, A.; Carmona, M.S.; Schwab, R.G.; Higueras, P. (2003). "Elimination of inorganic mercury from waste waters using crandallite type compounds" *Journal of Chemical Technology and Biotechnology78(4)*399–405.

[31] Velizarov, S.; Matos, C.; Oehmen, A.; Serra, S.; Reis, M.; Crespo, J. (2008). "Removal of inorganic charged micropollutants from drinking water supplies by hybrid ion exchange membrane processes" *Desalination* 223:85-90.

[32] (a)Biester, H.; Schuhmacher, P.; Muller, G. (2000). "Effectiveness of mossy tin filters to remove mercury from aqueous solution by Hg(0) reduction and Hg(0) amalgamation" *Water Research* 34 (7):2031-2036. (b) Ku, Y.; Wu, M-H.; Shen, Y-S. (2002). "Mercury removal from aqueous solutions by zinc cementation" *Waste Management*, 22(7)721-726.

[33] Robinson, J.B.; Tuovinen, O.H. (1984). "Mechanisms of microbial resistance and detoxification of mercury and organomercury compounds: physiological, biochemical, and genetic analyses" *Microbiology and Molecular Biology Reviews* 48:95-124.

[34] Gunasekara, A.S.; Donovan, J.A.; Xing. B. (2000). "Ground discarded tires remove naphthalene, toluene, and mercury from water" *Chemosphere*, 41: 1155-1160.

[35] Patterson, J. (1985). Wastewater Treatment Technology, 2nd Ed Ann Arbor, MI: *Ann Arbor Science*, 203-216.

[36] Patterson, J.W. (1985). Wastewater Treatment Technology, 2nd ed., Ann Arbor Science, Ann Arbor, Mi. b) Patterson, J.W. (1992). Toxicity reduction methodologies. In: *Toxicity Reduction, Evaluation, and Control.* Lancaster, PA, Technomic Publishing Co.

[37] Matlock, M.M.; Howerton, B.S.; Atwood, D.A. (2001). "Irreversible precipitation of mercury and lead." *Journal of Hazardous Materials* 84(1):73-82.

[38] Kazem; R.M.; Alireza; S.; Asieh, A. (2011). "Solid phase extraction of ultra traces mercury (II) using octadecyl silica membrane disks modified by 1,3-bis(2-ethoxyphenyl)triazene (EPT) ligand and determination by cold vapor atomic absorption spectrometry" *Journal of Hazardous Materials* 192:1358-1363.

[39] Yan, G.; Viraraghavan, T.(2001). "Heavy metal removal in a Biosorption column by immobilized M. rouxii biomass" *Bioresource Technology*, 78(3): 243-249.

[40] (a)Mohr, G.J.; Demuth, C.; Spichiger-Keller, U.E. (1998). "Application of chromogenic and fluorogenic reactands in the optical sensing of dissolved aliphatic amines" *Analytical Chemistry* 70:3868–3873. b) Mohr, G.J.; Tirelli, N.; Lohse, C.; Spichiger-Keller, U.E. (1998). "Development of chromogenic copolymers for optical detection of amines" *Advanced Materials* 10:1353–1356.

[41] (a)Middelhoek, S.; Noorlag, J. W. (1981-1982). "Three-dimensional representation of input and output transducers" *Sensors and Actuators* 2:29-41. . (b) El-Safty, S.A.; Ismail, A.A.; Shahat, A. (2011). "Optical Supermicrosensor Responses for Simple Recognition and Sensitive Removal of Cu (II) Ion Target" *Talanta* 83:1341–1351.

[42] (a)Potyrailo, R.A. (2004). "Sensors in Combinatorial Polymer Research" *Macromolecular Rapid Communications* 25(1):77–94. (b) Dickinson, T.A.; Walt, D.R.; White J.; Kauer, J.S. (1997). "Generating Sensor Diversity Through Combinatorial Polymer Synthesis," *Analytical Chemistry* 69(17): 3413-3418.

[43] (a)Turner, P.F. (2000), "Biosensors--Sense and Sensitivity" *Science* 290:1315-1317. b) Lei, B. Li, B. Zhang, H. Lu, S. Zheng, Z. Li, W. Wang Y. (2006). "Mesostructured Silica Chemically Doped with RuII as a Superior Optical Oxygen Sensor" *Advanced Functional Materials 16*:1883–1891.

[44] (a)McDonagh, C.; Burke, C.S.; MacCraith, B.D. (2008). "Optical chemical sensors" *Chemical Reviews* 108:400−422 (b) Orellana, G.; Moreno-Bondi, M. C. (2005). *"Frontiers in Chemical Sensors: NoVel Principles and Techniques"* Springer: New York, (c) Baldini, F.; Chester, A.N.; Homola, J.; Martellucci, S. (2006). *"Optical Chemical Sensors; NATO Science Series* 224; Springer: New York.

[45] (a)Wirnsberger, G.; Scott, B.J.; Stucky, G.D. (2001). "pH Sensing with mesoporous thin films" *Chemical Communications* 119-120. b) Nicole, L.; Boissière, C.; Grosso, D.; Hesemann, P.; Moreau, J.; Sanchez, C. (2004) "Advanced selective optical sensors based on periodically organized mesoporous hybrid silica thin films" *Chemical Communications* 2004, 2312-2313. c) Lee, S.J.; Lee, S.S.; Lee, J.Y.; Jung, J.H. (2006) "A Functionalized Inorganic Nanotube for the Selective Detection of Copper(II) Ion" *Chemistry of materials* 18 (20)4713–4715 d) Liu, J.W.; Lu, Y. (2004). "Optimization of a Pb2+ -directed gold nanoparticle/DNAzyme assembly and its application as a colorimetric biosensor for Pb2+" *Chemistry of materials* 16(17):3231–3238. e) Kalinina, M.A.; Golubev, N.V.; Raitman, O.A.; Selector, S.L.; Arslanov, V.V. (2006).

"A novel ultra-sensing composed Langmuir–Blodgett membrane for selective calcium determination in aqueous solutions" *Sensors and Actuators* B 114:19–27.

[46] (a)Zhang, C.; Suslick, K.S. (2005). "A Colorimetric Sensor Array for Organics in Water" *Journal of the American Chemical Society* 127:11548–11549 b) Mayr, T.; Igel, C.; Liebsch, G.; Klimant, I.; Wolfbeis, O.S. (2003) "Cross-Reactive Metal Ion Sensor Array in a Micro Titer Plate Format" *Analytical Chemistry* 75 (17), pp 4389–4396. DOI: 10.1021/ac020774t

[47] (a)Rodman, D.L.; Pan, H.; Clavier, C.W.; Feng, X.; Xue, Z.L. (2005). "Optical metal ion sensor based on diffusion followed by an immobilizing reaction. Quantitative analysis by a mesoporous monolith containing functional groups" *Analytical Chemistry* 77(10):3231-7. b) Potyrailo, R.A. (2006). "Polymeric Sensor Materials: Toward an Alliance of Combinatorial and Rational Design Tools?" *Angewandte Chemie International Edition* 45:702-723.

[48] Hammond, P.T. (2004). "Form and function in multilayer assembly: New applications at the nanoscale". *Advanced Materials* 16:1271-1293.

[49] (a)Descalzo, A.B.; Rurack, K.; Weisshoff, H.; Martinez-Manez, R.; Marcos, M. D.; Amoros, P.; Hoffmann, K.; Soto, J. (2005). "Rational Design of a Chromo- and Fluorogenic Hybrid Chemosensor Material for the Detection of Long-Chain Carboxylates" *Journal of the American Chemical Society* 127(1):184-200. b) Isik, S.; Schuhmann, W. (2006). "Detection of Nitric Oxide Release from Single Cells by Using Constant-Distance-Mode Scanning Electrochemical Microscopy" *Angewandte Chemie International Edition* 45:7451-7454.

[50] (a)Lu, Y.; Yang, Y.; Lu, M.; Huang, J.; Fan, H.; Haddad, R.; Lopez, G.; Burns, A.R.; Sasaki, D.Y.; Shelnutt, J.; Brinker, C.J. (2001). "Self-assembly of mesoscopically ordered chromatic polydiacetylene/silica nanocomposites" *Nature*, 410:913-917. (b) Trikalitis, P.N.; Rangan, K.K.; Bakas, T.; Kanatzidis, M.G. (2001). "Varied pore organization in mesostructured semiconductors based on the $[SnSe4]^{4-}$ anion" *Nature*, 410:671-675. c) Regan, B.C.; Aloni, S.; Ritchie, R.O.; Dahmen, U.; Zettl, A. (2004). "Carbon nanotubes as nanoscale mass conveyors" *Nature* 428:924-927. doi:10.1038/nature02496 (d) El-Safty, S.A.; Ismail, A.A.; Matsunaga, H.; Nanjo, H.; Mizukami, F. (2008). "Uniformly Mesocaged Cubic Fd3m Monoliths as Modal Carriers for Optical Chemosensors" *The Journal of Physical Chemistry C* 112(2008) 4825-4835. DOI: 10.1021/jp0764283

[51] (a) Xu, Z.; Wang, S-L.; Gao H-W. (2010). "Effects of nano-sized silicon dioxide on the structures and activities of three functional proteins" *Journal of Hazardous Materials* 180(1-3):375–383. doi:10.1016/j.jhazmat.2010.04.042 b) He, Q.; Zhang, Z.; Gao, F.; Li, Y.; Shi. J. (2011). "In vivo Biodistribution and Urinary Excretion of Mesoporous Silica Nanoparticles: Effects of Particle Size and PEGylation" *Small* 7(2):271–280. DOI: 10.1002/smll.201001459

[52] (a)Zhao, D.; Huo, Q.; Jianglin, F.; Chmelka, B.F.; Stucky, G.D. (1998). "Nonionic Triblock and Star Diblock Copolymer and Oligomeric Surfactant Syntheses of Highly Ordered, Hydrothermally Stable, Mesoporous Silica Structures" *Journal of the American Chemical Society* 120(24):6024–6036 DOI: 10.1021/ja974025i b) Tanev, P.T.; Pinnavaia, T.J. (1995). "A Neutral Templating Route to Mesoporous Molecular Sieves" *Science* 267(5199): 865-867. DOI: 10.1126/science.267.5199.865 (c) Attard, G.S.; Glyde, J.C.; Goltner, C.G. (1995). "Liquid-crystalline phases as templates for the

synthesis of mesoporous silica" *Nature* 378: 366 - 368 doi:10.1038/378366a0 d) El-Safty, S.A.; Hanaoka, T.; Mizukami, F. (2005). "Large Scale Design of Cubic Ia3d Mesoporous Silica Monoliths with High Order, Controlled Pores, and hydrothermal Stability" *Advanced Materials* 17(1):47-53. e) El-Safty, S.A.; Mizukami, F.; Hanaoka T. (2005). "General and Simple Approach for Control Cage and Cylindrical Mesopores, and Thermal/Hydrothermal Stable Frameworks" *The Journal of Physical Chemistry* 109(19):9255–9264. DOI: 10.1021/jp050304d

[53] (a)Kresge, C.T.; Leonowicz, M.E.; Roth, W.J.; Vartuli, J.C.; Beck, J.S. (1992) "Ordered mesoporous molecular sieves synthesized by a liquid-crystal template mechanism" *Nature* 359:710 – 712. doi:10.1038/359710a0 b) Che, S.; Garcia-Bennett, A. E.; Yokoi, T.; Sakamoto, K.; Kunieda, H.; Terasaki, O.; Tatsumi, T. (2003). "Mesoporous Silica of Novel Structures with Periodic Modulations Synthesized by Anionic Surfactant Templating Route" *Nature Materials* 2:801-805. c) Choi, M.; Cho, H.; Srivastava, R.; Venkatesan, C.; Choi D.; Ryoo, R. (2006). "Amphiphilic organosilane-directed synthesis of crystalline zeolite with tunable mesoporosity" *Nature Materials* 5:718-723.

[54] (a)Jaroniec, M. (2006). "Organosilica the conciliator" *Nature* 442:638-640 doi:10.1038/442638a b) Lu, Y.; Yang, Y.; Lu, M.; Huang, J.; Fan, H.; Haddad, R.; Lopez, G.; Burns, A.R.; Sasaki, D.Y.; Shelnutt, J.; Brinker, C.J. (2001). "Self-assembly of mesoscopically ordered chromatic polydiacetylene/silica nanocomposites" *Nature*, 410:913-917. c) Chen, P.; He, C. (2004). "Constructing Highly Sensitive and Selective Fluorescent Biosensors for Metal Ions by Using the MerR Family Proteins", *Journal of the American Chemical Society* 126:728-729 d) Stein, A. (2003)."Advances in Microporous and Mesoporous Solids--Highlights of Recent Progress" *Advanced Materials* 15(10):763-775. DOI: 10.1002/adma.200300007

[55] (a)Matos, J.R.; Kruk, M.; Mercuri, L.P.; Jaroniec, M.; Zhao, L.; Kamiyama, T.; Terasaki, O.; Pinnavaia, T.J.; Liu, Y. (2003). "Ordered Mesoporous Silica with Large Cage-Like Pores: Structural Identification and Pore Connectivity Design by Controlling the Synthesis Temperature and Time" *Journal of the American Chemical Society* 125 (3):821–829. DOI: 10.1021/ja0283347 b) Kleitz, F.; Liu, D.N.; Anilkumar, G.M.; Park, I.S.; Solovyov, L.A.; Shmakov, A.N.; Ryoo, R. (2003). "Large cage face-centered-cubic Fm3m mesoporous silica: Synthesis and structure" *The Journal of Physical Chemistry B* 107:14296-14300.

[56] Fukuoka, A.; Araki, H.; Sakamoto, Y.; Sugimoto, N.; Tsukada, H.; Kumai, Y.; Akimoto, Y.; Ichikawa, M. (2002). "Template Synthesis of Nanoparticle Arrays of Gold and Platinum in Mesoporous Silica Films" *Nano Letters* 2(7):793–795. DOI: 10.1021/nl0256107

[57] (a)Comes, M.; Marcos, M.D.; Sancenon, F.; Sato, J.; Villaescusa, L.A.; Amoros, P.; Beltran, D. (2004), "Chromogenic Discrimination of Primary Aliphatic Amines in Water with Functionalized Mesoporous Silica" *Advanced Materials* 16(20):1783–1786. DOI: 10.1002/adma.200400143 b) Metivier, R.; Leray, I.; Lebeau, B.D.; Valeur, B. (2005). "A mesoporous silica functionalized by a covalently bound calixarene-based fluoroionophore for selective optical sensing of mercury(II) in water" *Journal of Materials Chemistry* 15:2965- -2973 DOI: 10.1039/b501897h c) Ensafi, A.A.; Katiraei Far, A.; Meghdadi, S. (2008). "Highly selective optical sensor for mercury assay based on covalent immobilization of 4-hydroxy salophen on a triacetylcellulose membrane"

Sensors and Actuators B: Chemical 2008, 133, 84-90 d) Carrington, N.A.; Thomas, G.H.; Rodman, D.L.; Beach, D.B.; Xue, Z.-L. (2007) "Optical Determination of Cr(VI) Using Regenerable, Functionalized Sol-Gel Monoliths" *Analytica Chimica Acta* 581:232-240.

[58] (a)El-Safty, S.A.; Prabhakaran, D.; Ismail, A.A.; Matsunaga, H.; Muzukami, F. (2008). "Three-Dimensional Worm-like and Ordered Mesostructures and Their Applicability as Optically Ion-Sensitive Carriers" *Chemistry of materials* 20:2644-2654. b) El-Safty S.A. (2009). "Organic-Inorganic Hybrid Mesoporous Monoliths for Selective Discrimination and Sensitive Removal of Toxic Mercury Ions" J*ournal of Materials Scienc*e 44 (2009)6764–6774.

[59] (a)El-Safty, S.A.; Balaji, T.; Matsunaga, H.; Hanaoka, T.; Muzukami, F. (2006). "Optical Sensors-Based Nanostructured Cage Materials for Detection of Toxic Metal Ions" *Angewandte Chemie International Edition* 45:7202- 7208. b) Balaji, T.; Sasidharan, M.; Matsunaga, H. (2006). "Naked eye detection of cadmium using inorganic–organic hybrid mesoporous material" *Analytical and Bioanalytical Chemistry* 384(2):488-494, DOI: 10.1007/s00216-005-0187-2 c) El-Safty, S.A.; Prabhakaran, D.; Ismail, A.A.; Matsunaga, H.; Mizukami, F. (2007). "Nanosensor Design Packages: A Smart and Compact Development for Metal Ions Sensing Responses" *Advanced Functional Materials* 17:3731–3745 d) El-Safty, S.A.; Ismail, A.A.; Matsunaga, H.; Mizukami, F. (2007). "Optical nanosensor Design with Uniform Pore Geometry and Large Particle Morphology" *Chemistry - A European Journal* 13:9245 – 9255 e) El-Safty, S.A.; Ismail, A.A.; Matsunaga, H.; Hanaoka, T.; Mizukami, F. (2008). "Optical Nanoscale Pool-on-Surface Design for Control Sensing Recognition of Multiple Cations" *Advanced Functional Materials*18:1485–1500.

[60] Comes, M.; Lopez, G.R.; Marcos, M.D.; Manez, R.M.; Sancenon, F.; Soto, J.; Villaescusa, L. A.; Amoros, P.; Beltran, D. (2005). "Host Solids Containing Nanoscale Anion-Binding Pockets and Their Use in Selective Sensing Displacement Assays" *Angewandte Chemie International Edition* 44(19):2918–2922. DOI: 10.1002/anie.200461511

[61] Sandell, E.B. *"Colorimetric Determination of Traces of Metals"* 3th ed., Interscience Publisher, New York 1959, P326.

[62] (a)Li, H.; Li, Y.; Dang, Y.; Ma, L.; Wu, Y.; Hou, G.; Wu, L. (2009). "An easily prepared hypersensitive water-soluble fluorescent probe for mercury(II) ions" *Chemical Communications* 4453-4455. DOI: 10.1039/b907386h. b) Joshi, B. P.; Lohani, C. R.; Lee, K. H. (2010). "A highly sensitive and selective detection of Hg(II) in 100% aqueous solution with fluorescent labeled dimerized Cys residues" *Organic and Biomolecular Chemistry* 8(14):3220-3226. DOI: 10.1039/B925744F. c) Jana, A.; Kim, J. S.; Jung, H. S.; Bharadwaj, P. K. (2009). "A cryptand based chemodosimetric probe for naked-eye detection of mercury(II) ion in aqueous medium and its application in live cell imaging" *Chemical Communications* 4417-4419. DOI: 10.1039/b907646h. d) Ma, L. J.; Li, Y.; Li, L.; Sun, J.; Tian, C.; Wu, Y. (2008). "A protein-supported fluorescent reagent for the highly-sensitive and selective detection of mercury ions in aqueous solution and live cells" *Chemical Communications* 6345-6347. DOI: 10.1039/b815281k. e) Yang, M.H.; Lohani, C. R.; Cho, H.; Lee, K. H. (2011). "A methionine-based turn-on chemical sensor for selectively monitoring Hg2+ ions in

100% aqueous solution" *Organic and Biomolecular Chemistry*9:2350-2356. DOI: 10.1039/C0OB00780C

[63] (a)Kim, H.N.; Lee, M.H.; Kim, H.J.; Kim, J.S.; Yoon, J. (2008). "A new trend in rhodamine-based chemosensors: application of spirolactam ring-opening to sensing ions" *Chemical Society Reviews* 37(8):1465–1472. b) Lee, M.H.; Wu, J.S.; Lee, J.W.; Jung, J.H.; Kim, J.S. (2007). "Highly Sensitive and Selective Chemosensor for Hg2+ Based on the Rhodamine Fluorophore" *organic Letters* 9(13)"2501–2504. DOI: 10.1021/ol0708931

[64] Bhalla, V.; Tejpal, R.; Kuma, M. (2010). "Rhodamine appended terphenyl: A reversible "off–on" fluorescent chemosensor for mercury ions" *Sensors and Actuators* B 151:180–185

[65] Shi, W.; Sun, S.; Li X.; Ma, H. (2009). "Imaging Different Interactions of Mercury and Silver with Live Cells by a Designed Fluorescence Probe Rhodamine B Selenolactone" *Inorganic Chemistry* 49:1206–1210. DOI: 10.1021/ic902192a

[66] Wang, H-H.; Xue, L.; Yu, C-L.; Qian, Y-Y.; Jiang, H. (2011). "Rhodamine-based fluorescent sensor for mercury in buffer solution and living cells" *Dyes and Pigments* 91(3):350-355. doi:10.1016/j.dyepig.2011.04.00

[67] Ahamed, B.N.; Ghosh, P. (2011). "An integrated system of pyrene and rhodamine-6G for selective colorimetric and fluorometric sensing of mercury(II)" *Inorganica Chimica Acta* 372 (2011) 100–107. doi:10.1016/j.ica.2011.01.07.

[68] Soh, J.H.; Swamy, K.M.K.; Kim, S.K.; Kim, S.; Lee, S-H.; Yoon, J. (2007). "Rhodamine Urea Derivatives as Fluorescent Chemosensors for Hg2+" *Tetrahedron Letters* 48(34):5966-5969.

[69] (a)Clever, G.H.; Kaul, C.; Carell, T. (2007). "DNA–Metal Base Pairs" *Angewandte Chemie International Edition* 46(33):6226–6236. DOI: 10.1002/anie.200701185 b) Miyake, Y.; Togashi, H.; Tashiro, M.; Yamaguchi, H.; Oda, S.; Kudo, M.; Tanaka, Y.; Kondo, Y.; Sawa, R.; Fujimoto, T.; Machinami, T.; Ono, A. (2006). "MercuryII-Mediated Formation of Thymine−HgII−Thymine Base Pairs in DNA Duplexes" *Journal of the American Chemical Society* 128 (7):2172–2173. DOI: 10.1021/ ja056354d

[70] Tang, Y.; He, F.; Yu, M.; Feng, F.; An, L.; Sun, H.; Wang, S.; Li, Y.; Zhu, D. (2006). "A Reversible and Highly Selective Fluorescent Sensor for Mercury(II) Using Poly(thiophene)s that Contain Thymine Moieties" *Macromolecular Rapid Communications* 27(6):389–392. DOI: 10.1002/marc.200500837

[71] Wu, Y.; Zhan, S.; Xu, L.; Shi, W.; Xi, T.; Zhana, X.; Zhou, P. (2011). "A simple and label-free sensor for mercury(II) detection in aqueous solution by malachite green based on a resonance scattering spectral assay" *Chemical Communications* 47:6027-6029. DOI: 10.1039/C1CC10563A

[72] Zhu, Z.; Su, Y.; Li, J.; Li, D.; Zhang, J.; Song, S.; Zhao, Y.; Li, G.; Fan, C. (2009). "Highly Sensitive Electrochemical Sensor for Mercury(II) Ions by Using a Mercury-Specific Oligonucleotide Probe and Gold Nanoparticle-Based Amplification" *Analytical Chemistry* 81(18): 7660–7666. DOI: 10.1021/ac9010809.

[73] Tatay, S.; Gavina, P.; Coronado, E.; Palomares, E. (2006). "Optical Mercury Sensing Using a Benzothiazolium Hemicyanine Dye" *Organic Letters* 8(17: 3857–3860. DOI: 10.1021/ol0615580

[74] Zhang, Y.; Shi, W.; Feng, D.; Ma, H.; Liang, Y.; Zuo, J. (2011). "Application of rhodamine B thiolactone to fluorescence imaging of Hg 2+ in Arabidopsis thaliana" *Sensors and Actuators B* 153:261–265. doi:10.1016/j.snb.2010.10.00.

[75] Balamurugan, A.; Reddy, M.L.P.; Jayakannan, M. (2009). "Carboxylic Functionalized Water Soluble π-Conjugated Polymer: Highly Selective and Efficient Chemosensor for Mercury (II) ions" *Journal of Polymer Science Part A: Polymer Chemistry* 47:5144-5157. DOI: 10.1002/pola.23566.

[76] Yang, B.; Li, G.; Zhang, X.; Shu, X.; Wang, A.; Zhu, X.; Zhu J. (2011). "Hg2+ detection by aniline-based conjugated copolymers with high selectivity" *Polymer* 52(12):2537-2541

[77] Zhang, Q.Q.; Ge, J.F.; Xu, Q.F.; Yang, X.B.; Cao, X.Q.; Li, N.J.; Lu, J.M. "A selective, sensitive probe for mercury(II) ions based on oxazine-thione" *Tetrahedron letters* 52(5):595-597. DOI: 10.1016/j.tetlet.2010.11.130.

[78] Cheng, X.; Li, Q.; Li, C.; Qin, J.; Li, Z. (2011). "Azobenzene-Based Colorimetric Chemosensors for Rapid Naked-Eye Detection of Mercury(II*)" Chemistry - A European Journal* 17(26):7276–7281.

[79] Liu, X.; Tang, Y.; Wang, L.; Zhang, J.; Song, S.; Fan, C.; Wang, S. (2007). "Optical Detection of Mercury(II) in Aqueous Solutions by Using Conjugated Polymers and Label-Free Oligonucleotides" *Advanced Materials* 19(11):1471–1474. DOI: 10.1002/adma.200602578

[80] Huang, C.C.; Chang, H.T. (2007). "Parameters for selective colorimetric sensing of mercury(II) in aqueous solutions using mercaptopropionic acid-modified gold nanoparticles" *Chemical Communications* 1215-1217. DOI: 10.1039/b615383f

[81] Lee, J.-S.; Han, M.S.; Mirkin, C.A. (2007). "Colorimetric Detection of Mercuric Ion (Hg2+) in Aqueous Media Using DNA-Functionalized Gold Nanoparticles" *Angewandte Chemie International Edition* 46:4093-4096.

[82] Chang, H-Y.; Hsiung, T-M.; Huang, Y-F.; Huang, C-C. (2011). "Using Rhodamine 6G-Modified Gold Nanoparticles To Detect Organic Mercury Species in Highly Saline Solutions" *Environmental Science and Technology* 45 (4):1534–1539. DOI: 10.1021/es103369d

[83] Chai, F.; Wang, T.; Li, L.; Liu, H.; Zhang, L.; Su, Z.; Wang, C. (2010). "Fluorescent Gold Nanoprobes for the Sensitive and Selective Detection for Hg2+" *Nanoscale Research Letters* 5:1856–1860. DOI 10.1007/s11671-010-9730-y

[84] Darbha, G.K.; Singh, A.K.; Rai, U.S.; Yu, E.; Yu, H.; Ray, P.C. (2008). "Selective Detection of Mercury (II) Ion Using Nonlinear Optical Properties of Gold Nanoparticles" *Journal of the American Chemical Society* 130 (25):8038–8043. DOI: 10.1021/ja801412b

[85] Zheng, A.; Chen. J.; Wu, G.; Wei, H.; He, C.; Kai, X.; Wu, G.; Chen, Y. (2009). "Optimization of a sensitive method for the "switch-on" determination of mercury(II) in waters using Rhodamine B capped gold nanoparticles as a fluorescence sensor" *Microchim Acta* 164(1-2):17–27. DOI: 10.1007/s00604-008-0023-4

[86] (a)Sancenon, F.; Martınez-Manez, R.; Soto, J. (2001). "1,3,5-Triarylpent-2-en-1,5-diones for the colorimetric sensing of the mercuric cation" *Chemical Communications* 2262-2263. DOI: 10.1039/B106941C b) Yang, Y.-K.; Yook, K.-J.; Tae, J. (2005). "A Rhodamine-Based Fluorescent and Colorimetric Chemodosimeter for the Rapid Detection of Hg2+ Ions in Aqueous Media" *Journal of the American Chemical Society*

127 (48:16760–16761.DOI: 10.1021/ja054855t c) Nolan, E.M.; Racine M.E.; Lippard, S.J. "Selective Hg(II) Detection in Aqueous Solution with Thiol Derivatized Fluoresceins", *Inorganic Chemistry* 2006, 45, 2742-2749.

[87] (a)Zheng, H.; Qian, Z.H.; Xu, L.; Yuan, F.F.; Lan, L.D.; Xu, J.G. (2006). "Switching the Recognition Preference of Rhodamine B Spirolactam by Replacing One Atom: Design of Rhodamine B Thiohydrazide for Recognition of Hg(II) in Aqueous Solution" *Organic Letter* 8 (5):859–861.DOI: 10.1021/ol0529086 b) Chen, X.; Baek, K.; Kim, Y.; Kim, S.; Shin, I.; Yoon, J.A. (2010)"A Selenolactone-Based Fluorescent Chemodosimeter to Monitor Mercury/Methylmercury Species in vitro and in vivo" *Tetrahedron* 66(23):4016-4021.

[88] (a)Zhou, Y.; You, X.Y.; Fang, Y.; Li, J.Y.; Liu, K.; Yao, C.A. (2010). "A thiophen-thiooxorhodamine conjugate fluorescent probe for detecting mercury in aqueous media and living cells" *Organic and Biomolecular Chemistry* 8: 4819-4822 b) Lin, W.; Cao, X.; Ding, Y.; Yuan, L.; Yu, Q. (2010). "A reversible fluorescent Hg2+ chemosensor based on a receptor composed of a thiol atom and an alkene moiety for living cell fluorescence imaging" *Organic and Biomolecular Chemistry* 8:3618-3620 c) Lin, W.; Cao, X.; Ding, Y.; Yuan, L.; Long, L. (2010). "A highly selective and sensitive fluorescent probe for Hg2+ imaging in live cells based on a rhodamine–thioamide–alkyne scaffold" *Chemical Communications* 46:3529-353. DOI: 10.1039/b927373e

[89] (a)Palomares, E.; Vilar, R. Durrant, J.R. (2004). "Heterogeneous colorimetric sensor for mercuric salts" *Chemical Communications* 362-363.ISSN:1359-7345. b) Nazeeruddin, Md.K.; Di Censo, D.; Humphry-Baker, R.; Grätzel, M. (2006). "Highly Selective and Reversible Optical, Colorimetric and Electrochemical Detection of Mercury (II) by Amphiphilic Ruthenium complexes Anchored onto Mesoporous Oxide Films" *Advanced Functional Materials*, 16:189-194.

[90] (a)Ono, A.; Togashi, H. (2004). "Highly selective oligonucleotide-based sensor for mercury(II) in aqueous solutions" *Angewandte Chemie International Edition*43:4300–4302. b) Wang, J.; Qian, X. Cui, J. (2006). "Detecting Hg 2+ions with an ICT fluorescent sensor molecule: remarkable emission spectra shift and unique selectivity" *Journal of Organic Chemistry* 71:4308–4311. c) Nolan, E. M.; Lippard, S. J. (2007). "Turn-On and Ratiometric Mercury Sensing in Water with a Red-Emitting Probe" *Journal of the American Chemical Society* 129 (18):5910–5918. d) Nolan, E.M.; Lippard, S.J. (2008). "Tools and Tactics for the Optical Detection of Mercuric Ion," *Chemical Reviews* 108:3443-3480.

[91] Carrington, N.A.; Xue, Z.-L. (2007). "Inorganic Sensing Using Organofunctional Sol-Gel Materials" *Accounts of Chemical Research* 40: 343-350.

[92] (a)Liu, J.W.; Lu, Y. (2004). "Accelerated color change of gold nanoparticles assembled by DNAzymes for simple and fast colorimetric Pb2+ detection" *Journal of the American Chemical Society*126:12298–12305 b) Coronado, E.; Galán-Mascarós, J.R.; Martí-Gastaldo, C.; Palomares, E.; Durrant, J.R.; Vilar, R.; Gratzel, M.; Nazeeruddin, Md.K. J. (2005). "Reversible Colorimetric Probes for Mercury Sensing" *Journal of the American Chemical Society* 127:12351-12356. c) Hagleitner, C.; Hierlemann, A.; Brand, O.; Baltes, H. (2002). "CMOS Single Chip Gas Detection Systems — Part I" *Sensors Update* 11(1):101-155. d) Rex, M.; Hernandez, F. E.; Campiglia, A. D. (2006). "Pushing the Limits of Mercury Sensors with Gold Nanorods" *Analytical Chemistry* 78 (2): 445–451

[93] El-Safty, S.A.; Prabhakaran, D.; Kiyozumi, Y.; Mizukami, F. (2008). "Nanoscale Membrane Strips for Benign Sensing of Hg (II) ions: A Route to Commercial Waste Treatments" *Advanced Functional Materials* 18:1739-1750.
[94] Xiao, J.; Wei, J. "DIFFUSION MECHANISM OF HYDROCARBONS IN ZEOLITES-I, THEORY" *Chnntial Erzgimeering Sciemce*, 47(5)1123-1141, 1992.
[95] Sayari, A.; Hamoudi, S.; Yang, Y. (2005). "Applications of Pore-Expanded Mesoporous Silica. 1. Removal of Heavy Metal Cations and Organic Pollutants from Wastewater" *Chemistry of Materials*17 (1):212–216. DOI: 10.1021/cm048393e.

In: Mercury: Sources, Applications and Health Impacts
Editors: Ki-Hyun Kim and Richard J. C. Brown
ISBN: 978-1-62257-721-7

Chapter 3

ISOMERS IN THE CHEMISTRY OF MERCURY ORGANOMETALLIC AND HETEROMETALLIC COMPLEXES

M. Melnik and P. Mikuš

Department of Pharmaceutical Analysis and Nuclear Pharmacy, Faculty of Pharmacy, Comenius University in Bratislava, Bratislava, Slovak Republic

ABSTRACT

The mercury organometallics and heterometallic complexes cover huge fields, as shown by a survey covering the crystallographic and structural data of over five hundred examples about 14% of those complexes exist as isomers and are gathered and classified in this review. Included are distortion (by far prevails), ligand and linkage isomerism. These are discussed in terms of the coordination around the Hg atoms, and correlations are drawn between atoms, bond distances and bond angles. These isomers are discussed and summarized with those found in the chemistry of mercury coordination complexes. This review represents the first overall survey of the isomers in the chemistry of mercury.

Keywords: Structures, mercury, organometallics, heterometallics, isomers

ABBREVIATIONS

ad	adeninate
ala	alaninate
all	allopurinate
app	7-deaza-8-azaadenine
aza	7-azaindole
bcb	2-benzyl-o-carborane
c	cubic
$C_{10}H_{14}NCl_2$	1,1-dichloro-1-(2-methyl-5-tert-butyl-2H-pyrrole)

$C_{10}H_{15}$	camphenate
$C_{11}H_{11}O_2$	1-acetoxy-1-phenyl-2-methyl-a-cetylanate
$C_{12}H_6NS_2$	2-mercaptobenzothiazolate
$C_{12}H_8$	2,2′-biphenylylene
$C_{19}H_{14}F_3O_4$	3-fluoroacetoxy-2,3-diphenylcyclopopane-1-carboxylate
$C_{19}H_{40}O_2$	androsta-4,6-diene-3,17-dionate
$C_6H_{12}NCl$	(3-chloro-1-(n-propyl)prop-2-etyl)methylammonium
$C_9H_{15}NO$	2,2,6,6-tetramethyl-Δ^4-dehydropiperidine-1-oxyl
cp	cyclopentadienyl
cp*	pentamethylcyclopentadienyl
dppm	bis(diphenylphosphino)methane
en	ethylenediamine
Et_2dtc	N,N′-diethyldithiocarbamate
m	monoclinic
me	methyl
2-$MeOC_6H_{10}$	2-methoxycyclohexyl
Me_2phen	dimethyl-1,10-phenanthioline
Me_4phen	tetramethyl-1,10-phenanthroline
Methu	1-methyl-4-thiouracilate
or	orthorhombic
pen	D,L-penicillaminate
Ph	phenyl
phen	1,10-phenanthroline
Pr^i	iso-propyl
psit	2-phenylsulphonyl-imino-1,2-dihydrothiazole
py	pyridine
qu-8-th	quinoline-8-thiolate
quo	(ethylenedioxyl)-o-tolyl
sami	salicylalmethyliminate
tf	N,N-di-para-tolylformamide
thm	thymidinate
tr	triclinic
xan	xanthate

1. INTRODUCTION

A huge number of organomercury as well as heterometallic compounds are known and while many have useful physiological behaviours, for the most part they are harmful to life and the environment. There have been a number of environmental disasters attributable to mercury discharges. The actual agent involved has been shown to be the organometallic cation CH_3Hg^+. This water soluble ion can enter the body and induce complex irreversible disturbances to the central nervous system. The cation is formed by biological methylation of sludge metallic mercury [1].

Organomercury as well as heterometallic mercury compounds have been extensively studied from a structural point of views, with over three hundred and twenty organometallic [2] and over two hundred heterometallic mercury compounds [3] with over sixty isomeric examples noted. In this review we analyze and classify these examples.

Stereoselectivity in complex compounds is very often related to important stereospecificity of biological systems, catalysis and stereochemical effects in technical processes. Isomers can be broadly classed into two major categories, structural and stereoisomers. The former can be divided into: ionisation, hydrate, coordination number, linkage and polymerization sub-categories, and the latter can be divided into geometric (cis-trans, fac-mer), optical, distortion and ligand isomerism.

Analyzed of mercury coordination compounds showed [4] that there are four isomeric types: distortion (73%), polymerization (20.6%), coordination number (3.2%) and ligand (3.2%) isomers.

The aim of this review is to discuss the factors which could lead to a better understanding of stereochemical interactions within the inner coordination sphere. A comprehensive view on isomers of mercury complexes is given and discussed.

2. Mercury Organometallics

There are over fifty examples of mercury organomethallic isomers which are analyze and classify.

2.1. Distortion Isomers

The coexistence of two or more species even within the same crystal, differ by degree of distortion is typical of general class of distortion isomerism [5]. There are ten examples which exist in two isomeric forms, and twenty six examples which contain crystallographically independent molecules within the same crystal.

2.1.1. Pair of Isomers

There are ten derivatives which exist in two isomeric forms and their crystallographic and structural data are gathered in Table 1. In two colourless $Hg(CF_3)_2$ cubic [6] and orthorhombic [7] each Hg(II) atom is linearly coordinated by two CF_3^- groups with the mean HgC bond distances of 2.109 and 2.139Å, respectively. The orthorhombic was studied at ambient temperature and at 99K. The crystallographic parameters as well as Hg-C bond distance reduced with the lowering temperature. In another six pairs of isomers, monoclinic α- and orthorhombic β-$Hg(CF_3)(N_3)$ [8]; monoclinic α-Hg(Me)(DL-ala) [9] and orthorhombic β-Hg(Me)(DL-ala) [10]; monoclinic α- Hg(Me)(ad) [11] and orthorhombic β-Hg(Me)(ad) [12]; monoclinic α- and orthorhombic β-Hg(Me)(thm) [13]; and two monoclinic α- and β-Hg(Me)(all) [14]; two monodentate ligands created an almost linear coordination about each Hg(II) atom (HgNC). The mean Hg-N and Hg-C bond distances α- vs. β- isomer are 2.075 and 2.04Å vs. 2.10 and 2.085Å. This indicated that the inner coordination sphere in α-isomers is somewhat more tightly than in β-isomers. The N-Hg-C bond angles in the α-isomers range from 171.8(7) to 178.3(6)° (av. 176°) and in the β-isomers from 173(1) to 178.5(7)° (av. 176°). β-Hg(Me)(thm) contains two crystallographically independent molecules with the Hg-N and Hg-C bond distances (molecule 1 vs. molecule 2) are: 2.10(1) and 2.07(2)Å vs. 2.10(1) and2.03(2)Å which indicate that the inner coordination sphere in molecule 2 is somewhat more tighter than in molecule 1. The inner coordination sphere in the molecules is less crowded than that in the α-Hg(Me)(thm) [13] (Table 1). In two colourless triclinic α- and β-Hg(2-MeOC_6H_{10})Cl isomers [15] two monodentate ligands about each Hg(II) atom form almost linear C-Hg-Cl 178° in α- and linear 180° in β- isomer. The Hg-C and Hg-Cl bond distances (α vs. β) are 2.34 and 2.53Å vs. 2.15 and 2.50Å the β-isomer is somewhat more crowded than α-isomer. For another two pairs of the α- and β-Hg(2-MeOC_6H_{10})X (X=Br or I) [15] only unit cell dimensions are available in the original paper.

Table 1. Crystallographic and Structural Data for Monomeric Mercury Organometallics-Distortion Isomers[a]

COMPLEX (colour)	Crys.cl Sp.Grp Z	a [Å] b [Å] c [Å]	α [°] ß [°] γ [°]	Chromo - phore	Hg-L [Å]	L-Hg-L [°]	Ref
α-Hg(CF_3)$_2$ (colourless)	c Pa3 4	8.127(2)		HgC_2	C[b] 2.109(16)	C,C[b] 180	6
β-Hg(CF_3)$_2$ (colourless)	or Pbca 4	8.118 8.115 8.113		HgC_2	C 2.139	C,C notgiven	7
(at 99K)	or Pbca 4	7.957 7.957 7.957		HgC_2	C 1.998	C,C notgiven	7
α-Hg(CF_3) (N_3) (colourless)	m $P2_{1/c}$ 4	9.720(3) 5.721(1) 9.938(2)	116.05(3)	HgNC	N 2.02(2) C 1.96(2)	N,C 177.3(9)	8
β-Hg(CF_3)(N_3) (colourless)	or Pbca 8	13.68(5) 6.81(3) 10.50(4)		HgNC			8
α-Hg(Me)(DL-ala) (colourless)	m $P2_{1/c}$ 4	9.460(2) 8.794(2) 8.723(2)	97.49(2)	HgNC	N 2.14(2) C 2.04(3)	N,C 175.6(9)	9
β-Hg(Me)(L-ala) (colourless)	or P2,2,2, 4	5.763(3) 6.851(6) 18.75(2)		HgNC	N 2.10(3) C 2.08(3)	N,C 173(I)	10
α-[Hg(Me)(ad)]H_2O (colourless)	m C2/c 8	19.796(6) 7.119(3) 16.471(4)	128.83(2)	HgNC	N 2.07(1) C 2.06	N,C 176.8(7)	11
β-Hg(Me)(ad) (colourless)	or Pbca 8	14.658(8) 8.407(5) 13.006(9)		HgNC	N 2.070(7) C 2.076(9)	N,C 176.5(3)	12
α-Hg(Me)(thm) (colourless)	m $P2_1$ 2	4.798(6) 14.270(8) 10.390(4)	102.74(9)	HgNC	N 2.07(1) C 2.01(2)	N,C 178.3(6)	13
β-[Hg(Me)(thm)]. 0.65H_2O[c] (colourless)	or P2,2,2, 8	10.484(3) 14.633(3) 18.538(5)		HgNC	N 2.10(1) C 2.07(2)	N,C 178.5(7)	13
α-Hg(Me)(all) (white)	m $P2_{1/c}$ 4	11.303(2) 10.550(1) 7.271(1)	96.56(2)	HgNC	N2.08(1) C2.08(3)	N,C 171.8(6)	14
β-[Hg(Me)(all)]. $2H_2O$ (white)	m C2/c 8	19.838(3) 7.058(2) 17.869(3)	119.32(1)	HgNC	N 2.13(1) C 2.14(6)	N,C 175.1(6)	14

COMPLEX (colour)	Crys.cl Sp.Grp Z	a [Å] b [Å] c [Å]	α [o] ß [o] γ [o]	Chromo - phore	Hg-L [Å]	L-Hg-L [o]	Ref
α-Hg(2-MeOC$_6$H$_{10}$)Cl (colourless)	tr P$\bar{1}$ 2	8.20(1) 10.58(2) 6.07(1)	83.6(1) 116.2(1) 102.5(1)|	HgCCl	C 2.34 Cl 2.53	C,Cl 178	15
β-(Hg(2-MeOC$_6$H$_{10}$)Cl (colourless)	tr P$\bar{1}$ 2	7.88(1) 9.29(1) 6.371(1)	88.3(1) 102.2(1) 92.6(1)	HgCCl	C 2.15 Cl 2.50	C,Cl 180	15
α-Hg(2-MeOC6H$_{10}$)Br	tr P$\bar{1}$ 2	8.33(2) 1.73(3) 6.09(1)	83.4(2) 115.3(2) 104.1(2)	HgCBr			
β-Hg(2-MeOC$_6$H$_{10}$)Br	tr P$\bar{1}$ 2	7.96(2) 9.41(2) 6.46(2)	86.2(2) 102.0(2) 93.4(2)	HgCBr			15
α-Hg(2-MeOC$_6$H$_{10}$)I	tr P$\bar{1}$ 4	8.58(2) 19.87(3) 6.12(2)	84.0(2) 115.4(2) 102.5(2)	HgCI			15
β- Hg(2-MeOC$_6$H$_{10}$)I	tr P$\bar{1}$ 2	8.18(2) 9.63(2) 6.57(2)	84.3(2) 102.5(2) 94.5(2)	HgCI			15
α-Hg(Ph)(η^2quo)	or Pnam 8	23.12(3) 16.59(2) 6.62(1)		HgONC HgONC	O 2.19(4) N 2.29(4) C 2.18(6) O 2.15(4) N 2.40(4) C 2.14(5)	O,N 74(1) O,C 142(2) N,C 144(2) O,N 75(1) O,C 154(2) N,C 1.31(2)	16
β- Hg(Ph)(η^2quo)	or P2,2,2, 4	5.584(2) 9.902(3) 22.073(9)		HgONC	O 2.06(2) N 2.57(2) C 2.05(2)	O,N 72.1(5) O,C175.0(7) N,C 112.3(7)	16

a. Where more than one chemically equivalent distance or angle is present, the mean value is tabulated. The first number in parenthesis is the e.s.d., and the second is the maximum deviation from the mean.

b.The chemical identity of the coordinated atom or ligand is specified in these columns.

In two orthorhombic α- and β-Hg(η^2-quo)(Ph) [16] the ligands created about each Hg(II) atom very distorted trigonal plane (HgONC). α-isomer contains two crystallographically independent molecules. The sum of Hg-L(x3) bond distances are 6.66Å (α-molecule 1), 6.69Å (α-molecule 2) and 6.68Å (β-isomer).

2.1.2. Independent Molecules – Monomers

There are twenty eight derivatives which contain crystallographically independent molecules within the same crystal (one contains five [17], another one four [18], two derivatives contain three [19,20] and remainders [18,21-42] two independent molecules) (Table 2).

In monoclinic colourless Hg($C_9H_{15}NO$)Cl [17] which contains five crystallographically independent molecules, two monodentate ligands form about the Hg(II) atom an almost linear coordination (HgCCl), with the mean C-Hg-Cl bond angle of 177.9°.

Another colourless orthorhombic Hg(Ph)$_2$(Me$_2$-phen) [18] contains four crystallographically independent molecules unfortunately only mean structural data are available.

Two phenyl groups about the Hg(II) atoms form a diagonal coordination (HgC$_2$) with C-Hg-C bond angle of 180°. Two N atoms of Me$_2$phen molecule are interact with the Hg(II) atom (av. Hg-N is 2.90Å).

Table 2. Crystallographic and Structural Data for Monomeric Mercury Complexes – Independent Molecules[a]

COMPOUND (colour)	Crys.cl Sp.Grp Z	a [Å] b [Å] c [Å]	α [°] β [°] γ [°]	Chrom o- phore	Hg-L [Å]	L-Hg-L [°]	Ref
$Hg(C_9H_{15}NO)Cl$ (colourless)	m $P2_1/n$ 20	16.349(1) 18.931(3) 20.105(4)	 106.70(4)	HgCCl HgCCl HgCCl HgCCl HgCCl	C[b] 2.05(2) Cl 2.32(6) C 2.07(2) Cl 2.324(6) C 2.02(2) Cl 2.315(6) C 2.04(3) Cl 2.324(6) C 2.04(3) Cl 2.322(6)	C,Cl[b] 177.1(7) C,Cl 178.2(6) C,Cl 178.1(7) C,Cl 177.2(8) C,Cl 178.8(8)	17
$Hg(Ph)_2(\eta^2$- $Me_2phen)$[c] (colourless)	or Pbca 4	15.42 14.28 14.25		HgC_2 (N_2)	PhC 2.10 N 2.90(18)	C,C 180	18
$Hg(Ph)(Pr^iOCS_2)$ (colourless)	m $P2_{1/a}$ 12	13.678(5) 21.347(7) 14.570(6)	 114.99(2)	HgCS HgCS HgCS	C 2.03(1) S 2.374(4) C 2.03(1) S 2.394(4) C 2.10(1) S 2.394(4)	C,S 176.6(4) C,S 175.6(3) C,S 174.3(3)	19
Hg(aza)(Me) (colourless)	tr $P\bar{1}$ 6	10.926(10) 11.333(8) 11.647(10)	92.13(8) 104.83(9) 101.86(7)	HgNC HgNC HgNC	N 2.11(2) MeC 2.11(3) N 1.96(2) MeC 2.07(3) N 2.05(2) MeC 2.08(3)	N,C 173.8(11) N,C 177.0(9) N,C 176.0(10)	20
Hg(Me)(bcb) (colourless)	m $P2_{1/n}$ 8	20.44(1) 7.580(5) 22.07(1)	 108.79(4)	HgC_2 HgC_2	Me,C 2.07(2) bcbC2.11(2) MeC 2.09(2) bcbC2.11(2)	C,C 174.4(6) C,C 173.0(6)	21
$Hg(MesPCH_2)_2$ (colourless)	m $P2_{1/c}$ 4	13.8900(7) 14.4928(6) 19.8504(5)	 124.49(1)	HgC_2 HgC_2	C 2.042(11,0) C 2.079(10,0)	C,C 180 C,C 180	22
$Hg(ClCH_2CC)_2$ (phen))	tr P1 4	21.66(1) 9.486(5) 8.814(4)	91.13(4) 101.14(5) 103.90(7)	HgC_2 HgC_2	C 2.07(4,2) C 2.03(4,1)	C,C 165 C,C 168	23
$Hg(Ph)_2$ (Me_4phen)	m Cm 4	21.09 14.54 14.65	 20.34	HgC_2 HgC_2	C 2.10 C 2.13	C,C 180 C,C 180	18
Hg(Me) (Methu). $O.5NaNO_3$ (colourless)	or Pnma 8	11.088(3) 6.785(2) 27.31(2)		HgNC HgNC	N 2.092(8) MeC 2.058(10) N 2.100(8) MeC 2.072(10)	N,C 175.9(3) N,C 174.4(3)	24
Hg(4-Mep)(2NO- 4-Mep) (dark red)	tr $P\bar{1}$ 4	10.26(2) 22.71(3) 5.61(3)	97.2(2) 93.1(2) 91.1(2)	HgNC HgNC	N2.08(3) C 2.07(4) N 2.20(4) C 2.01(5)	N,C174 N,C 176	25
Hg(Ph)(psit) (colourless) (at 153K)	m $P2_{1/c}$ 8	15.516(2) 10.534(2) 22.877(3)	 105.29(1)	HgNC HgNC	N 2.07(1) PhC 2.03(1) N 2.08(1) PhC 2.03(1)	N,C 170.3(4) N,C 176.6(4)	26

COMPOUND (colour)	Crys.cl Sp.Grp Z	a [Å] b [Å] c [Å]	α [o] β [o] γ [o]	Chromo- phore	Hg-L [Å]	L-Hg-L [o]	Ref
Hg(Ph)(tf) (colourless)	m $P2_{1/c}$ 8	23.586(4) 21.324(3) 7.240(1)	91.29(2)	HgNC HgNC	N 2.13(3) PhC 2.08(4) N 2.02(3) PhC1.96(4)	N,C 176(1) N,C 174(1)	27
[Hg(Ph)(py)]. F_3CCO_2 (colourless)	or $Pb2_1$ 8	34.43(3) 7.404(7) 10.64(1)		HgNC HgNC	N 2.19(4) PhC 2.09(4) N 2.11(4) PhC 2.01(4)	N,C 178 N,C 167(2)	28
$Hg(C_{19}H_{14}F_3O_4)Cl$ (colourless)	m $P2_{1/c}$ 8	17.595(2) 11.045(1) 20.933(1)	97.281(8)	HgCCl HgCCl	C 2.12(4) Cl 2.30(1) C 2.12(3) Cl 2.32(1)	C,Cl 178.4(7) C,Cl 178.7(7)	29
$Hg\{Ph(py)_2\}$ (pale yellow)	tr $P2_{1/n}$ 8	10.334(2) 7.484(1) 27.052(6)	97.60(2)	HgCCl HgCCl	C 2.069(13) Cl 2.314(4) C 2.070(13) Cl 2.335(4)	C,Cl 178.1(4) C,Cl 176.9(4)	30
$Hg(C_{11}H_{11}O_2)Cl$ (colourless)	m $P2_{1/n}$ 8	17.815 15.6725) 9.299(1)	111.25	HgCCl HgCCl	not given not given	C,Cl 175.9 C,Cl 176.4	31
$Hg(C_{19}H_{40}O_2)Cl$. Me_2CO (pale yellow) (at 153K)	m $P2_{1/a}$ 4	12.074(3) 9.453(2) 15.817(7)	105.09(3)	HgCCl HgCCl	C 2.00(3) Cl 2.372(7) C 2.11(3) Cl 2.375(12)	C,Cl 177.0(10) C,Cl 174.5(10)	32
Hg(cp*)Cl (yellow)	tr $P\bar{1}$ 4	8.315(2) 11.859(2) 12.458(5)	92.80(2) 101.24(2) 109.34(2)	HgCCl HgCCl	C 2.08(2) Cl 2.350(6) C 2.10(2) Cl 2.361(5)	C,Cl 176.8(7) C,Cl 17.39(5)	33
$Hg(C_{11}H_{11}O_2)Cl$	m $P2_{1/n}$ 8	17.815 15.672 9.299	111.25	HgCCl HgCCl			34
$[ClHg(C_6H_{12}NCl)]$. $[HgCl_3]$	m Pc 4	10.532(6) 9.866(2) 14.093(7)	99.23(6)	HgCC HgCl $HgCl_3$ $HgCl_3$	C 2.03(3) Cl 2.37(1) C 2.09(4) Cl 2.28(1) Cl 2.1(1,6) Cl 2.41(1,13)	C,Cl 175.3(9) C,Cl 173.6(9) ClCl 90.1(4,7.1) 129.0(4,8.8) Cl,Cl 90.4(4,14.7) 128.1(4,14.1)	35
$Hg(C_{10}H_{14}NCl_2)Cl$ (colourless)	m $P2_{1/c}$ 8	16.549(4) 11.484(4) 16.242(3)	113.9(2)	HgCCl HgCCl	C 2.05 Cl 2.34 C 2.09 Cl2.30	C,Cl 166 C,Cl 173	36
$Hg(Me)(C_{12}H_6NS_2)$ (white)	tr $P\bar{1}$ 4	8.009(4) 10.042(4) 13.074(3)	101.25(2) 102.61(3) 101.42(3)	HgCS HgCS	C 2.06(2) S 2.369(6) C 2.10(3) S 2.375(6)	C,S 177.7(7) C,S 178.8(6)	37
$[Hg(Me)(pen)].H_2O$ (colourless)	m $C\bar{1}$ 16	23.154(33) 9.819(14) 19.321(27)	106.4(1)	HgCS HgCS	C 2.07(6) S 2.38(1) C 2.09(5) S 2.36(1)	C,S 175(2) C,S 175(2)	38

Table 2. (Continued)

COMPOUND (colour)	Crys.cl Sp.Grp Z	a [Å] b [Å] c [Å]	α [o] β [o] γ [o]	Chromo-phore	Hg-L [Å]	L-Hg-L [o]	Ref
$Hg(Ph)(Et_2dtc)$ (colourless)	tr P1̄ 8	9.959(2) 12.359(4) 13.098(2)	65.53(2) 65.81(2)	HgCS HgCS	C 2.041(9) S 2.387(2)) C 2.06(1) S 2.388(2)	C,S 175(2) N,C 167(2)	18
Hg(Ph)(qu-8-th) (colourless) at 153K	m $P2_{1/c}$ 8	6.732(3)) 27.43(2) 16.811(8)	95.68(4))	HgCS (N) HgCS (N)	C 2.07(3) S 2.402(7) N 2.46(2) C 2.10(3) S 2.40(7) N 2.46(2)	C,S 160.9(8) C,S 162.3(7)	39
$Hg(Cl_3CH)Br$	or Pbnm 8	18.28 10.66 6.56		HgCBr HgCB	C not given Br 2.36(2) C not given) Br 2.43(2)	C,Br not given C,Br not given	40
$Hg(C_{12}H_{21}OSi)I$ (yellow)	m $P2_{1/n}$ 8	10.715(3) 9.539(2) 33.235(7)	95.43(2)	HgCI HgCI	C 2.10(2) I 2.604(2) C 2.08(2) I 2.594(2)	C,I 173.5(7) C,I 172.8(7)	41
Hg(Ph)(sami) (colourless)	tr P1̄ 4	10.148(2) 11.195(2) 14.386(2)	99.08(5) 109.56(5) 104.16(5)	HgNC (O) HgNC (O)	C 2.02(4) N 2.15(3) O 2.44(3) C 1.93(4) N 2.09(3) O 2.49(3)	N,C 167(1) N,C 166(1)	42

a. Where more than one chemically equivalent distance or angle is present, the mean value is tabulated. The first number in parenthesis is the e.s.d., and the second is the maximum deviation from the mean value.

b.The chemical identity of the coordinated atom or ligand is specified in these columns.

c.There are four crystalloqraphically independent molecules.

Two colourless, monoclinic $Hg(Ph)(Pr^iOCS_2)$ [19] and triclinic Hg(Me)(aza) [20] contain three such molecules. In the former each Hg(II) atom is diagonally coordinated (HgCS) with the sum of Hg-C plus Hg-S bond distances and C-Hg-S bond angles of 4.40Å and 176.6(4)° (molecule 1), 4.42Å and 175.6(3)° (molecule 2), and 4.49Å and 174.3(3)° (molecule 3). In the latter [20] also each Hg(II) atom has diagonal coordination (HgNC). The sum of Hg-N plus Hg-C bond distance increases with the deviation of N-Hg-C bond angle from the linearity (180°) as in the former derivative [19]. The values are 4.03Å and 177.0° (molecule 1), 4.13Å and 176.0° (molecule 2), and 4.22Å and 173.8° (molecule 3).

Remainders, twenty four derivatives [18,21-42] contain two crystallographically independent molecules (Table 2). Each Hg(II) atom has diagonal coordination with the following chromophores (HgC_2 [18,21-23], HgNC [24-28], HgCCl [29-36], HgCS [18,37-39], HgCBr [40], HgCI [41] and HgNC(O) [42]. The mean value of the sum of Hg-L(x2) bond distance (more crowded molecule vs. less crowded molecule) increases with the covalent radii of the coordinate atoms in the sequence: 4.10 vs. 4.20Å (HgNC,1.52Å) < 4.13 vs. 4.19Å (HgC_2, 1.54Å) < 4.38.vs 4.43Å (HgCCl, 1.76Å) < 4.45 vs. 4.45Å (HgCS, 1.79Å) < 4.67 vs. 4.70Å (HgCI, 2.10Å).

Inspection of the data in Tables 1 and 2 reveals that the mean value of Hg-L bond distance (more crowded molecule vs. less crowded molecule) elongated with the covalent radius of coordinated atom, except C atom, in the order: 2.07 vs. 2.13Å (N, 0.75Å) < 2.06 vs. 2.09Å (C, 0.77Å) < 2.35 vs. 2.37Å (Cl, 0.99Å) < 2.36 vs. 2.38Å (S, 1.02Å) < 2.36 vs. 2.43Å (Br, 1.14Å) < 2.59 vs. 2.61Å (I, 1.33Å).

2.1.3. Independent Molecules – Di-, Tri-, Tetra- and Polymeric

Table 3. Crystallographic and Structural parameters of di-, tri-, tetra- and Polymeric Mercury organometallics independent molecules[a]

COMPLEX (colour)	Crys.cl Sp.Gr Z	a [Å] b [Å] c [Å]	α [°] β [°] γ [°]	Chromo-phore	Hg-L [Å]	Hg-Hg [Å] Hg-L-Hg [°]	L-Hg-L [°]	Ref
[{Hg(Me)}$_2$(μ-app)]. NO_3.H_2O (colourless)	m $P2_1/_c$ 8	13.419(3) 13.676(6) 14.571	90.68(2)	HgNC HgNC	N[b] not given C -“- not given	3.39..2(1) 3.388(1)	N,C[b] not given N,C not given	43
[{Hg(Me)}$_{2(}$(μ-ad)]. EtOH (colourless)	m $P2_1/_c$ 8	7.999(3)) 24.056(8) 14.625(2)	90.56(2)	HgNC	N 2.060(19,3) C 2.046(22,59)		N,C 174.7(8,1.7)	44
[{Hg(Me)}$_3$($μ_3$-O)]. NO_3 (colourless)	m $P2_{1/c}$ 12	12.404 20.029 14.047	115.97	HgOC	$μ_3$O 2.08 C 2.15	3.44 115	O,C 175	45
[{Hg(NO_3)}$_3$($μ_3$-CCOOH)]HNO_3 (colourless)	m $P2_{1/c}$ 4	11.555(6) 7.670(3) 27.004(9)	90.23(1)	HgOC HgOC	O 2.14(4,1) $μ_3$C 2.07(5,1) O 2.09(5,6) $μ_3$C 2.07(6,2)	3.359(4,32) 109(2,2) 3.351(4,95) 108(2,5)	O,C 175(2,3) O,C 173(1,3)	46
[Hg($C_{12}H_8$)]$_3$ (colourless)	tr P$\bar{1}$ 4	11.983(5) 12.947(6) 20.068(8)	82.13(4) 80.28(4) 77.22(4)	HgC_2 HgC_2	C 2.116(16,29) C 2.124(16,11)	3.419(1,22) 3.422(1,28)	C,C 177.2(6,6) C,C 176.8(7,1.8)	47
[{Hg(Me)}$_3$. (μ-MeguH)]. NO_3 (colourless)	tr P$\bar{1}$ 4	11.101(3) 15.746(5) 11.042(5)	90.51(5) 91.62(4) 73.12(4)	HgNC HgNC	N 2.09(2,2) C 2.05(3,7) N 2.07(2,1) C 2.05(3,5)		N,C 176(1,3) N,C 176(1,1)	48
[{Hg(Me)}$_3$.(μ-ad)]. $(ClO_4)_2$ (colourless)	tr P$\bar{1}$ 4	6.304(7) 11.925(4) 25.946(4)	85.76(2) 89.78(5) 83.57(6)	HgNC	N 2.08(3) C 2.05(4)	3.290(2,20)	N,C 177(1)	49
(PPh_4) [{Hg_2(o-C_6H_4Cl)$_2$}$_2$($μ_4$-Cl)]. 0.5CH_2Cl_2 (colourless)	tr P$\bar{1}$ 4	13.351(8) 16.420(8) 20.200(17)	108.43(6) 94.48(6) 105.41(6)	$HgCl_2C$ $HgCl_2C$	Cl 2.326(4,38) $μ_4$Cl 3.076(4,259) C 2.056(14,45) Cl 2.334(4,24) $μ_4$Cl 3.017(4,107) C 2.062(13,12)	81.9(1,10.9) 157.7(1) 81.1(1,11.6) 159.0(1)	$μ_4$Cl,Cl 90.9(1,4.6) $μ_4$Cl,C 96.5(4,1.6) Cl,C 167.3(4) $μ_4$Cl,Cl 91.3(4,5.3) $μ_4$Cl,C 96.5(4,1.6) Cl,C 172.0(4,4.9)	50
[{Hg(Me)}$_4$(μ-xan)]. NO_3 (white)	m C2/c 16	24.55(2) 17.016(4) 19.108(5)	110.26(5)	HgNC	N 2.07(4,5) C 2.06(7,3)	3.353(4) 3.639(4)	N,C 175(2,5)	51
2[Hg_3(μ-O)(μ-CCHO)]. $(NO_3)_2$.HNO_3 (colourless)	or Cmc2_1 4	6.853(2) 12.043(3) 18.508(1)		HgOC HgOC	μO 2.17(-,1) μC 2.08(-,3) μO 2.15(-,2) μC2.11(-,1)	O 108.1(-,4) C 110.4(-,3) O 109.3(,1.0) C107.1(-,1.3)	O,C 178.3(-,1.0) O,C 175.8(-,9)	52

a. Where more than one chemically equivalent distance or angle is present, the mean value is tabulated. The first number in arenthesis is the e.s.d., and the second is the maximum deviation from the mean value.

b.The chemical identity of the coordinated atom or ligand is specified in these columns.

There are ten derivatives (two di- [43,44], five tri- [45-49], two tetra- [50,51] and one polymeric [52]) which contain two crystallographically independent molecules within the same crystal (Table 3). In two colourless monoclinic dimers [43,44] both Hg(II) atoms sit on a N-heterocyclic ligand (7-diaza-8-azaadenine [43] (Fig 1) or adenine [44] and the second coordination site is occupied by a methyl group (HgNC). Unfortunately, in the formed derivative only Hg ... Hg separations are available, with the values of 3.392(1)Å in dimer 1 and 3.388(1)Å in dimer 2 [43]. In the latter dimer [44] only average values of Hg-N and Hg-C bond distances are available.

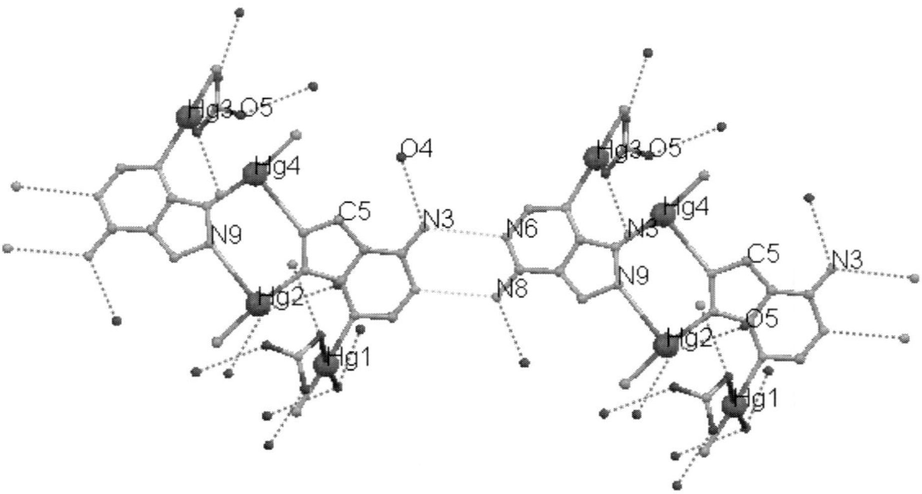

Figure 1. Structure of [{Hg(Me)}$_2$(μ-app)] [43].

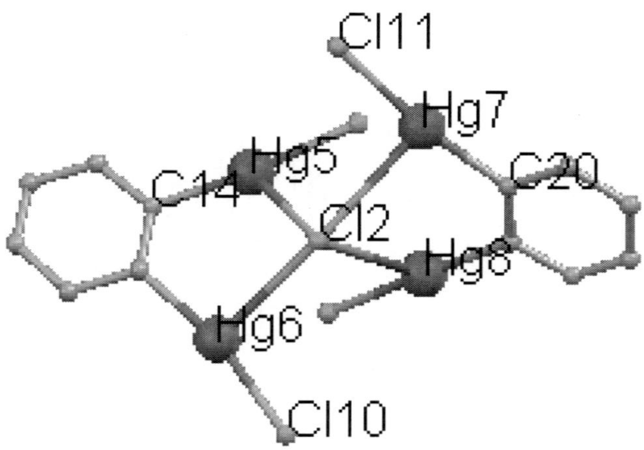

Figure 2. Structure of {Hg2(o-C6H4Cl)2}2(μ4-Cl)]- [50].

Structure of colourless monoclinic trimer [45] consists of well separated [{Hg(Me)}$_3$(μ_3-O)] cations and NO_3^- anions. In complex cation three crystallographically equivalent $(HgMe)_3$ units are symmetrically linked by a triply bridging O atom. In another colourless monoclinic trimer [46] three {Hg(ONO_2)}$_3$ units are symmetrically linked by a triply bridging C atom of

CCOOH ligand. Each Hg(II) atom is diagonally coordinated (HgOC). The mean values of the sum of Hg-O plus Hg-C bond distances are 4.16Å (dimer 1) and 4.21Å (dimer 2).

Another colourless triclinic trimer $[Hg(C_{12}H_8)]_3$ [47] has a fifteen – membered ring where three mercury (II) atoms are bridged by 2,2′-biphenylene groups. Each Hg(II) atoms has diagonal coordination (HgC_2), with the mean Hg-C bond distances of 2.116 and 2.124Å, respectively.

In another colourless triclinic trimers [48,49] all three mercury (II) atoms are sitting on N-heterocyclic ligands (4-MequH [48], and [49]) and the diagonal coordination is complete by a methyl group on each Hg(II) atom (HgNC).

Structure of colourless triclinic tetramer [50] consists of well separated PPh_4^+ cations, $[\{Hg_2(o\text{-}C_6H_4Cl_2)\}_2(\mu_4\text{-}Cl)]^-$ anions and 0.5 CH_2Cl_2 molecules. In the complex anion the real center is a chlorine atom (Fig. 2) with mean Hg-μ_4Cl distances of 3.017 and 3.076Å, respectively. In white monoclinic tetramer [51] all four mercury (II) atoms are sitting on purine base with the mean Hg-N bond distance of 2.07Å. Diagonal coordination about the mercury is completed by methyl group with a mean Hg-C bond distance of 2.06Å and mean N-Hg-C angle of 175°.

In $\{Hg_3(\mu\text{-}O)(\mu\text{-}CCHO)\}_n$ [52] diagonal coordinated Hg(II) atoms are linked via O and C donor ligands to form infinite chain polymer (Fig. 3).

Inspection of the data in Table 3 reveals that the mercury atoms are mostly diagonally coordinated HgNC [43,44,48,49,51], HgOC [45,46,52] and HgC_2 [47], and in one example trigonally, $HgCl_2C$ [50]. The mean Hg-L bond distance in the series of diagonal coordination elongated in the order: 2.07Å (NL, CL, μ_3Cl) < 2.08Å (μ_3O) < 2.10Å (μ-CL) <2.11Å (OL) < 2.16Å (μOL). The Hg...Hg separation range from 3.25 to 3.45Å (av. 3.40Å) excluded a direct bond.

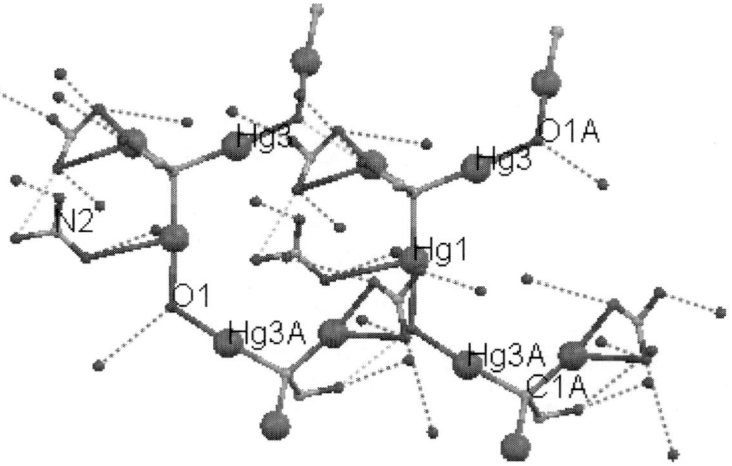

Figure 3. Structure of $\{Hg_3(\mu\text{-}O)(\mu\text{-}CCHO)\}_n$ [52].

2.2. Ligand Isomers

There are two colourless monoclinic examples $Hg(o\text{-}tolyl)_2$ [53] and $Hg(\eta\text{-}tolyl)_2$ [54] which are isostructural. Each mercury (II) atom is diagonally coordinated (HgC_2) with the mean Hg-C bond distances of 2.08Å [53] and 2.09Å [54]. The former is exactly linear (C-Hg-C, 180°) and the latter is almost linear, 178.0(4)°.

These are example of ligand isomerism (Table 4).

Table 4. Crystallographic and Structural Data of ligand isomers and temperature evoke isomerism[a]

COMPLEX (colourless)	Crys.cl Sp.Grp Z	a [Å] b [Å] c [Å]	α [°] β [°] γ [°]	Chromo- phore	Hg-L [Å]	L-Hg-L [°]	Ref
A: Ligand isomers							
$Hg(p\text{-}tolyl)_2$ (colourless)	m $P2_1/c$ 2	10.12(1) 5.07(1) 11.65(2)	 91.8(3)	HgC_2[b]	C 2.08(2,0)	C,C[b] 180	53
$Hg(o\text{-}tolyl)_2$ (colourless)	m C2/c 4	10.97(2) 10.448(3) 11.409(3)	115.48(2)	HgC_2	C 2.09(1,0)	C,C 178.0(4)	54
B: Temperature evoke isomers							
$Hg(CF_3)_2$ (colourless)	or Pbca 4	8.118 8.115 8.113		HgC_2	C 2.139	C,C not given	7
$Hg(CF_3)_2$ (colourless) (at 99K)	or Pbca 4	7.957 7.957 7.957		HgC_2	C 1.998	C,C notgiven	7
$Hg(C_{10}H_{15})Cl$ (colourless)	or $P2_12_12_1$ 4	14.928(1) 11.784(1) 6.1514(7)		HgCCl	C 2.09(2) Cl 2.352(2)	C,Cl 177.7(3)	55
$Hg(C_{10}H_{15})Cl$ (colourless) (at 153K)	m P21 2	6.943(2) 11.915(4) 6.831(2)	 106.89(2)	HgCCl	C 2.08(2) Cl 2.35(2)	C,Cl 175	56

a. Where more than one chemically equivalent distance or angle is present, the mean value is tabulated. The first number in parenthesis is the e.s.d., and the second is the maximum deviation from the mean.

b.The chemical identity of the coordinated atom or ligand is specified in these columns.

2.3. Isomers Evoke by Temperature

There are two derivatives $Hg(CF_3)_2$ [7] and $Hg(C_{10}H_{15})Cl$ [55,56] which were studied at two different temperature, the former at ambient and at 99K, and the latter at ambient [55] and at 153K [56].

The colourless $Hg(CF_3)_2$ at both temperatures is orthorhombic, when temperature decreases, the unit cell parameters decreases as well as the Hg-C bond distance. The Hg(II) atoms are diagonally coordinated (HgC_2) with the mean Hg-C bond distances of 2.139Å (at ambient temperature) and 1.998Å (at 99K).

X-ray analysis of colourless $Hg(C_{10}H_{15})Cl$ at ambient temperature shows [55] that the complex is orthorhombic and Hg(II) atom has diagonal coordination (HgCCl) with C-HgCl bond angle of 177.7(3)°. However, at 153K the complex become monoclinic and Hg(II) atom is also diagonally coordinated (HgCCl) with C-Hg-Cl bond angle of 175° [56].

Both derivatives [7,55,56] are classical examples of distortion isomerism.

3. Mercury Heterometallic Complexes

There are nine examples of distortion isomers and two examples of linkage isomer.

3.1. Distortion Isomers

The [η^5-cp)$_2$MoHg$_2$(SEt)$_2$] [57] exists in two isomeric forms, orange and red and both crystallized in the monoclinic crystal class. In both the molybdenum is a central core with two other „external" mercury atoms (Hg-Mo-Hg) with the mean Hg-Mo bond distances of 2.676Å (orange) and 2.677Å (red). The Hg-Mo-Hg bond angles are 73.1° and 70.8°, respectively. Each of the SEt group terminally coordinated to the mercury atom with the mean Hg-S bond distance of 2.43Å. Each central molybdenum atom is also sandwiched between two cyclopentadienyl molecules, with cp-Mo-cp (centroid) angles of 140.5°. The S-Hg-Mo bond angles are 171.2° (orange) and 167.6° (red) (Table 5).

Crystal of yellow triclinic derivative [58] contains three crystallographically independent $HgCl_4^{2-}$ anions and two $[Cr(\eta^2\text{-en})_3]^{3+}$ cations. Each Hg(II) atom is tetrahedrally coordinated with the mean Hg-Cl bond distances of 2.472, 2.513 and 2.517Å.

There are six derivatives HgMo [59], $HgPt_2$ [60], $HgCo_2$ [61], $HgCr_2$ [62], Hg_2Fe_2 [63] and $HgCo_3$ [64], which contain two crystallographically independent molecules within the same crystal (Table 5). In tridinic yellow HgMo dimer [59] two units (Ph_3P)Hg and Mo(COI$_3$(η^5-cp) are held together by direct Hg-Mo bond with the distance of 2.685(1)Å (molecule 1) and 2.687(1)Å (molecule 2). The mercury atoms are diagonally coordinated (HgPMo) with Hg-P bond distances and P-Hg-Mo bond angles (molecule 1 vs. molecule 2) of 2.490(2)Å and 169.7(1)° vs. 2.476(2)Å and 160.8(1)°.

In triclinic orange $HgPt_2$ trimer [60] the metal atoms form a triangle with Pt-Hg-Pt angles of 59.96(2) and 60.45(2)°. Two chlorine atoms are coordinated to the mercury atom and completed highly distorted tetrahedron about each mercury atom ($HgCl_2Pt_2$). The mean Hg-Cl and Hg-Pt bond distances (trimer 1 vs trimer 2) are 2.429 and 2.714Å vs 2.447 and 2.704Å. The dppm ligand serves as a bridge between platinum atoms by its P atoms and in addition chlorine atom is terminally coordinated to each Pt atom. Each platinum atom is five coordinated ($PtP_2ClHgPt$). The sum of all four [Hg-Cl(x2) plus Hg-Pt(x2)] and five [Pt-P(x2) plus Pt-Cl, Pt-Hg and P-Pt] bond distances of 10.28 and 12.36Å in trimer 1 are somewhat smaller than those in trimer 2 (10.30 and 12.38Å) indicates that the trimer 1 is somewhat more crowded than trimer 2.

The structure of [{(cp)Co(P(O)(OEt)$_2$)$_3$}$_2$Hg] [61] contains two crystallographically independent molecules. It consists of a linear Co...Hg....Co arrangement bridged by diethylphosphanate groups, with Co...Hg separations of 4.065(1)Å in one molecule and 4.078(2)Å in the other. The Hg atoms are surrounded by six phosphanate oxygen atoms forming a tetragonal distortion about Hg with Hg-O bond lengths of 2.270 and 2.292Å, respectively. Each Co atom has CoC_5P_3 inner coordination sphere. The sum of Hg-O(x6) and Co-C(x5) plus Co-P(x3) bond distances (trimer 1 vs trimer 2) are 13.62 and 16.90Å vs 13.74 and 16.97Å indicated that the trimer 1 is somewhat more tighter than the trimer 2.

Table 5. Crystallographic and Structural Data for Heterometallic Mercury Complexes – Distortion and Linkage Isomers[a]

COMPOUND (colour)	Crys .cl Sp. Gr Z	a [Å] b [Å] c [Å]	α [°] β [°] γ [°]	Chromo-phore	M-L [Å]	Hg-M [Å]	L-M-L [°]	Ref
A: Distortion isomers								
$[(\eta^5\text{-cp})_2Mo.Hg_2.$ $(SEt)_2]$ (orange)	m C2/c 2	31.503(8) 7.986(3) 20.843(1)	97.87(3)	HgSMo MoC_{10}	S^b 2.43(2.1) η^5C not given	Mo2.676(6,6)	Mo,S^b 171.2(1) cp,cp 140.5 Hg,Hg 73.1	57
$[(\eta^5\text{-cp})_2Mo.Hg_2.$ $(SEt)_2]$ (red)	m $P2_1/c$ 4	10.342(5) 17.757(10) 10.193(8)	 113.39(4)	HgSMo MoC_{10}	S 2.43(2,2) η^5C not given	Mo2.677(6,13)	S,Mo 167.6 cp,cp 140.5 Hg,Hg 70.8	57
$[HgCl_4]_3.$ $[Cr(\eta^2\text{-en})_3]_2$ (yellow)	tr $P\bar{1}$ 2	11.250(2) 11.245(1) 17.421(4)	81.66(1) 72.25(2) 90.01(2)	$HgCl_4$ $HgCl_4$ $HgCl_4$ CrN_6	Cl 2.472(4,6) Cl 2.513(4,99) Cl 2.515(4,97) N 2.074(9,22)		Cl,Cl 109.5(1,1.1) Cl,Cl 109.8(1,26.7) Cl,Cl 109.7(1,25.9) N,N 170.6(4,2.6)	58
$[(Ph_3P)HgMe.$ $(CO)_3(\eta^5\text{-cp})].$ BF_4 (yellow)	tr $P\bar{1}$ 4	11.294(1) 12.147(1) 20.401(1)	98.813(6) 102.615(6) 91.362(7)	HgPMo MoC_8Hg HgPMe MoC_8Hg	Ph_3P 2.476(2) OC 1.963(8,9) cpC 2.32(3) Ph_3P 2.490(2) OC 1.963(9,40) cpC 2.31(2)	Mo 2.687(1) Mo 2.685(1)	P,Mo 160.8(1) C,C 76.4(4,2) 110.3(3) P,Mo 169.7(1) C,C 78.4(4,7) 106.1(3)	59
$Cl_2HgPt_2Cl_2(\eta^2\text{-dppm})_2$ (orange)	tr $P\bar{1}$ 4	11.587(3) 22.156(5) 23.036(5)	65.53(1) 76.53(2) 84.30(2)	$HgCl_2Pt_2$ PtP_2ClHg Pt $HgCl_2Pt_2$ PtP_2ClHg Pt	Cl 2.429(4,12) P 2.301(4,10) Cl 2.334(4,4) Cl 2.447(5,2) P 2.300(4,11) Cl 2.343(4,7)	Pt 2.714(1,2) Pt 2.712(1) Pt 2.704(1,5) Pt2.736(1)	Cl,Cl 106.0(1) Pt,Pt 59.96(2) Cl,Pt 122.5(1,20.0) P,P 173.3(1,1.1) P,Cl 88.6(1,2.4) Cl,Cl 103.7(1) Pt,Pt 60.78(2) Cl,Pt 123.0(1,16.1) P,P 174.3(1,1) P,Cl 87.4(1,1.7)	60
$[\{\eta^5\text{-cp})Co(\mu\text{-}\eta^1\text{:}\eta^1\text{-}$ $P(O).$ $OEt)_2)_3\}_2Hg]$	tr $P\bar{1}$ 1	12.042(4) 20.359(11) 12.694(6)	101.51(4) 118.51(2) 92.20(4)	HgO_6 CoC_5P_3 HgO_6 CoC_5P_3	O 2.292(12,24) cpC 2.09(2,1) P 2.174(6,2) O 2.27(12,14) cpC 2.07(2,3) P 2.185(8,25)	Co 407.8(2) Co 4.065(2)	O,O 85.7(4,1.1) P,P 92.3(2,6) O,O 86.3(5,5) P,P 91.3(3,3)	61
$[\{(OC)_3Cr(\mu\text{-}Ph)\}_2Hg$ (at 153K)	m $P2_{1/c}$ 4	11.143(3) 14.191(6) 12.195(4)	 116.56(2)	HgC_2 CrC_4 HgC_2 CrC_4	Ph,C 2.074(6) OC 1.851(8,7) PhC 2.228(8,15) PhC 2.066(7) OC 1.848(8,4) PhC 2.224(8,19)		C,C 180 C,C not given C,C 180 C,C not given	62
$[\{(\eta^5\text{-cp})Fe(\mu\text{-}\eta^5\text{:}\eta\text{-}$ $C_5H_3CH_2NMe_2)Hg\}_2]$ (orange) (at 163K)	tr $P\bar{1}$ 2	10.062(4) 11.684(4) 12.517(4)	114.46(3) 111.33(3) 96.17(3)	HgC_2 FeC_{10} HgC_2 FeC_{10}	η^1C 2.053(7,12) η^5C 2.063(7,27) η^1C 2.060(6,5) η^5C 2.054(7,29)	Hg 3.323(2) Hg 3.362(2)	C,C 177.1(3) C,C not given C,C 176.6(3) C,C not given	63

COMPOUND (colour)	**Crys .cl Sp. Gr Z**	**a [Å] b [Å] c [Å]**	**α [o] ß [o] γ [o]**	**Chromo-phore**	**M-L [Å]**	**Hg-M [Å]**	**L-M-L [o]**	**Ref**
$(NMe_4)[Hg\{Co(CO)_4\}_3$ (not given)	m $P2_{1/a}$ 8	16.286(3) 16.460(3) 19.213(4)	 96.84(1)	$HgCo_3$ CoC_4 $HgCo_3$ CoC_4	 OC 1.771(24,63) OC 1.753(24,43)	Co 2.686 (2,26) Co 2.686(2,2)	Co,Co120.0(1,1. 6) C,C 80.9(6,7.1) 173.6(7,2.2) Co,Co120.0(1,4. 8) C,C 80.8(6,8.3) 171.8(7,4.7)	64
B: Linkage isomers								
$[HgAu.\{Ph_2(S).$ $PCH_2\}_2]PF_6$ (light yellow)	m C2/c 8	29.761(7) 8.824(2) 27.422(2)	 125.39(2)	HgCS AuCS	C 2.09(2) S 2.336(5) C 2.10(2) S 2.337(5)	Au 2.989(1)	C,S 174.8(7) C,S 175.2(7)	65
$[HgAu\{Ph_2(S).$ $PCH_2\}_2]PF_6$ (yellow)	m P2/c 8	30.356(6) 14.028(5)) 16.491(5)	 121.48(2)	HgC_2 AuS_2	C 2.14(2,1) S 2.306(6,6)	Au 3.088(1)	C,C 175.0(7) S,S 175.0(2)	65

a. Where more than one chemically equivalent distance or angle is present, the mean value is tabulated. The first number in parenthesis is the e.s.d., and the second is the maximum deviation from the mean. b.The coordinated atom or ligand is specified in these columns.

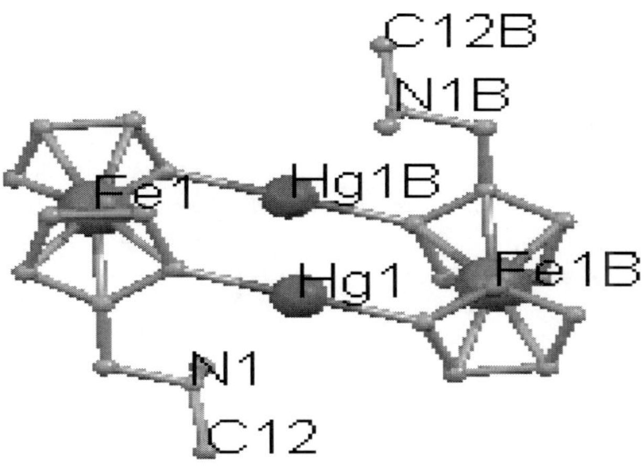

Figure 4. Structure of $[\{(cp)Fe(\mu\text{-}C_5H_3CH_2NMe_2)\}_2Hg_2]$ [63].

In monoclinic $HgCr_2$ trimer [62] a almost linear Cr...Hg...Cr arrangement is bridged by phenyl groups. The Hg atom is linearly coordinated C-Hg-C, 180°, with the mean Hg-C bond distances of 2.066Å (trimer 1) and 2.074Å (trimer 2). Three CO groups about each Cr atom completed four coordination Cr atoms (CrC_4). The sum of Hg-C(x2) and Cr-C(x4) bond distances (trimer 1 vs trimer 2) are 4.13 and 7.77Å vs 4.15 and 7.78Å.

Structure of triclinic orange Hg_2Fe_2 tetramer [63] is shown in Figure 4. As can be see two substituted cyclopentadienyls serve as bridges. Each mercury atom has almost linear coordination (HgC_2). The Hg....Hg separations are 3.323(2) and 3.362(2)Å, respectively. Each iron atom is sandwiched (FeC_{10}).

In monoclinic $HgCo_3$ tetramer [64] the mercury atom is coordinated by three $Co(CO)_4$ units with a mean Hg-Co bond distance of 2.686Å and Co-Hg-Co angles of 120.0°

3.2. Linkage Isomers

The monoclinic $[HgAu\{Ph_2(S)PCH_2\}_2]PF_6$ exists in two isomeric forms a light yellow and yellow [65]. In light yellow form one $Ph_2(S)PCH_2$ anion is bonded to Hg via C and the other anion via S (HgCS) and correspondingly to the Au atom (AuSC). In yellow form two $Ph_2(S)PCH_2$ anions are bonded to the Hg atom via C atoms (HgC_2) and to the Au atom via S atoms (AuS_2) (Fig. 5).The pair of derivatives are typical examples of linkage isomerism [5].

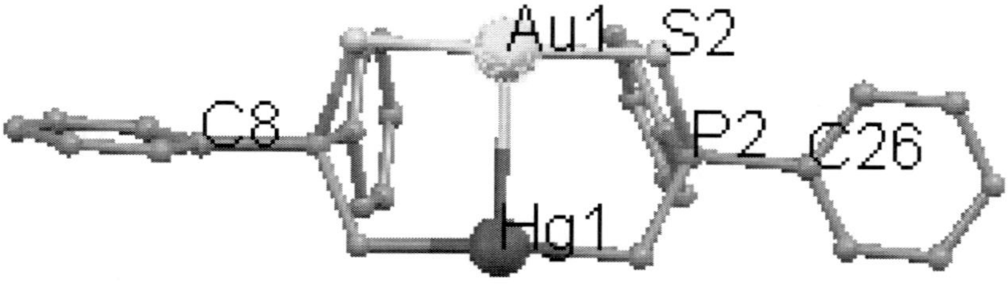

Figure 5. Structure of $HgAu\{Ph_2(s)PCH_2\}_2]^+$ [65].

Conclusion

This review summarizes the crystallographic and structural data for over sixty isomers of mercury organometalics and heterometallics. These isomers are classifies there are three categories of the isomers, namely distortion, ligand and linkage. The distortion isomers by far prevails (88.5%). The distortion isomers can be divided into several subgroups. Ten derivatives [6-16] exist in two isomeric forms. In five derivatives, the isomers differ from each other not only by degree of distortion, but also by crystal class, one isomer is cubic [6] and another is orthorhombic [7], in the remaining four derivatives, one isomer is monoclinic [8,9,11,13] and the other is orthorhombic [8,10,12,13]. In another five derivatives, both isomers belong to the homo-crystal classes: monoclinic [14], triclinic (3 pairs) [15] and orthorhombic [16].

There are two pairs of isomers, $Hg(CF_3)_2$ both are orthorhombic [7] and $Hg(C_{10}H_{15})Cl$ orthorhombic [55] and monoclinic [56], where temperature evokes distortion isomers (Table 4B).

Remainders derivatives contain crystallographically independent molecules within the same crystal; differ mostly by degree of distortion. In one example [17] exists five such molecules, in another one [18] four, in three derivatives exist three such molecules [19,20,58], and the forty five derivatives (Tables 2,3,5) contain two such molecules. These examples exist in four crystal classes monoclinic (x33), triclinic (x22), orthorhombic (x16) and cubic (x1).

Mercury atoms are two, three, four and even six coordinated. The summary of $Hg\text{-}L_n$ (n=2,3,4 or 6) bond distances as well as L-Hg-L bond angles (only for two coordinated derivatives) is given in Table 6. As can be see the sum of the $Hg\text{-}L_n$ bond distance increases with the covalent radii of the respective donor atoms, as well as with the coordination number, as expected. The sums of the more crowded isomers are some what smaller than for their partners.

There are two monoclinic derivatives Hg(p-tolyl)$_2$ [53] and Hg(o-tolyl)$_2$ [54] which are examples of ligand isomerism. The [HgAu{$Ph_2(S)PCH_2$}$_2$]PF_6 which exists in two monoclinic forms are examples of linkage isomerism, with the chromophores HgC_2 and AuS_2 in one isomer and HgCS and AuCS in another one.

Table 6. Summary of the mean Hg-L$_n$ bond distances and L-Hg-L bond angles of distortion isomers more vs less crowded

CROWDED CHROMOPHORE	MORE		LESS	
	Hg-L[Å]	L-Hg-L[°]	Hg-L[Å]	L-Hg-L[°]
HgNC	4.14(3.98-4.18)	173.6(166-178.3)	4.17(3.98-4.28)	175.2(167-178.5)
HgOC	4.20(4.16-4.25)	175.6(173-178)	4.24(4.21-4.26)	174.4(175-175.8)
HgC_2	4.21(3.99-5.00)	177.7(168-180)	4.27(4.12-5.00)	176.6(165-180)
HgCCl	4.42(4.33-4.65)	175.9(166-180)	4.48(4.39-4.487)	176.1(172-178.7)
HgCS	4.44(4.40-4.47)	172.4(160.9-177.7)	4.48(4.45-4.50)	172.5(162.3-198.8)
HgCI	4.67	172.8	4.70	173.5
HgSMo	5.10	171.2	5.11	167.6
HgPMo	5.16	166.8	5.18	169.7
HgONC	6.58	not given	6.69	not given
$HgCl_2C$	7.41	not given	7.46	not given
$HgCo_3$	8.06	not given	8.06	not given
$HgCl_4$	9.89	not given	10.06	not given
$HgCl_2P_2$	10.28	not given	10.30	not given
HgO_6	13.62	not given	13.75	not given

This review together with its precursor [4], represents the first overall survey of the structural data of mercury complexes. The mercury complexes exist in the following types of isomers: distortion (by far prevails), polymerization, coordination number, ligand and linkage isomerism. It is hoped that review, will help to focus attention on areas of mercury chemistry, that could be enhanced by further study, and assist in the predicting the behavior of the mercury atom in a variety of important biological and environmental situations. In gathering the data it became evident that some of it is buried in the literature and effectively invisible to automated data retrieval searches for a variety of reasons in other cases incorrect information had been promulgated in the literature. This review sets out to systematize and correct where necessary, the information and provide a base upon which new information can grow.

ACKNOWLEDGMENTS

The author wish to thank those who gave permission for reproduction of figures, the Ministry of Education of the Slovak Republic for financial support.

REFERENCES

[1] N.D. Madsen, in R.M. Hochster and J.H. Quastel (eds.) *Metabolic Inhibition, Academic Press*, New York,Vol. II, 1963, Chapter 21.

[2] C.E. Holloway and M. Melník, *J. Organomet. Chem.*, 495 (1995) 1.

[3] C.E. Holloway and M. Melník, *Main Group Met. Chem.,* 17 (1994) 741.

[4] M. Melník, *Cent. Eur. J. Chem.,* 8 (2010) 469.

[5] M. Melník, *Coord. Chem. Rev.*, 47 (1982) 239

[6] D.J. Brauer, H. Burger and R. Eujen, *J. Organomet. Chem.*, 135 (1977) 281.

[7] D. Grdenic, B. Kamenar and Z. Dodic, Izv. Jug. *Cent. Krist., Ser. A*, 11 (1976) 28.

[8] D.J. Brauer, H. Burger, G. Pawelke, K.H. Flegler and A. Haas, *J. Organomet. Chem.*, 160 (1978) 389.

[9] M.C. Corbeil, A.L. Beauchamp, S. Alex and R.Savoie, *Can. J. Chem.*, 64 (1986) 1876.

[10] M.C. Corbeil and A.L. Beauchamp, *J. Crystallogr. Spectr. Res.*, 19 (1989) 123.

[11] L. Prizant, M.J. Olivier, R. Rivest and A.L. Beauchamp., *Can. J. Chem.*, 59 (1981) 1311.

[12] J.P. Charland and A.L. Beauchamp, *Croat. Chem. Acta*, 57 (1984) 679.

[13] F. Guay and A.L. Beauchamp, *Can. J. Chem.,* 63 (1965) 3456.

[14] W.S. Sheldrick and P. Bell, *Inorg. Chim. Acta*, 137 (1987) 181.

[15] A.G. Brook and G.F. Wright, *Acta Crystallogr.*, 4 (1951) 50.

[16] C.L. Raston, B.W. Skelton and A.H. White, *Aust. J. Chem.*, 31 (1978) 537.

[17] L.G. Kuzmina, L.A. Litvinov and M.A. Poraj – Koshits, *Koord. Klum.*, 15 (1989) 902; *Engl. Ed.*, p. 539

[18] A.J. Carty and B.L. Gatehouse, *Acta Crystallogr.*, Sect. B, 28 (1972) 1872

[19] E.R. Tiekink, *J.Organomet. Chem.*, 322 (1987) 1.

[20] P. Dufour, Z. Dartiguenave, M. Dartiguenave, N. Dufour,A.M. Lebuis, F. Belanger-Gariepy and A.L.Beauchamp, *Can.J. Chem.*, 68 (1990) 193.

[21] N.G. Formanova, Y.T. Struchkov, V.N. Kalanin, B.Y.Finelshtein and L.I. Zakharin, Zh. Strukt. Khim., 21 (1980) 96; *Engl. Ed.*, p. 640.

[22] S.J. Goede, H.P. van Shaik,F. Bickelhaupt,H. Kooijman and A.L. Spek, *Organometallics*, 11 (1992) 3844.

[23] E. Gutierrez –Puebla, A. Vegas and S. Garcia-Blamco, *Cryst. Struct. Commun.*, 8 (1979) 861.

[24] F. Guay and A.L. Beauchamp, *Inorg. Chim. Acta*, 66 (1982) 57.

[25] Y. Kobayashi, Y. Iitaka and Y. Kido, *Bull. Chem. Soc. Jpn.,* 43 (1970) 3070.

[26] L.G. Kuzmina, Y.T.Struchkov, D.N. Kravisov and L.S. Kravtsvov, *Zh. Strukt. Khim.,* 23 (1982) 102.

[27] L.G. Kuzmina, N.G. Bokii, Y.T. Struchkov, V.I. Minkin, L.P. Olekhnovich and I.E. Mikhailov, *Zh. Strukt. Khim.*, 18 (1977) 122; Engl. Ed., p.96

[28] J. Halfpenny and R.W.H. Small, *Acta Crystallogr.*, Sect.B, 36 (1980) 138.

[29] L.G. Kuzmina,Y.T. Struchkov, A.M. Chernov,V.R. Karpatsov, N.V.Malisova and E.V. Skorobogatova, *Zh. Strukt. Khim.*, 28 (1987) 109; Engl. Ed., p. 726.

[30] E.C. Constable, A.M.W. Cargill Thompson, T.A. Leese, D.G.F.Reese and D.A. Tocher, *Inorg. Chim. Acta*, 182 (1991) 93.

[31] L.G. Kuzmina and M.A.Poraj-Koshits, *Koord. Khim.*, 15 (1989) 185.

[32] M.G.B. Drew, J. Mann and B. Pietziak, *J. Chem. Soc.*, Perkin Trans. I, (1985) 1049.

[33] J. Lorberth, T.F. Berlitz and W. Massa, *Angew. Chem. Int. Ed. Engl.*, 28 (1989) 611.

[34] V.R. Karpashov, T.N. Sokolova, E.V. Skorobogatova, Y.K. Greshin, D.V. Bashenov, A.N. Chernov and N.S. Zefirov, *Zh. Org. Khim.*, 24 (1988)1684.

[35] R.C. Larock, L.D. Burns, S.Varaprath, C.E. Russell, J.W. Richardson, Jr., M.N. Janakiraman and R.A. Jacobson, *Organometllics*, 6 (1987) 1780.

[36] A.Gambacorta,R. Nicolettis, S. Cerrini, W. Fedeli and G. Gavuzzo, *Tetrahedron*, 36 (1980) 1367.

[37] J. Bravo,J. Casas, M.V. Castano, M.Gayoso,Y.P. Mascarenhas, A. Sanchez, C. de O.P. Santos and J. Sordo, *Inorg. Chem.*, 24 (1985) 3435.

[38] Yau Shing Wong, C. Chieh and A.J. Carty, J. Chem. Soc., Chem. *Commun.*, (1973) 741; Yau Shing Wong, A.J. Carty and C. Chieh, *J. Chem. Soc., Dalton Trans.,* (1977) 1801.

[39] L.G. Kuzmina,Y.T. Struchkov, E.M. Rokhlina and D.N. Kravtsov, *Zh. Strukt., Khim.*, 24 (1983) 130; *Engl. Ed.*, p. 764.

[40] T.A. Babushkina, E.V. Bryukhova, F.K. Velichko, V.I. Pakhomov and G.K. Semin, *Zh. Strukt. Khim.*, 9 (1968) 207; *Engl. Ed.*, p. 153.

[41] M.A. Bonaventura, J. Drouin, F. Theobald and N. Rodier, *Bull. Soc. Chim. Fr.*, (1987) 1006.

[42] L.G. Kuzmina, N.G. Bokii, Y.T. Struchkov, V.I. Miakin, L.P. Olekhnovich and I.E. Mikhailov, *Zh. Strukt. Khim.,* 15 (1974) 659; *Engl. Ed.*, p. 561.

[43] W.S. Sheldrick, P.Bell and H.J. Hausler, *Inorg. Chim. Actra*, 163 (1989) 181.

[44] J.P. Charland, *Inorg. Chim. Acta,* 135 (1987) 191.

[45] M. Bruvo, D. Grdenic, B.Kamenar and S. Popov, *Izv. Jug. Cent. Krist., Ser. A,* 11 (1976) 24.

[46] D. Grdenic, M. Sikirica and D. Matkovic-Calogovic, *J. Organomet. Chem.*, 306 (1986)1.

[47] K. Stender, W. Hinriche, J. Kopf and G. Klar, *Cryst. Struct. Commun.,* 10 (1981) 613.

[48] W.S. Sheldrick and P. Gross, *Inorg. Chim. Acta,* 153 (1988) 247.

[49] J. Hubert and A.L. Beauchamp, *Can. J. Chem.*, 58 (1980)1439.

[50] . J.D. Wuest and B. Zacharie, Organometallics, 4 (1985) 410; A.L. Beauchamp, M. J. Olivier, J.D. Wuest and B. Zacharie, *J. Amer. Chem. Soc.*, 108 (1986) 73.

[51] F. Allaire and A.L. Beauchamp, *Can. J. Chem.,* 62 (1984) 2249.

[52] D.Grdenic, D.Matkovic-Calogovic and M. Sikirica, *J. Organomet. Chem.*, 319 (1987) 1.

[53] M. Mathew and N.R. Kunchur, *Can. J. Chem.*, 47 (1969) 429.

[54] D. Liptak, W.H. Iisley, M.D. Glick and J.P. Oliver, *J.Organomet. Chem.*,191 (1980) 11.

[55] V.G. Andrianov, Y.T. Struchkov, V.A. Blinova and I.I. Krickaya, *Izv. Akad. Nauk SSSR, Ser. Khim,* (1979) 2021.

[56] I.V. Shirina-Einzorn, L.G. Kuzmina, A.G. Makarovskaja, Y.T. Struchkov and I.I. Krickaya, *Izv. Akad. Nauk SSSR, Ser. Khim.*, (1983) 374.

[57] M.M. Kubicki, R. Kergoat, J.E. Guerchais, I. Bkouche-Waksman, C. Bois and P. L'Haridon, *J. Organomet. Chem.*, 219 (1981) 329.

[58] D.A. House, V. McKee and W.T. Robinson, *Inorg. Chim. Acta,* 157 (1989) 15.

[59] D.F. Mullica, S.L. Gűpson, L.C. Snell,E.L. Sappenfield and D.H. Leschnitzer, *Polyhedron*, 11 (1992) 2265.

[60] P.R. Sharp, *Inorg. Chem.*, 25 (1986) 4185.

[61] E.Kamenar, R. Hergold-Brundic and M. Bruwo, *Z. Kristallogr.*, 184 (1988) 103.

[62] L.G. Kuzmina and Y.T. Struchkov, *Koord. Khim.*, 8 (1982) 1542.

[63] L.G. Kuzmina, Y.T. Struchkov, D.A. Lemenovskii, I.F. Urazovsky, I.E. Nifant'ev and E.G. Perevalova, *Koord. Khim.*, 9 (1983) 1212; *Engl., Ed.*, p. 694.

[64] J.M. Brulitch, J.M. Ragosta and M.C.Van Derveer, *Acta Crystallogr., Sect. C*, 40 (1984) 1549.

[65] S. Wang and J.P. Fackler, Jr., Organometallics, 7 (1968) 2415; *Inorg. Chem.*, 28 (1989) 2615.

In: Mercury: Sources, Applications and Health Impacts
Editors: Ki-Hyun Kim and Richard J. C. Brown
ISBN: 978-1-62257-721-7

Chapter 4

GENERATION OF SUPEROXIDE BY THE INTERACTION BETWEEN MERCURY IONS AND ENDOGENOUS THIOLS

Margarita E. Aliaga[1], **Alexis Vásquez-Arce**[1], **Camilo López-Alarcón**[1], **Claudio Olea-Azar**[2], **Germán Barriga**[2] and **Hernán Speisky**[2,3]*

[1]Facultad de Química, Pontificia Universidad Católica de Chile, Santiago, Chile
[2]Faculty of Chemical and Pharmaceutical Sciences, University of Chile, Santiago, Chile
[3]Nutrition and Food Technology Institute, University of Chile, Macul, Santiago, Chile

ABSTRACT

The interaction between Hg^{2+} ions and endogenous thiols (RSH), such as glutathione (GSH), cysteine (Cys), homocysteine (Hcy), cysteinyl-glycine (CysGly) and γ-glutamyl-cysteine (γ-GluCys), leads to the swift formation of the Hg(II)-thiol complexes. We investigated the potential capacity of Hg(II)-[RSH] and Hg(II)-$[RSH]_2$ complexes to reduce oxygen into superoxide.

Among these complexes, only Hg(II)-$[GSH]_2$, Hg(II)-$[Cys]_2$ and Hg(II)-[Hcy] were able to reduce oxygen in a concentration-dependent manner. The generation of the superoxide induced by these complexes was confirmed by EPR spin-trapping. Further evidence on the ability of the complexes to generate superoxide was attained through the demonstration of their capacities to reduce cytochrome *c* and/or to oxidize dyhydroethidium (two superoxide-susceptible probes), both effects being SOD-inhibitable.

In the case of the solutions containing either Hg(II)-$[CysGly]_n$ or Hg(II)-$[\gamma\text{-GluCys}]_n$ (n = 1 or 2) no redox activity toward molecular oxygen was observed.

* Corresponding author: Facultad de Química, Pontificia Universidad Católica de Chile, Av. Vicuña Mackenna 4860, Santiago 6094411, Chile.

The incubation of the complexes Hg(II)-$[GSH]_2$, Hg(II)-$[Cys]_2$ and Hg(II)-[Hcy] with oxygen during 60 min led to the hydrogen peroxide-dependent oxidation of acetaminophen (inhibited by catalase) and an increment in the formation of their oxidized thiols (oxidized glutathione, cystine and homocystine, respectively).

Thus, the ability of the complexes to generate superoxide and, thereby, to deleteriously affect superoxide-susceptible biological targets warrants further studies.

1. INTRODUCTION

Mercury is a heavy transition metal that possess mostly three distinct oxidation states 0, +1 and +2, although interestingly, quantum chemical predictions have also suggested the existence of mercury in its oxidation state +4 [1]. The mercuric state (Hg^{2+}) can form stable organic compounds by attaching one or two carbon atoms (i.e. CH_3Hg^+, $C_2H_5Hg^+$ and $(CH_3)_2Hg$) [2-3].

Hg is considered one of the most toxic and dangerous heavy metal elements; specifically in its state +2 has a very high affinity for sulfhydryl groups especially those present in proteins, enzymes and endogenous thiols (RSH; Scheme 1). In the latter case, formation constants for mercury bonding to thiolate anions are of the order of $\geq 10^{40}$ [4-6]. In an excess of an endogenous thiol, mercuric ions (Hg^{2+}) have a high propensity toward the formation of compounds with linear geometry and the coordination number 2 [7]. Thus, the interaction between Hg^{2+} and RSH leads to the favorable generation of the Hg(II)-$[RSH]_2$ complexes.

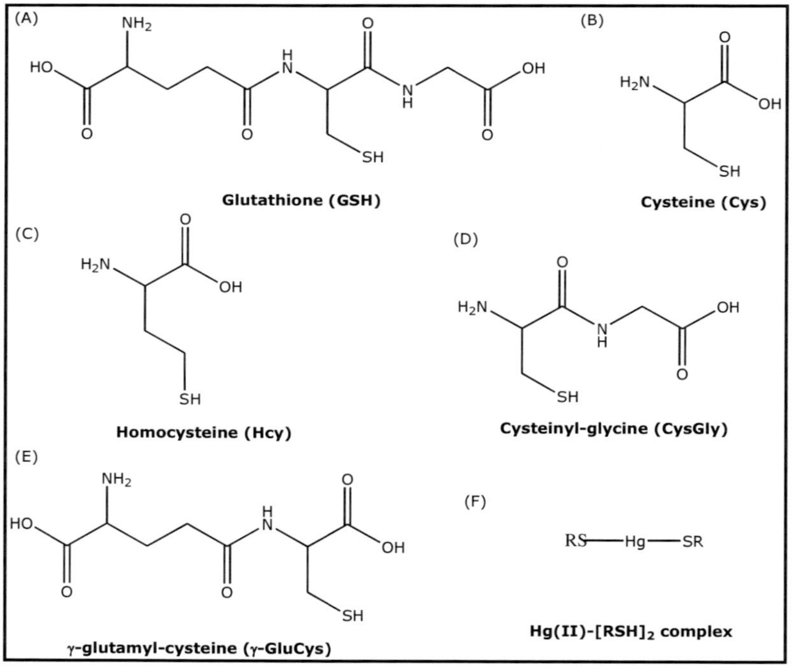

Scheme 1. Structures of endogenous thiols (RSH): (A) glutathione (GSH), (B) cysteine (Cys), (C) homocysteine (Hcy), (D) cysteinyl-glycine (CysGly), (E) γ-glutamyl-cysteine (γ-GluCys) and (F) the Hg(II)-$[RSH]_2$ complex.

Studies conducted in non-cellular systems, using ^{13}C-NMR spectroscopy [7,8] and electrospray ionization and mass spectrometry [9] among others, have demonstrated that in mixtures containing Hg^{2+} and RSH in a 1:2 molar ratio, respectively, a Hg(II)-$[RSH]_2$ complex is the major species in solution, with the sulfhydryl group as the only coordination site [10].

Interestingly, others authors [6] have also reported, in such conditions, the presence of relatively low concentrations of 1:1 Hg(II)-thiol complexes.

Studies conducted in mammalian epithelial cells [10] have postulated that certain mercuric *S*-conjugates of endogenous thiols -cysteine (Cys), homocysteine (Hcy) or glutathione (GSH)- can act as functional or structural mimic of essential molecules such as cystine, homocystine, glutathione disulfide (GSSG), or methionine. In this context the formation of the some Hg-thiol complexes would facility their entry into certain target cells by serving as molecular mimics of essential molecules.

Previous studies conducted by Divine *et al.* [11] have demonstrated that mercury cytotoxicity in LLC-PK1 cells is dependent on Hg(II) concentration, the ligand (RSH) and the presence of a cysteine source for the cells.

On the other hand, a recent work reported by us [12] provided direct evidence that a preformed Hg(II)-$[GSH]_2$ complex is able to concentration-dependently reduce oxygen into superoxide and we postulated that as a result of this, the complex can induce either the oxidation or the reduction of molecules that are redox-susceptible towards superoxide. Thus, in view of the biological role of the Hg-thiol complexes and the potential toxicological importance of the ability of the Hg(II)-$[GSH]_2$ complex to generate superoxide, we present here results relative to the redox activity of the latter complex. In addition we extend our investigation to evaluate the capacity of other Hg-endogenous thiols complexes -cysteine (Cys), homocysteine (Hcy), cysteinyl-glycine (CysGly) and γ-glutamyl-cysteine (γ-GluCys)- to generate reactive oxygen species. Results from the present work are expected to contribute –primarily from a chemical point of view- to the mechanism that explains the toxicity induced by mercury ions after the formation of the above mentioned mercury-containing thiol complexes.

2. Preparation of the Mercury-Containing Thiol Complexes

The Hg(II)-RSH complexes (RSH= GSH, Cys, Hcy, CysGly and γ-GluCys) were prepared as previously described [6] by mixing $HgCl_2$ and RSH in a 1:1 molar ratio, respectively.

The Hg(II)-$[RSH]_2$ complexes (RSH= GSH, Cys, Hcy, CysGly and γ-GluCys) were prepared by direct mixing of $HgCl_2$ and RSH in a 1:2 molar ratio as previously described [8].

Whenever stating a given concentration of each complex, this refers to the concentration of mercury used in its preparations. Unless indicated otherwise, all complexes were freshly prepared prior to use.

3. Capacity of the Mercury-Containing Thiol Complexes to Generate Superoxide

In the present report, we have collected information consistent with the view that some Hg(II)-[RSH] and Hg(II)-$[RSH]_2$ complexes are indeed redox-active towards molecular oxygen. In fact, we provide evidence that, when present in aqueous solutions, some of these complexes react with molecular oxygen to generate superoxide, and that as a result of this behavior; these redox-active complexes can induce hydrogen peroxide production.

3.1. EPR Studies

To evaluate whether the mercury-containing thiol complexes (Hg(II)-$[RSH]_n$, n being 1 or 2) are capable of generating superoxide, the spin trap DMPO (100 mM) was added to such solutions.

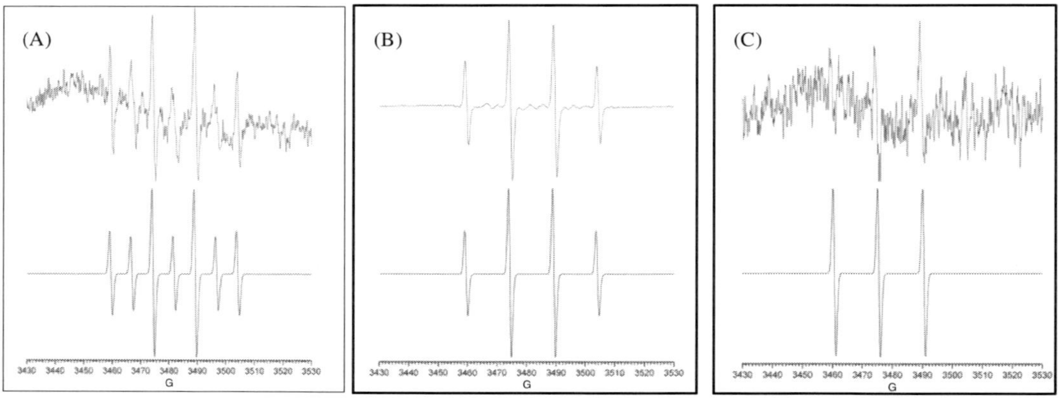

Figure 1. EPR spectra of DMPO-radical adducts formed from solutions containing a preformed Hg(II)-$[GSH]_2$ complex and SOD-treated complex. (A) Spectrum of a solution containing the preformed Hg(II)-$[GSH]_2$ complex (1.5 mM) (top) and simulation of the same spectrum (bottom). (B) Spectrum of DMPO-OH radical adduct, generated by mixing $FeSO_4$ (1 mM) with H_2O_2 (1%) (Fenton's reaction) (top) and simulation of the same spectrum (bottom). (C) Spectrum obtained 1 h after the addition of SOD (1000 U/mL) to a solution containing the Hg(II)-$[GSH]_2$ complex (arising from in A) (top) and simulation of the same spectrum (bottom). Hyperfine splitting for DMPO-OH: $a_H = a_N = 14.99$ G (Spectrum (A) and (B)) and DMPO-X: $a_N = 14.66$ G (Spectrum (C)). Data were taken from reference [12].

Figure 1(A) depicts the EPR spectrum, recently reported by us [12], resulting from adding DMPO (100 mM) to a preformed Hg(II)-$[GSH]_2$ complex (1.5 mM). The spectrum obtained for such mixture reveals the presence of a typical signal for the DMPO-OH adduct (Fig. 1(B)), suggesting that rather than superoxide, hydroxyl radicals [13-15] have been generated. It should be noted, however, that the formation of such adduct originates, most likely, from the decomposition of a DMPO-OOH adduct [13-15] formed previously from the interaction between DMPO and $O_2^{\bullet -}$. Noteworthy, the signal observed in Figure 1(A) was almost totally prevented in the presence of the superoxide dismutase (SOD) enzyme (Figure 1(C)), thus confirming the generation of superoxide by the Hg(II)-$[GSH]_2$ complex.

Computer simulations of the EPR spectra are presented in Fig. 1(A), (B) and (C) (bottom) and were obtained using the hyper-fine splitting constants resulting from the analysis of the experimental spectra. Such simulations, showed a characteristic hyper-fine pattern for the oxidation of DMPO/OH$^•$ (minor signal in Fig. 1(A) and (C)), confirmed the presence of a DMPO-OH adduct (Fig. 1(A)) and the disappearance of signals of such adduct in the presence of SOD (Fig. 1(C)) [12].

When a solution containing the Hg(II)-[Cys]$_2$ complex was evaluated, similar signals (associated to the DMPO-OOH adduct) were detected. However, the EPR spectra obtained for solutions containing either the Hg(II)-[CysGly]$_n$ or Hg(II)-[γ-GluCys]$_n$ (n = 1 or 2) complexes did not present such signals (not shown).

3.2. Dihydroethidium Oxidation Experiments

The postulated ability of the Hg(II)-[RSH]$_n$ complexes to generate superoxide was also assessed by measuring the increase in fluorescence that arises from oxidizing dihydroethidium (DHE), a probe widely used to make evident the formation of such radicals [16].

As shown in Figure 2, the incubation of DHE (50 μM) with Hg(II)-[GSH]$_2$, Hg(II)-[Cys]$_2$ or Hg(II)-[Hcy] complexes (10 μM), led to an increment in the DHE oxidation (expressed as an increment in the RFU, λ_{exc} = 470nm and λ_{em} = 590nm). The addition of SOD (250 U/well) to such mixtures completely blocked such effect. Thus, the DHE oxidation induced by Hg(II)-[GSH]$_2$, Hg(II)-[Cys]$_2$ or Hg(II)-[Hcy] is clearly mediated by superoxide.

Interestingly, as seen in Figure 2(A) the addition of DHE to the Hg(II)-[GSH] complex, prepared by mixing directly Hg^{2+} ions and GSH in a 1:1 molar ratio, led to a low but SOD-inhibitable fraction of DHE oxidation.

According to a previous report by Oram *et al.* [6], solutions containing a Hg(II)-GSH ratio of exactly 1:1 show as principal species the 1:1 Hg:complexes but also relatively low concentrations of the 1:2 complexes. Thus, such the minor fraction of DHE oxidation observed would be a consequence of the minor presence of the redox-active Hg(II)-[GSH]$_2$ complex in the above referred 1:1 mixtures. The redox activity of the latter complex has been confirmed by EPR studies (Figure 1) [12].

In the case of the Hg(II)-[Hcy]$_n$ (n=1 or 2) complexes, the capacity of these compounds to oxidize DHE (Figure 2C) could only be attributed to the Hg(II)-[Hcy] complex. This contention is supported by data obtained from HPLC-DAD, using the experimental conditions previously reported by other authors [17] for the study of mixtures containing mercury and endogenous thiols. The latter results demonstrate that in a solution containing a Hg(II)-Hcy 1:1 the principal species present is Hg(II)-[Hcy] (data not shown). However, when solutions containing a 1:2 molar ratio of Hg(II)-Hcy were assessed, a mixture of Hg(II)-[Hcy] plus Hg(II)-[Hcy]$_2$ was observed. Presumably, the lower oxidation of DHE observed for a molar ratio of 1:2 in our system (Fig. 2C) could be associated to the minori presence of the Hg(II)-[Hcy] complex in such mixtures.

On the other hand, controls conducted with the endogenous thiols, cysteinyl-glycine (CysGly) and γ-glutamyl-cysteine (γ-GluCys) (in absence of the metal), show that under the conditions of the DHE oxidation assay these thiols themselves were able to oxidize the probe. Thus, the evaluation of their complexes is not possible using the DHE oxidation assay.

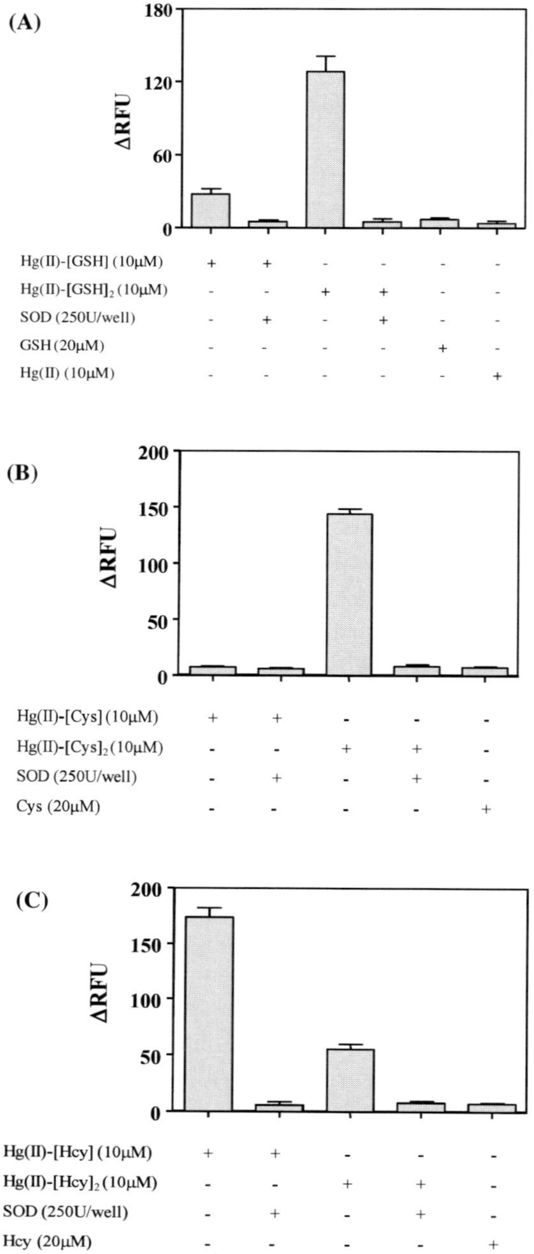

Figure 2. DHE oxidation induced by the Hg(II)-$[RSH]_n$ complexes. DHE (50 μM) was added to solutions containing either (A) Hg(II)-$[GSH]_n$ (n= 1 or 2), (B) Hg(II)-$[Cys]_n$ (n= 1 or 2) or (C) or Hg(II)-$[Hcy]_n$ (n= 1 or 2). The increase in fluorescence ($\lambda_{exc.}$ 470 nm and $\lambda_{em.}$ 590 nm) resulting from DHE oxidation was registered 30 min after, at 30 °C and the results expressed as ΔRFU ($RFU_{mixture}$-$RFU_{DHE\ alone}$). Controls were carried out using only $HgCl_2$ (10 μM) and RSH alone (20 μM).

3.3. Cytochrome *c* Reduction Experiments

The reduction of Cyt *c*, a highly sensitive superoxide-reducible probe [18], was also used to evaluate the capacity of the mercury-containing thiol complexes to generate superoxide. As shown in Figures 3(A) and 3(B), the addition of the preformed Hg(II)-$[GSH]_2$ or Hg(II)-[Hcy] complexes, to a solution containing Cyt *c* led to a substantial reduction of the cytochrome (tested as an increment in the OD_{550nm}). The sole addition of GSH (Figure 3(A)) or Hcy (Figure 3(B)) had no Cyt *c*-reduction effect.

In addition, Figures 3(A) and 3(B) depict the results from adding SOD to a complexes-containing mixture. As shown in such Figures the Hg(II)-$[GSH]_2$ and Hg(II)-[Hcy] complexes were able to reduce Cyt *c* and such effect was totally prevented by SOD (250 U/well). Thus, the Cyt *c* reduction process induced by these complexes is clearly mediated by superoxide.

Likewise, as seen in the EPR studies and DHE experiments, Cyt *c* reduction was also found to be susceptible to be inhibited by SOD for the Hg(II)-$[GSH]_2$ and Hg(II)-[Hcy] complexes. Controls conducted with the endogenous thiols cysteine (Cys), cysteinyl-glycine (CysGly) and γ-glutamyl-cysteine (γ-GluCys) (in the absence of the metal) show that these thiols by themselves are able to reduce Cyt *c*. Thus, the evaluation of their complexes is not possible using the Cyt *c* reduction assay.

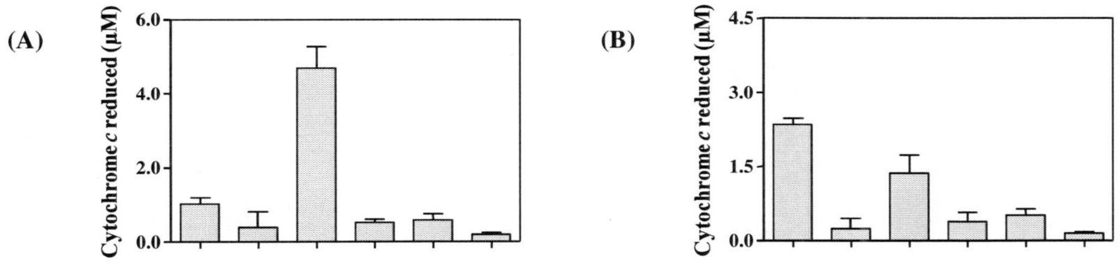

(A)						
Hg(II)-[GSH] (10µM)	+	+	-	-	-	-
Hg(II)-[GSH]$_2$ (10µM)	-	-	+	+	-	-
SOD (250U/well)	-	+	-	+	-	-
GSH (20µM)	-	-	-	-	+	-
Hg(II) (10µM)	-	-	-	-	-	+

(B)						
Hg(II)-[Hcy] (10µM)	+	+	-	-	-	-
Hg(II)-[Hcy]$_2$ (10µM)	-	-	+	+	-	-
SOD (250U/well)	-	+	-	+	-	-
Hcy (20µM)	-	-	-	-	+	-
Hg(II) (10µM)	-	-	-	-	-	+

Figure 3. Cyt *c* reduction induced by the Hg(II)-$[RSH]_n$ complexes. Cyt *c* (50 µM) was added to solutions containing either (A) Hg(II)-$[GSH]_n$ (n= 1 or 2) or (B) Hg(II)-$[Hcy]_n$ (n= 1 or 2). The increase in the optical density (OD_{550nm}) resulting from Cyt *c* reduction was registered 2 min after, at 30 °C and the results expressed as concentration of reduced Cyt *c* calculated using a ε value of 21.1 $mM^{-1}cm^{-1}$. Controls were carried out using $HgCl_2$ (10 µM) and RSH alone (20 µM).

However, considering all results obtained, using EPR, DHE-oxidation and Cyt c-reduction assays, we postulate that the Hg(II)-$[GSH]_2$, Hg(II)-$[Cys]_2$ and Hg(II)-[Hcy] complexes are able to generate superoxide (as shown in Rx. 1-3).

$$\text{Hg(II)-[GSH]}_2 + O_2 \rightarrow \text{Oxidized Complex} + O_2^{\bullet -} \quad \text{(Rx. 1)}$$

$$\text{Hg(II)-[Cys]}_2 + O_2 \rightarrow \text{Oxidized Complex} + O_2^{\bullet -} \quad \text{(Rx. 2)}$$

$$\text{Hg(II)-[Hcy]} + O_2 \rightarrow \text{Oxidized Complex} + O_2^{\bullet -} \quad \text{(Rx. 3)}$$

To further characterize the ability of redox-active complexes to generate superoxide, we recently reported [12] for the preformed Hg(II)-$[GSH]_2$ complex the existence of a possible relationship between increasing concentrations of this complex and its capacity to reduce Cyt *c*. When instead of Hg(II)-$[GSH]_2$, the Hg(II)-[Hcy] complex was evaluated, a similar biphasic behavior was obtained as a function of its concentration (Figure 4).

Figure 4 depicts that the Cyt *c* reduction increased linearly as the concentration of the complex increased from 5 to 10 μM. Concentrations greater than 10 μM (and up to 50 μM) were associated with proportionally lower degrees of Cyt *c* reduction. For concentrations greater than 10 μM, Cyt *c* reduction started to decline. The latter is, presumably, because the greater flux of superoxide expected to be generated by the higher concentrations of the Hg(II)-[Hcy] complex favor the dismutation of superoxide (Eq. 1; k_1=2.4×10^5 M^{-1}s^{-1}) [19] over its reaction with Cyt *c* (Eq. 2; k_2=2.6×10^5 M^{-1}s^{-1}) [20].

$$v_{dismut}. := k_1\,[O_2^{\bullet -}][O_2^{\bullet -}] \qquad \text{(Eq. 1)}$$

$$v_{Cyt\,c\,red}. := k_2\,[\text{Cyt } c][O_2^{\bullet -}] \qquad \text{(Eq. 2)}$$

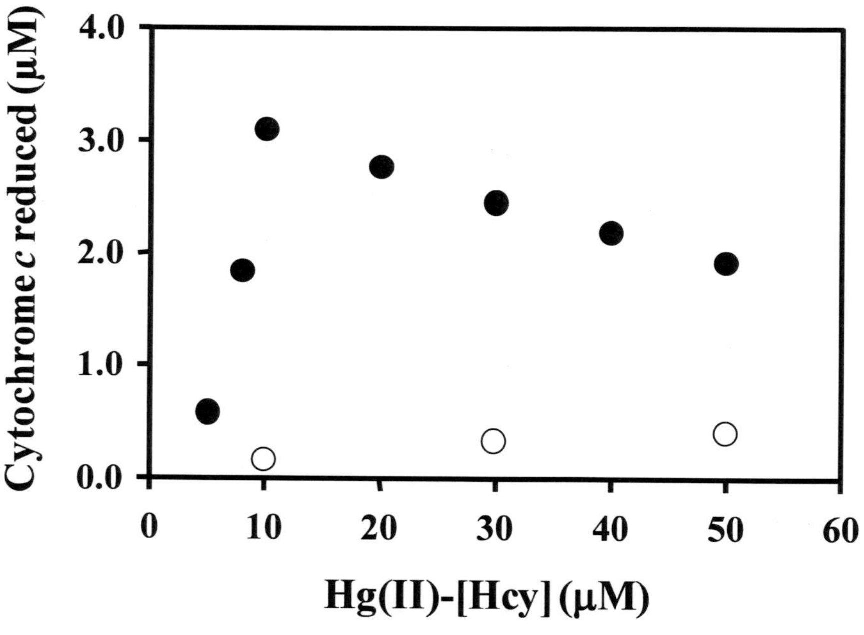

Figure 4. Effect of increasing concentration of the Hg(II)-[Hcy] complex on the Cyt *c* reduction. Cyt *c* (50 μM) was added to solutions containing increasing concentration of Hg(II)-[Hcy] (5-50 μM) in the presence (•) or absence (o) of SOD (250 U/well). Results were expressed as concentration of reduced Cyt *c* calculated using a ε value of 21.1 mM^{-1}cm^{-1}.

Thus, higher concentrations of the Hg(II)-[Hcy] complex would be associated with a higher rate of production and subsequent dismutation of superoxide.

4. Capacity of the Mercury-Containing Thiol Complexes to Generate Hydrogen Peroxide

4.1. Acetaminophen Oxidation Experiments

The interpretation that an increment in the production of superoxide by the redox-active complexes results in an increase in the rate of dismutation was confirmed when studying the generation of hydrogen peroxide (H_2O_2) by the above referred redox-active complexes.

Hydrogen peroxide (H_2O_2) was quantified as described [21], employing acetaminophen as hydrogen peroxide reacting agent. The fluorescence obtained (excitation and emission wavelengths were 298 nm and 333 nm, respectively), associated with the oxidation of acetaminophen in acidic medium, was monitored at 30 °C in a 96-well plate using a Multi-Mode Microplate Reader (Synergy^{TM} HT).

The generation of hydrogen peroxide (H_2O_2) by the Hg(II)-[Hcy] complex was assessed from the interaction between increasing concentrations of this complex and molecular oxygen in a Cyt *c*- or DHE-free medium by the acetaminophen oxidation methodology [21]. Figure 5 shows that, by employing this assay increasing concentrations of the Hg(II)-[Hcy] complex were associated with the generation of correspondingly increasing concentrations of H_2O_2. Furthermore, the estimation of H_2O_2 was totally inhibited by CAT (50 U/mL). Similar results have been observed for Hg(II)-[GSH]$_2$ [12] and Hg(II)-[Cys]$_2$.

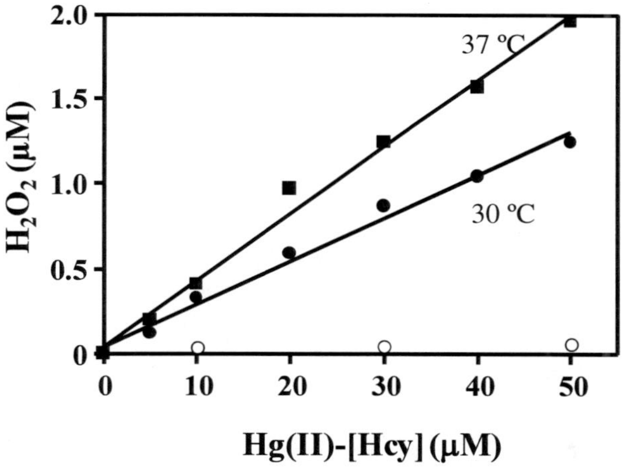

Figure 5. Relationship between increasing concentrations of the Hg(II)-[Hcy] complex (5-50 μM) and the formation of hydrogen peroxide (H_2O_2), in the presence (○) or absence (●) of CAT (50 U/mL) at 30 °C or in the absence (■) of CAT (50 U/mL) at 37 °C of co-incubation temperature.

As shown in Figure 5, the generation of H_2O_2 was markedly increased when the solution containing the Hg(II)-[Hcy] complex was co-incubated at 37 °C, instead of 30 °C. The latter could be explained by a temperature-dependent increment in the rate at which the superoxide generated by the complex undergo spontaneous dismutation. Such behavior was also observed when the solution containing the Hg(II)-[Hcy] complex was co-incubated at 30 °C, but in the presence of Tempol, a SOD-mimmetic compound, (i.e. formation of 0.45 μM of

H_2O_2 from 10 μM of Hg(II)-[Hcy]). This effect would be attributable to the superoxide dismutase-mimetic properties of Tempol [22].

On the other hand, the incubation of Hg(II)-$[CysGly]_n$ or Hg(II)-$[\gamma\text{-GluCys}]_n$ (n = 1 or 2) complexes with oxygen during 60 min did not give place to the hydrogen peroxide-dependent oxidation of acetaminophen.

5. Capacity of the Mercury-Containing Thiol Complexes to Generate Oxidized Thiols

To investigate whether the interaction between the redox-active complexes and oxygen involves an oxidative consumption of the thiols present in the complexes, the formation of oxidized thiols (i.e. oxidized glutathione (GSSG), cystine (Cys-S-S-Cys) and homocystine (Hcy-S-S-Hcy)) was measured in a DHE- or Cyt *c*-free system. The formation of these compounds was determined by HPLC with UV-vis detection, using a C18 column with a mobile phase of 0.1% TFA/H_2O:0.1% TFA/acetonitrile (95:5) at a flow rate of 0.5 mL min^{-1} at 202 nm [17].

Figure 6 shows that the incubation of a solution containing the Hg(II)-$[GSH]_2$ complex (1 mM) during 120 minutes led to the formation of oxidized glutathione (GSSG). In the case of the Hg(II)-$[Cys]_2$ and Hg(II)-[Hcy] complexes the formation of Cys-S-S-Cys and Hcy-S-S-Hcy, respectively show the same behavior in comparison to Figure 6 but was assessed using Diferential Pulse Cyclic Voltammetry.

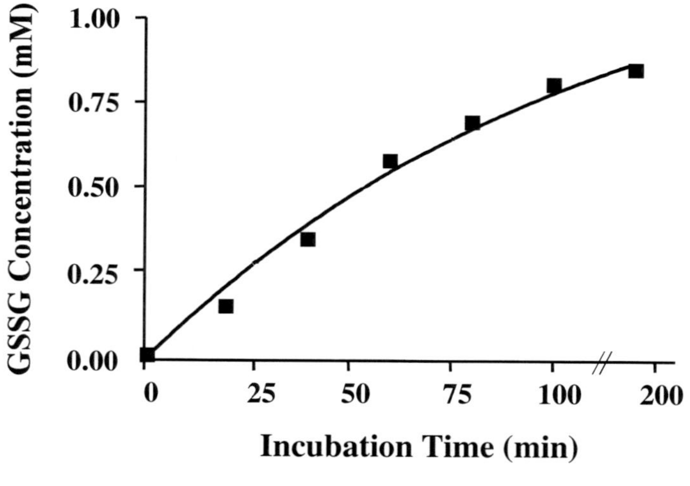

Figure 6. Recovery of GSSG from solutions containing the Hg(II)-$[GSH]_2$ complex (1 mM) incubated at different times and at 30 °C. GSSG was assessed using HPLC-DAD, as described in [17]. Results are expressed as mM GSSG concentration.

Thus, the present study addresses the ability of five endogenous thiols -glutathione (GSH), cysteine (Cys), homocysteine (Hcy), cysteinyl-glycine (CysGly) and γ-glutamyl-cysteine (γ-GluCys)- bound to mercury (Scheme 1) to generate reactive oxygen species. The results show that, Hg(II)-$[CysGly]_n$ and Hg(II)-$[\gamma\text{-GluCys}]_n$ (n = 1 or 2) totally failed to reduce Cyt *c* and to oxidize DHE and acetaminophen. These results reveal that neither of such complexes would be able to generate superoxide and hydrogen peroxide. Therefore, they suggest that these complexes would lack redox-activity towards molecular oxygen. In contrast, an important reduction of Cyt *c* and oxidation of DHE and acetaminophen was evident when the Hg(II)-$[GSH]_2$, Hg(II)-$[Cys]_2$ and Hg(II)-[Hcy] complexes were evaluated. The latter results imply that a significant part of the superoxides generated by the complexes undergo dismutation leading to the accumulation of hydrogen peroxide in the media. A potential implication of the latter is that, when redox-active metal ions are present, the co-occurrence of hydrogen peroxide and superoxide can easily lead to the formation of hydroxyl radicals, the single most reactive species in biological systems [23].

Interestingly, similar results of redox activity were obtained for other metal-containing complex (*i.e.* the Cu(I)-$[GSH]_2$ complex) and redox-inactivity for the Cu(I)-$[CysGly]_n$ and Cu(I)-$[\gamma\text{-GluCys}]_n$ complexes [24-26]. Thus, the presence of the dipeptides (CysGly and γ-GluCys) in a complex containing copper or mercury ions would lead to the formation of stable complexes. The opposite effect would occur in the case of glutathione- containing complexes, which can induce either the oxidation or the reduction of molecules that are redox-susceptible towards superoxide through of the decomposition of such complexes.

Conclusion

Based in our results we conclude that solutions containing the Hg(II)-$[GSH]_2$, Hg(II)-$[Cys]_2$ and Hg(II)-[Hcy] complexes are able to reduce molecular oxygen into superoxide. Removal of superoxide from these solutions, spontaneously or by the addition of superoxide interceptors, led to the formation and accumulation of hydrogen peroxide and of their oxidized thiols.

Acknowledgments

This work was supported by FONDECYT Grants #11090115 and #1110018.

References

[1] Wang, X.; Andrews, L.; Riedel, S.; Kaupp, M. *Angew. Chem.* 2007, 119, 8523-8527.

[2] Diez, S. (2009). *Human Health Effects of Methylmercury Exposure. Reviews of Environmental Contamination and Toxicology*, vol. 198, pp 111–132, Springer, New York.

[3] Countera, S. A.; Buchananb, L. H. *Toxicol. and Appl. Pharmacol.* 2004, 198, 209-230.

[4] Ballatori, N. *Drug Metab. Rev.* 1991, 23, 83-132.

[5] Stricks, W.; Kolthoff, I. *J. Am. Chem. Soc.* 1953, 75, 5673-5680.

[6] Oram, P. D.; Fang, X.; Fernando, Q.; Letkeman, P.; Letkeman, D. *Chem. Res. Toxicol.* 1996, 9, 709-712.

[7] Rabenstein, D. L. (1989). *Metal complexes of glutathione and their biological significance, in Glutathione: Chemical, Biochemical and Medical Aspects: Coenzymes and Cofactors*, Dolphin, D.; Avramovic, O. and Poulson, R. (Eds.) vol. 3, pp 147-186, John Wiley and Sons, New York.

[8] Fuhr, B. J.; Rabenstein, D. L. *J. Am. Chem. Soc.* 1973, 95, 6944-6950.

[9] Rubino, F. M.; Verduci, C.; Giampiccolo, R.; Pulvirenti, S.; Brambilla, G.; Colombi, A. *J. Am. Soc. Mass Spectrom.* 2004, 15, 288-300.

[10] Bridges, C. C.; Zalups, R. K. *Toxicol. Appl. Pharmacol.* 2005, 204, 274-308. And references described therein.

[11] Divine, K. K.; Ayala-Fierro, F.; Barber, D. S.; Carter, D. E. *J. Toxicol. Environ. Health A.* 1999, 57, 489-505.

[12] Aliaga, M. E.; López-Alarcón, C.; Barriga, G.; Olea-Azar, C.; Speisky, H. *J. Inorg. Biochem.* 2010, 104, 1084-1090.

[13] Finkelstein, E.; Rosen, G. M.; Rauckman, E. J. *J. Am. Chem. Soc.* 1980, 102, 4994-4999.

[14] Villamena, F. A. *J. Phys. Chem. A* 2009, 113, 6398-6403.

[15] Buettner, G. R. *Free Radical Res. Commun.* 1993, 19, S79-S87.

[16] Benov, L.; Sztejnberg, L.; Fridovich, I. *Free Radical Biol. Med.* 1998, 25, 826-831.

[17] Dago, A.; González-García, O.; Ariño, C.; Díaz-Cruz, J. M.; Esteban, M.; *J. Chromatogr. A* 2009, 1216, 6752-6757.

[18] Van Gelder, B.; Slater, E. C. *Biochim. Biophys. Acta* 1962, 58, 593-595.

[19] Zhao, H.; Kalivendi, S.; Zhang, H.; Joseph, J.; Nithipatikom, K.; Vásquez-Vivar, J; Kalyanaraman, B. *Free Radical Biol. Med.* 2003, 34, 1359-1368.

[20] Butler, J.; Koppenol, W. H.; Margoliash, E. *J. Biol. Chem.* 1982, 257, 10747-10750.

[21] Jie, N.; Yang, J.; Huang, X.; Zhang, R.; Song, Z. *Talanta* 1995, 42, 1575-1579.

[22] Soule, B. P.; Hyodo, F.; Matsumoto, K.; Simone, N. L.; Cook, J. A.; Krishna, M. C.; Mitchell, J. B. *Free Radical Biol. Med.* 2007, 42, 1632-1650.

[23] Haliwell, B.; Gutteridge, J. M. C. (2007) *Free Radical Biol. Med.* In 4th Ed, pp 30-78; University Press: Oxford.

[24] Carrasco-Pozo, C.; Aliaga, M.E.; Olea-Azar, C.; Speisky, H. *Bioorg. Med. Chem.* 2008, 16, 9795-9803.

[25] Speisky, H.; Gómez, M.; Carrasco-Pozo, C.; Pastene, E.; Lopez-Alarcón, C.; Olea-Azar, C. *Bioorg. Med. Chem.* 2008, 16, 6568-6574.

[26] Speisky, H.; Gómez, M.; Burgos-Bravo, F.; López-Alarcón, C.; Jullian, C.; Olea-Azar, C. Aliaga, M. E. *Bioorg. Med. Chem.* 2009, 17, 1803-1810.

In: Mercury: Sources, Applications and Health Impacts ISBN: 978-1-62257-721-7
Editors: Ki-Hyun Kim and Richard J. C. Brown

Chapter 5

EFFECTS OF ACUTE MERCURY EXPOSITION ON THE EXPRESSION AND FUNCTION OF ORGANIC ANION TRANSPORTERS IN THE KIDNEYS

Adriana M. Torres
Area Farmacología, Facultad de Ciencias Bioquímicas y Farmacéuticas, CONICET,
Universidad Nacional de Rosario, Rosario, Argentina

ABSTRACT

All living organisms are exposed to mercury (Hg) due to its widespread presence in the environment, but exposure leading to health hazards are generally related to specific human activities and originate from many sources including: unintentional occupational exposure, ingestion of Hg as an ingredient in folk remedies, amalgam fillings, food and as a preservative in vaccines.

In order to develop better pharmacological strategies for the treatment of intoxicated and exposed people, it is very important to understand how the diverse chemical species of this metal arrives and affects target organs.. Mercury accumulates in the brain, intestine, liver, kidneys and placenta. Recent studies have demonstrated that accumulation levels of mercury in the kidneys are determined by the activity of membrane transporters, mainly organic anion transporters (For example: Oat1, Oat3, Oat5 and Mrp2). Indeed, this is relevant to the study of the expression, function and regulation of drug transporters in the design and development of new therapeutics strategies. The following topics will be discussed in this chapter:

i) the effects of acute mercury exposure on the expression, function and regulation of organic anion transporters in the liver and kidneys,
ii) the implications of modulation in the expression and function of these transporters in the pharmacotherapeutics of mercuric intoxicants.
iii) the validation of urinary excretion of Oat5 as biomarker of mercury-induced nephrotoxicity.

Introduction

Mercury is a prominent environmental contaminant that causes detrimental effects to human health. This heavy metal has been used medically and commercially for centuries and it was a common component of many medications. It is still used in thermometers and blood-pressure cuffs, in batteries, switches and fluorescent light bulbs. Metallic mercury is also employed in the electrolytic production of chlorine and sodium hydroxide from saline [1,2].

All living organisms are exposed to mercury due to its ubiquitous presence in the environment, but exposure leading to health hazards are generally related to specific human activities and originate from many sources including: unintentional occupational exposure, ingestion of Hg as an ingredient in folk remedies, amalgam fillings, food and as a preservative in vaccines. The main exposure of the general population comes from three major sources: fish consumption, dental amalgams and vaccines [1-4].

Chemical Forms of Mercury

Three main chemical forms of mercury must be distinguished:

- elemental mercury
- inorganic mercury salts
- organic mercury compounds

Elemental mercury is the most volatile of the inorganic forms of mercury. Exposure to mercury vapour is mainly occupational but it can also be released from dental amalgams [1-4].

Salts of mercury exist as monovalent mercurous salts or as divalent mercuric salts. Mercurous chloride (calomel) has been used in some skin creams as an antiseptic and was employed as a diuretic and cathartic. Mercuric salts are still used in industry, and the industrial discharge into rivers has contaminated the environment the world over [1-4].

Organic mercury compounds have been used as fungicides (methylmercury), and now fish are the main source of human exposure to methylmercury. Ethylmercury is the active ingredient of the preservative thimerosal used in vaccines [1-4].

Pharmacokinetics of Mercury

Elemental mercury has a low absorption from the gastrointestinal tract. Inhaled mercury vapour is completely absorbed by the lungs and then is oxidized to the divalent mercuric cation by catalase erythrocytes. Mercury vapour enters the brain before its oxidation and induces central nervous system toxicity [1-5].

Following ingestion, 10% of soluble inorganic mercuric salts (Hg^{2+}) are absorbed in the gastrointestinal tract. The highest concentration of Hg^{2+} is found in the kidneys. Inorganic mercury compounds do not pass through the placenta and the blood-brain barrier. They are excreted in the urine and faeces. The half-life of inorganic mercury is about 60 days [1-5].

Organic mercury compounds are absorbed efficiently from the gastrointestinal tract since they are lipid soluble. Around 90 % of methylmercury is absorbed from the gastrointestinal tract, across the placenta and blood-brain barrier and produces neurological and teratogenic effects. The carbon-mercury bond of organic mercury compounds is cleaved after absorption and inorganic mercury species are produced.

Methylmercury is mainly excreted in the faeces in the form of a conjugate with glutathione and around 10 % of any ingested appears in urine. The half-life of methylmercury is between 40 and 105 days [1-5].

Mercury forms covalent bonds with sulphur, a characteristic that accounts for most of the biological properties of the metal. It binds to endogenous molecules containing thiols, such as glutathione, cysteine, homocysteine, N-acetylcysteine, metallothionein and albumin [1-5].

Even in low concentrations organic mercury compounds are capable of inactivating sulfhydryl enzymes and thus interfering with cellular metabolism and function. The affinity of mercury for thiols provides the basis for the treatment of mercury poisoning with such agents as dimercaprol and penicillamine. Mercury also combines with several important ligands, including phosphoryl, carboxyl, amide and amine groups [1-5].

MERCURY AND THE KIDNEYS

Following exposure to elemental or inorganic forms of mercury, the kidneys are the primary sites of mercury accumulation. Organic forms of mercury also accumulate in the kidneys [5-7].

Mercury accumulation causes acute renal failure. This is characterized by profound renal vasoconstriction, reduction in the glomerular filtration rate, histological damage, and activation of the renin-angiotensin system [5-7].

Membrane Transporters

Mercury accumulates in the renal cortex and outer stripe of the outer medulla, almost exclusively along the convoluted and straight segments of the proximal tubule. Complex factors may account for this susceptibility, including the selective presence of specific transport systems involved in the uptake of mercuric ions by target cell populations [5-7].

Data from in vitro studies indicate that there are luminal and basolateral mechanisms involved in the uptake of mercuric ions by proximal tubular epithelial cells. Some GSH-S-conjugates of Hg^{2+} are filtered at the glomerulus and delivered into the lumen of proximal tubules, where they are degraded rapidly and sequentially by gɤ--glutamyltransferase and cysteinylglycinase to yield Cys S-conjugates of Hg^{2+} (Cys-S-Hg-S-Cys). It was determined that the luminal uptake of Cys-S-Hg-S-Cys involves at least one Na^{+}-dependent and one Na+-independent amino acid carrier. A likely candidate for the Na^{+}-independent transport of Cys-S-Hg-S-Cys is system $b^{o,+}$ [8-12].

It has also been recently postulated that the Multidrug Resistance Protein 2 (Mrp2), is involved in the renal export of mercuric ions from the tubular cell to the tubular lumen [13].

Oat1 and Oat3

Transport of Hg^{2+} from peritubular blood into the intracellular compartment of proximal tubular cells accounts for 40-60 % of proximal tubular uptake of Hg^{2+}. Experimental evidence indicating that mercury conjugates of cysteine (Cys), Homocysteine (Hcy) and N-acetyl-cysteine (NAC) are taken up by the Organic Anion Transporter 1 (Oat1) and/or the Organic Anion Transporter 3 (Oat3), which have both been localized in the basolateral plasma membrane of proximal tubular epithelial cells [5, 8-12].

Oat1 and Oat3 are involved in the organic anion transport into the renal cell at the basolateral membrane. Oat1 is detected exclusively in the proximal tubules and Oat3 is localized in the proximal tubule, cortical and medullary thick ascending limb of the loop of Henleé, connecting tubules, and cortical and medullary collecting ducts [14].

Both Oat1 and Oat3 support organic anion/α-ketoglutarate exchange [15-17]. It has been reported that Oat1 mediates renal tubular uptake of para-aminohippuric acid (PAH) and several pharmacological agents, such as methotrexate, β-lactam antibiotics, antiviral, antineoplasic and non-steroidal anti-inflammatory drugs. Oat3 appears to efficiently mediate estrone-3-sulfate transport with high affinity, in addition to transporting several anionic compounds such as PAH, methotrexate, prostaglandin E2, and cyclic adenosine monophosphate [15-17]. We have described modifications in the expression and function of Oat1 and Oat3 in the presence of acute renal failure induced by ischemia [18] and by ureteral obstruction [19-21], in chronic renal failure [22] and in the presence of extrahepatic cholestasis [23-27].

Numerous studies on experimental animals treated with nephrotoxic doses of mercury and other heavy metals have described shortening and focal loss of microvilli of brush border membranes, fragmentation and loss of invagination of basolateral membranes, and inhibition of transporters. These alterations may explain the well-known impairment of reabsorptive and secretory functions of the proximal tubule [28,29]. Our laboratory has pioneered the study of the modifications induced in vivo by $HgCl_2$ in the expression and function of some of its own renal transporters such as Oat1 and Oat3.

We have induced an acute renal failure by the administration of a nephrotoxic dose of $HgCl_2$ (5 mg/kg b.w., i.p.) to male adult Wistar rats, 18 h before the studies. In this experimental model of acute nephrotoxicity, the Oat1 protein expression significantly increased in the homogenates and decreased in the basolateral membranes from the kidneys of treated animals. The exposure to $HgCl_2$ also decreased the expression of Oat1 mRNA. The increase of Oat1 protein expression in homogenates suggests alterations of the mechanisms for its degradation, maybe mediated by the mercury toxicity-dependent inhibition of some specific enzyme for Oat1 [5,30].

In the kidneys of rats treated with $HgCl_2$, the expression of both Oat3 protein (in homogenates and in basolateral membranes, BLM) and Oat3 mRNA was significantly reduced, indicating a down-regulation at transcriptional level. A higher decrease in Oat3 mRNA levels (60%) was observed in comparison with the Oat3 protein decrease in homogenates (30%), which might suggest an alteration in some pathway of its degradation.

To evaluate the functional activity of Oat1 and Oat3, PAH uptake in renal basolateral membrane (BLM) vesicles prepared from control and $HgCl_2$ rats was measured. Mercury treatment caused a marked inhibition of PAH uptake due to a reduction in Vmax (capacity). The capacity of a transfer is determined by the total number of active carriers presented in

BLM. The difference in Vmax indicates that a lower number of functional carrier units exist in BLM vesicles from $HgCl_2$ rats, which is in agreement with the lower expression of Oat1 and Oat3 in BLM. Additionally we studied whether the $HgCl_2$-induced modification in PAH uptake in BLM was supported by the systemic clearance of this organic anion. Rats treated with $HgCl_2$ exhibited an increase in the area under the PAH plasma concentration-time curve and, therefore, a statistically different decrease in total body clearance. In renal basolateral membranes, Oat1 and Oat3 protein abundances decreased by 30 and 24%, respectively, which might account for the 60% decrease in the capacity of PAH uptake observed in $HgCl_2$-treated rats. The systemic clearance of PAH was still greatly reduced (80%) in animals' exposure to mercury. The decrease in the systemic clearance of PAH might be justified, on one hand, by the decreased PAH uptake in renal basolateral membranes, and on the other hand, by the fact that Oat1 and Oat3 mediate the uptake of a number of mercuric-thiol conjugates [12, 31]. As low molecular weight thiols (such as Cys and Hcy) conjugated with mercuric ions are present in plasma from mercury treated animals [5], they could compete with PAH for Oat1 and Oat3, producing additional reductions in the renal tubular uptake of PAH. Our results demonstrated that inorganic $HgCl_2$ administered to rats induces significant modifications in critical steps of the transport processes involved in the cell excretion of mercury-conjugated compounds. $HgCl_2$ induces, in vivo, a down-regulation of Oat1 and Oat3 protein expression in the BLM. The lower protein levels of Oat1 and Oat3 at the BLM explain the lower uptake capacity for PAH and the reduction in the systemic clearance of this organic anion. It has been described that, in case of mercury intoxication, effects favoring cell survival [28,29] could be activated by the proximal tubule cells if they are not immediately killed by apoptosis or necrosis, such as the increase of the intracellular concentration of protective thiols (GSH and metallothioneins) and up-regulation of genes encoding for specific efflux pumps capable of removing Hg-GSH conjugates from the cells (MRP1 and MRP2). Aleo et al [28] have demonstrated, in vitro, that $HgCl_2$ induces the upregulation of Mrp1 and Mrp2 in MDCK cells.

Oat1 and Oat3 mediate mercury compounds' access to the proximal tubule cells. Therefore their down-regulation, demonstrated in vivo in our laboratory, might be another defensive mechanism developed by the cell to protect itself against mercury damage. These aspects may be useful for medicinal purposes to provide a biological protection against toxic effects exerted by mercury or by specific forms of mercury. The pharmacological modulation of the expression and/or the function of Oat1 and Oat3 might be an effective therapeutic strategy for reducing the nephrotoxicity of mercury.

The mercury-induced renal injury has also been investigated in *Oat1* knock out mice by intraperitoneal administration of 4 mg/kg of body weight of $HgCl_2$ [32]. The results showed that, although the rodents are susceptible to mercury, mice deficient in *Oat1* due to genetic deletion are largely protected from mercury-induced kidney injury. For example, following $HgCl_2$ treatment the observed increase in kidney weight (presumably due to edema) and blood urea nitrogen levels seen in wild-type mice wereas reduced dramatically in the *Oat1* knock out mice. Moreover, following $HgCl_2$ treatment, histological parameters were largely unchanged in the *Oat1* knock out kidneys. These differences in renal injury were striking and indicated that Oat1 is the main transporter responsible for $HgCl_2$-induced renal injury. This is also one of the most impressive phenotypes reported in the *Oat1* knock-out. This was consistent with the notion that Oat1 is the major transporter mediating cellular uptake of mercuric conjugates and provided direct genetic evidence implicating Oat1 in renal injury caused by $HgCl_2$ by mediating its transport into the proximal tubular cells.

Oat5

The Organic Anion Transporter 5 (Oat5) is another organic anion transporter recently cloned and characterized [33-35] that is exclusively expressed in the kidneys and is localized in the apical membranes of the proximal tubule straight segment (S3) in the outer stripe [33,34]. Oat5 transports several compounds, including ochratoxin A, estrone-3-sulphate, dehydroepiandrosterone sulphate, dicarboxylates (α-ketoglutarate and succinate) and interacts with nonsteroidal anti-inflammatory drugs, diuretics, bromosulfophthalein and penicillin G [33]. Our group has pioneered the detection of Oat5 in urine [36].

We have recently evaluated Oat5 renal expression and urinary excretion in rats exposed to different doses of $HgCl_2$, in comparison with traditional markers of renal injury (such as histology, creatinine and urea plasma levels and urinary alkaline phosphatase activity) [37]. We planned to determine whether Oat5 represents a more sensitive and early biomarker of mercuric-induced acute kidney injury when compared to traditional, routinely used biomarkers of nephrotoxicity. A $HgCl_2$ dose–response study was performed, employing $HgCl_2$ doses of 0.2, 1, and 5 mg/kg b.w., s.c. 18 h before the studies. The kidney/body weight ratio, urea and creatinine plasma levels and alkaline phosphatase urinary excretion were significantly elevated above control values only at the highest $HgCl_2$ dose of 5 mg/kg s.c. Histological alterations were detected in the Hg1 group and dramatic damage was observed in Hg5 rats. In contrast, urinary Oat5 protein levels increased in a dose-related manner after injection of $HgCl_2$. Urinary Oat5 excretion increased by 89, 146, and 232% over control urinary Oat5 for $HgCl_2$ doses of 0.2, 1, and 5 mg/kg b.w., s.c., respectively. Urinary Oat5 excretion was elevated at a dose of 0.2 mg/kg b.w., s.c., predicting a perturbation before the manifestation of histopathological change. A down-regulation of Oat5 protein expression in homogenates and of Oat5 mRNA in a dose-related way was observed, suggesting a decrease in the synthesis of the carrier. On the other hand, Oat5 protein expression in the apical membrane only decreased at the highest dose of mercury. The ratio between the abundance of Oat5 in apical membranes to the abundance of Oat5 in the homogenates increased by 54, 77, and 61% with respect to the control for $HgCl_2$ doses of 0.2, 1, and 5 mg/kg b.w., s.c., respectively. This suggests a redistribution of Oat5 from the cytoplasm to the apical membrane after mercury exposition. Several relatively new biomarkers of nephrotoxic acute kidney injury are being investigated. 24 h after treatment using $HgCl_2$ doses of 0.25 and 0.5 mg/kg b.w., i.v, kidney injury molecule-1 (Kim-1) werewas found in urine at concentrations 14- and 25-fold higher than in the control. In addition, Kim-1 was usually found at concentrations higher than in the control at Hg doses of 1 mg/kg b.w., i.v.. At $HgCl_2$ doses of 0.5 and 1 mg/kg b.w., i.v. urinary N-acetyl-βb-D-glucosaminidase (NAG) increased 3- and 6-fold compared to control [38]. Results obtained in our laboratory showed a statistically significant elevation of Oat5 abundance in urine at $HgCl_2$ doses as low as 0.2 mg/kg b.w., s.c. 18 h after treatment, far preceding the increase or the detection in urine of traditional and novel biomarkers of renal dysfunction.

Further to these investigations, we have also recently postulated the urinary excretion of Oat5 as an early biomarker of proximal tubule damage in ischemic renal failure [36]. Hence, urinary Oat5 excretion may serve as potentially valuable biomarker of ischemic and mercury induced acute kidney injury. Oat5's possible role as a selective biomarker of S3 segments of proximal tubule injury deserves to be investigated further. On the basis of our results, we suggest that further studies using animal experimental models of acute renal failure and clinical studies of patients at high risk of developing acute renal failure must be undertaken to explore the full utility of this potential marker.

Studies aimed at evaluating the renal expression of Oat1, Oat3 and Oat5 and the urinary excretion of Oat5 in rats under chronic exposition to $HgCl_2$ and to methylmercury are currently being performed in our laboratory.

ACKNOWLEDGMENTS

The present work was supported by the following grants: Agencia Nacional de Promoción Científica y Tecnológica (ANPCyT) de la República Argentina, PICT-2007 Nro: 00966; Consejo Nacional de Investigaciones Científicas y Técnicas (CONICET) de la República Argentina, PIP Nro. 01665. The authors thank Prof. H. Endou and Dr N. Anzai (Department of Pharmacology and Toxicology, Kyorin University School of Medicine, Tokyo, Japan) for kindly providing Oat1, Oat3 and Oat5 specific antibodies.

REFERENCES

[1] Clarkson T. W. (1997). The toxicology of mercury. *Crit Rev Clin Lab Sci* 34, 369-403.

[2] Clarkson T. W., Magos L. and Myers G. J. (2003). The toxicology of mercury - Current exposures and clinical manifestations. *N Engl J Med* 349, 1731-1737.

[3] Counter S. A. and Buchanan L. H. (2004). Mercury exposure in children: a review. *Toxicol Appl Pharmacol* 198, 209-230.

[4] Magos L. and Clarkson T. W. (2006). Overview of the clinical toxicity of mercury. *Ann Clin Biochem* 43, 254-268.

[5] Zalups R. K. (2000). Molecular interactions with mercury in the kidney. *Pharmacol Rev* 52, 113-143.

[6] Stacchiotti A., Borsani E., Rodella L., Rezzani R., Bianchi R. and Lavazza A. (2003). Dose-dependent mercuric chloride tubular injury in rat kidney. *Ultrastruct Pathol* 27, 253-259.

[7] Stacchiotti A., Lavazza A., Rezzani R., Borsani E., Rodella L. and Bianchi R. (2004). Mercuric chloride-induced alterations in stress protein distribution in rat kidney. *Histol Histopathol* 19, 1209-1218.

[8] Aslamkhan A. G., Han Y. H., Yang X. P., Zalups R. K. and Pritchard J. B. (2003). Human renal organic anion transporter 1-dependent uptake and toxicity of mercuric-thiol conjugates in Madin-Darby canine kidney cells. *Mol Pharmacol* 63, 590-596.

[9] Bridges C. C. and Zalups R. K. (2010). Transport of inorganic mercury and methylmercury in target tissues and organs. *J Toxicol Environ Health Part B* 13, 385-410.

[10] Giacomini K. M. and Sugiyama Y. (2006). Membrane transporters and drugs response. In *"Goodman and Gilman´s The Pharmacological Basis of Therapeutics"* (Ed. Brunton L. L., Lazo J. S., Parker K. L., Goodman L. S. and Gilman A. G.). pp 41-70. 11a edition, The Mc Graw-Hill companies, Inc.

[11] Zalups R. K., Aslamkhan A. G. and Ahmad S. (2004). Human organic anion transporter 1 mediates cellular uptake of cysteine-S conjugates of inorganic mercury. *Kidney Int* 66, 251-261.

[12] Zalups R. K. and Ahmad S. (2004). Homocysteine and the renal epithelial transport and toxicity of inorganic mercury: role of basolateral transporter organic anion transporter 1. *J Am Soc Nephrol* 15, 2023-2031.

[13] Bridges C. C., Joshee L. and Zalups R. K. (2011). MRP2 and the handling of mercuric ions in rats exposed acutely to inorganic and organic species of mercury. *Toxicol Appl Pharmacol* 251, 50-58.

[14] Kojima R., Sekine T., Kawachi M., Cha S. H., Suzuki Y., and Endou H. (2002). Immunolocalization of multispecific organic anion transporters, OAT1, OAT2, and OAT3, in rat kidney. *J Am Soc Nephrol* 13, 848-857.

[15] Anzai N., Kanai Y. and Endou H. (2006). Organic anion transporter family: current knowledge. *J Pharmacol Sci* 100, 411-426.

[16] Sekine T., Miyasaki H. and Endou H. (2006). Molecular physiology of renal organic anions transporters. *Am J Physiol* 290, F251-F261.

[17] Srimaroeng C., Perry J. L. and Pritchard J. B. (2008). Physiology, structure and regulation of the cloned organic anion transporters. *Xenobiotica* 38, 889-935.

[18] Di Giusto G., Anzai N., Endou H. and Torres A. M. (2008). Elimination of organic anions in response to an early stage of renal ischemia-reperfusion in the rat. Role of basolateral plasma membrane transporters and cortical renal blood flow. *Pharmacology* 81, 127-136.

[19] Villar S. R., Brandoni A., Quaglia N. B. and Torres A. M. (2004). Renal elimination of organic anions in rats with bilateral ureteral obstruction. *Biochim Biophys Acta* 1688, 204-209.

[20] Villar S. R., Brandoni A., Anzai N., Endou H. and Torres A. M. (2005). Altered expresión of rat cortical OAT1 and OAT3 in response to bilateral ureteral obstruction. *Kidney Int* 68, 2704-2713.

[21] Villar S. R., Brandoni A. and Torres A. M. (2008). Time course of organic anion excretion in rats with bilateral ureteral obstruction. Role of organic anion transporters (Oat1 and Oat3). *Nephron Physiol.* 110, 45-56.

[22] Torres A. M., Mac Laughlin M., Muller A., Brandoni A., Anzai N. and Endou H. (2005). Altered renal elimination of organic anions in rats with chronic renal failure. *Biochim. Biophys Acta* 1740, 29-37.

[23] Brandoni A., Quaglia N. B. and Torres A. M. (2003). Compensation increase in organic anions excretion in rats with acute biliary obstruction. Role of the renal organic anion transporter 1 (OAT1). *Pharmacology* 68, 57-63.

[24] Brandoni A., Anzai N., Kanai Y., Endou H. and Torres A. M. (2006a). Renal elimination of p-aminohippurate (PAH) in response of three days of biliary obstruction in the rat. The role of OAT1 and OAT3. *Biochim Biophys Acta* 1762, 673-682.

[25] Brandoni A., Villar S. R., Picena J. C., Anzai N., Endou H. and Torres A. M. (2006b). Expression of rat cortical OAT1 and OAT3 in response to acute biliary obstruction. *Hepatology* 43, 1092-1100.

[26] Torres A. M. (2008). Renal elimination of organic anions in cholestasis. *World J Gastroenterol* 14, 6616-6621.

[27] Brandoni A, and Torres A.M. (2010). Expression and function of renal organic anion transporters (Oats) in health and disease. *Current Topics in Pharmacology* 14:1-9.

[28] Aleo M. F., Morandini F., Bettoni F., Giuliani R., Rovetta F., Steimberg N., Apostoli P., Parrinello G. and Mazzoleni G. (2005). Endogenous thiols and MRP2 transporters contribute to Hg^{2+} efflux in $HgCl_2$-treated tubular MDCK cells. *Toxicology* 206, 137-151.

[29] Sabolic I. (2006). Common mechanisms in nephropathy induced by toxic metals. *Nephron Physiol* 104, 107-114.

[30] Di Giusto G., Anzai N., Ruiz M. L., Endou H. and Torres A. M. (2009). Expression and function of Oat1 and Oat3 in rat kidney exposed to mercuric chloride. *Arch Toxicol* 83, 887-897.

[31] Bridges C. C. and Zalups R. K. (2005). Molecular and ionic mimicry and the transport of toxic metals. *Toxicol Appl Pharmacol* 204, 274-308.

[32] Torres A. M., Dnyanmote A. V., Bush K. T., Wu W. and Nigam S. K. (2011). Deletion of multispecific organic anion transporter Oat1/Slc22a6 protects against mercury-induced kidney injury. *J Biol Chem* 286, 26391-26395.

[33] Anzai N., Jutabha P., Enomoto A., Yokoyama H., Nonoguchi H., Hirata T., Shiraya K., He X., Cha S.H., Takeda M., Miyazaki H., Sakata T.,Tomita K., Igarashi T., Kanai Y. and Endou H. (2005). Functional characterization of rat organic anion transporter 5 (Slc22a19) at the apical membrane of renal proximal tubules. *J Pharmacol Exp Ther* 315:534–544.

[34] Youngblood G.L. and Sweet D.H. (2004). Identification and functional assessment of the novel murine organic anion transporter Oat5 (SLC22a19) expressed in kidney. *Am J Physiol* 287, F236–244.

[35] Kwak J-O, Kim H-W, Oh K-J, Ko C-B, Park H, Cha SH (2005). Characterization of mouse organic anion transporter 5 as a renal steroid sulfate transporter. *J Steroid Biochem Mol Biol* 97:369–375.

[36] NaDC1 protein abundance in kidney and urine after renal ischemic reperfusion injury. *J Histochem Cytochem* 57,17–27.

[37] Di Giusto G. and Torres A. M. (2010). Organic anion transporter 5 renal expression and urinary excretion in rats exposed to mercuric chloride: a potential biomarker of mercury-induced nephropathy. *Arch Toxicol* 84, 741-749.

[38] Zhou Y., Vaidya V.S., Brown R.P., Zhang J., Rosenzweig B.A., Thompson K.L., Miller T.J., Bonventre J.V.and Goering P.L. (2008) Comparison of kidney injury molecule-1 and other nephrotoxicity biomarkers in urine and kidney following acute exposure to gentamicin, mercury and chromium. *Toxicol Sci* 101,159–170.

In: Mercury: Sources, Applications and Health Impacts ISBN: 978-1-62257-721-7
Editors: Ki-Hyun Kim and Richard J. C. Brown

Chapter 6

MERCURY: ENVIRONMENTAL CONCERNS AND MANAGEMENT TECHNOLOGIES

I. Padilla*, H. Tayibi, O. Rodríguez and A. López-Delgado
National Centre for Metallurgical Research, CSIC. Madrid, Spain

ABSTRACT

Mercury is considered a priority hazardous substance due to its adverse effects on human health and the environment. A world-wide common effort is underway to reduce both the supply and demand of mercury. In order to solve the environmental concern of mercury, a number of technologies have been developed to reduce its mobility towards the environment such as thermal treatments, amalgamation, stabilization/solidification, vitrification and encapsulation. This chapter aims to review the most common technologies that allow the immobilization of mercury (solid-waste and elemental liquid Hg) and the disposal or safe storage of these wastes without risk for the environment.

1. INTRODUCTION

Mercury is a metallic element that enters in the environment due to release from natural and anthropogenic sources. Mercury exists in three oxidation states, Hg^0, Hg(I) and Hg(II). Elemental mercury (Hg^0) a liquid metal at ambient-temperature, has a low but significant vapor pressure which produces direct toxicity (gaseous mercury) and also gives rise to the possibility of natural reactions producing other toxic species. Monovalent mercury, (Hg(I)), can form mercurous oxide (Hg_2O) and mercurous chloride (Hg_2Cl_2). Mercurous oxide (HgO) is unstable and its dismutation to elemental and divalent mercury occurs spontaneously. Divalent mercury, Hg(II), includes mercuric chloride, mercuric oxide and mercuric sulfide. HgS is the most stable compound of mercury in both polymorphs, the red -cinnabar- and the black -meta cinnabar-. Concerning the organic compounds, methylmercury, an

* National Centre for Metallurgical Research. CSIC. Avda. Gregorio del Amo, 8. 28040. Madrid. Spain. isapadilla@cenim.csic.es.

organometallic cation with the formula $[CH_3Hg]^+$, it is found to be one of the most important concerns to global environmental level, because it is a bioaccumulative potent neurotoxic. Methylmercury is formed, principally, from inorganic mercury by the action of anaerobic organisms that live in aquatic systems by means of a methylation process. This complex consists of the transformation of inorganic mercury to methylmercury in the natural environment.

The United Nations Environment Program (UNEP) established the Global Mercury Program with the purpose of delivering activities on mercury and supporting negotiations of international measures for its control. Thus a series of Governing Council Decisions have been established from 2001 which include prevention, minimization and management of mercury waste and also its long-term storage and disposal.

European Directive 2000/60/EC considers mercury as a priority hazardous substance due to its adverse effects on human health and the environment. Indeed, a world-wide common effort is now under way to reduce both, the supply and demand of mercury. EU Council and European Parliament Regulation 1102/2008 (EC) set 2011 as the due date for industrial use of Hg and the banning of mercury exports and set also the issue of safe storage of metallic mercury.

Different processes have been described for the recovery of mercury from waste including thermal, pyrometallurgical, hydrometallurgical and electrometallurgical technologies. In the same manner, a number of technologies have been developed to reduce the mobility of mercury towards the environment, such as roasting/retorting, amalgamation, stabilization/solidification, vitrification and encapsulation. This chapter aims to review the most sustainable technologies that allow the immobilization of mercury (wastes and elemental Hg) and the disposal of these wastes without risk for the environment.

2. MERCURY AND ITS COMPOUNDS

Mercury is one of the oldest known elements since it occurs either as a native metal or in cinnabar, corderoite, livingstonite and other minerals. There are three categories of mercury: metallic, liquid or elemental mercury, inorganic and organic mercury compounds. Mercury exists in three oxidation stages: Hg(0), Hg(I) and Hg(II). Elemental mercury is the only liquid metal at room temperature and it is the most volatile at the same temperature. Mercury has a high surface tension and density and a poor electrical conductivity of electric current at room temperature, not being so when it is close to absolute zero, or at high temperature. Mercury does not dissolve in non-oxidizing acids but dissolve in dilute or concentrated nitric acid, concentrated and hot sulfuric acid, in lipids and pentane. Mercury, in the oxidation state (I), forms mercurous compounds. This oxidation state is apparent and really consists in two ions Hg(II) linked by a bond Hg^+-Hg^+. This species is stable only in acidic medium because it is dismutated easily at higher pH to form metallic mercury and mercuric ion. Mercury ion Hg^{2+} forms mercuric compounds. The most important inorganic mercury compound is mercury (II) sulfide. It occurs as two crystalline forms, the cubic metacinnabar (black) and the hexagonal cinnabar (red). Both metacinnabar and cinnabar are insoluble in water, alcohol and mineral acids, but they are soluble in *aqua regia*. Cinnabar is the most abundant ore in mercury mines [1, 2].

Mercury combines with carbon based compounds to form mercury organic compounds. These compounds are especially hazardous to life. The most common organic mercury compound in the environment is methylmercury. It is formed in the environment from metallic mercury in a gaseous state and inorganic compounds of Hg(II) and certain microorganisms. Exposure to methylmercury can cause subtle and not-so-subtle neurological deficits, especially in fetuses and children [3].

3. The Cycle of Mercury in Nature and its Sources

The main sources of mercury emissions into the environment are natural sources (volcanic activity, fires, movement of rivers, lakes and oceanic, and biological processes) and anthropogenic sources, such as the combustion of coal and refined petroleum products, the extraction of metals from ore and the use and disposal of consumer products containing mercury (batteries, light bulbs, etc.). These activities have contributed to a wide dispersion of mercury and its compounds around the Earth surface. When mercury enters in the environment, it can be converted to various compounds and/or forms. Several mercury species can be present in the air, including gaseous elemental mercury (GEM), reactive gaseous mercury (RGM) and particulated mercury (PM), while mercury that reaches the water and soil may come from wet- and dry- deposition of atmospheric mercury species, as well as soluble inorganic or organic compounds present in the soil (and transferred to water), or even formed in the soil as a consequence of physico-chemical and biological reactions involving some of the species present in the soil as a consequence of the geology of the area or as a consequence of the before mentioned deposition processes. Mercury can be transformed into a highly toxic compound, methylmercury. Fish and marine mammals are the largest receptors and accumulators of methylmercury in the environment; this compound is accumulated in the tissues and consequently, its concentration biomagnifies at concentrations thousands of times greater than that found in the immediate environment. This is the form of mercury to which humans are most often exposed, primarily through consumption of fish [4, 5]. The background level of mercury in the environment is constantly growing, even regions with no significant mercury releases, such as the Arctic, are adversely contaminated due to the transcontinental transport effect of mercury [6]. Figure 1 shows the cycle of mercury in the nature and its sources [7, 8].

Exposure to elemental mercury generally occurs through inhaling mercury vapors, by non-occupational or occupational pathways. The last one is the most important because of both the high concentration of mercury vapor and the long exposure times. Workers from elemental mercury related industries present an exposure level very high. In the case of female workers, when pregnant, fetus can exposed to elemental mercury through placenta. For general population, one of the most important exposures to mercury vapor comes from the use of dental amalgam (non-occupational pathways). But mercury also releases as metallic ions or fine particles. The vapor mercury can be inhaled into lungs during tooth filling and metallic ions and particles can be retained in the saliva and reach the gastrointestinal tract. Concerning inorganic mercury compounds, population has been also exposed during decades to compounds such as Hg_2Cl_2, $HgCl_2$, HgO, HgI_2, among other, which were used as antiseptic, bactericidal, fungicidal, etc.

In relation to organic mercury, three types of compounds can be found: aryl, short chain alkyl and long chain alkyl, which are absorbed in the gastrointestinal tract. The most toxic organic mercury is methylmercury, which is ingested by general population, principally through the consumption of fish, aquatic mammals and waterfowl. The mercury concentration in food increases as one ascends the food-chain. The highest concentrations of mercury are found in the liver and kidney [9-12, 15].

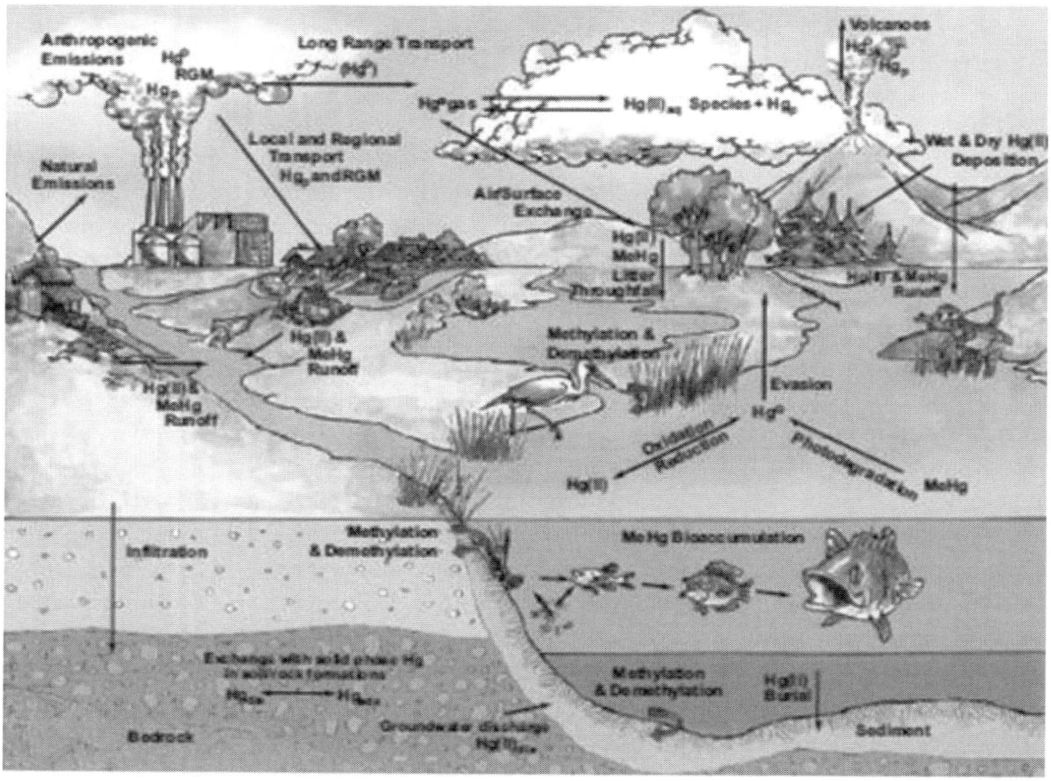

Source: Tetra Tech, Inc. 2010.

Figure 1. The mercury cycle (RGM:reactive gaseous mercury; Hg_p:particulated mercury; MeHg: methylmercury) [7, 8].

4. Effects of Mercury on Health

The adverse effects of mercury on the health are more important when longer is the exposure time and/or greater is the exposure concentration. The main routes of mercury poisoning are inhalation of elemental mercury vapor, oral ingestion of inorganic ($HgCl_2$, Hg_2Cl_2) and organic mercury (methylmercury) compounds and finally absorption of inorganic mercury through the skin.

Once inhaled mercury vapor is mostly converted to Hg^{2+} by catalase in the erythrocytes. This form present similar property to inorganic mercury compounds. Small amounts of the inhaled elemental mercury persist and penetrate in central nervous system (CNS), causing tremors, emotional lability, insomnia, memory loss, neuromuscular changes, headaches, polyneuropathy and performance deficits in test of cognitive and motor function (hydrargyrism). These symptoms can be improved when the patient is removal from the source of exposure but some of them can be irreversible. Exposure to high levels of elemental mercury in a short time causes respiratory, cardiovascular, gastrointestinal, hepatic, renal, and dermal problems [3, 13-17].

Inorganic mercury compounds such as Hg_2Cl_2 o $HgCl_2$ are absorbed through the gastrointestinal tract and skin. Due to the poor lipid solubility they accumulate mostly in the kidney, causing renal damage (kidney failure) and several gastrointestinal effects. The low solubility in lipids limits their presence in CNS, but the slow elimination and long-term exposure results in a significant accumulation of Hg^{2+}, causing neurological problems. Ingestion of these compounds causes also respiratory, cardiovascular, gastrointestinal, hepatic, renal, and dermal problems [3, 13-16, 18].

Methylmercury is readily absorbed in the gastrointestinal tract because of its high lipid solubility. It is accumulated in the brain, kidney, liver, hair, among others. In the case of pregnant women, the methylmercury can cross the placenta into the fetus and it is accumulated in the fetal brain and other tissues. Also it can be transferred to infants through breast milk. Methylmercury is a potent neurotoxin able to alter the neurological development of the fetuses and the young children and damage seriously the central nervous system of adults (Minamata disease) [19]. Other complications include visual, respiratory, dermal problems have been described. Severe and acute poisoning with metylmercury can lead to coma and death [3, 13-16, 20].

5. Legislation on Mercury

The European Union launched a series of measures to face the problems generated by mercury, thus the European Directive 2000/60/EC [21] which considers mercury as a priority hazardous substance due to its adverse effects on human health and the environment. These measures include reduction of mercury emissions strategies, cut supply and demand of mercury and the protection of population from mercury exposure, especially from methylmercury contained in fish. To protect the environment and the global population, a directive restricting the sale of measuring devices containing mercury has been dictated and the European Union banned exports of mercury from 2011 according to the regulation adopted in 2008. The strategy also establishes the conditions of mercury storage requirements [22]. Furthermore, the existing regulatory is complemented by restrictions on the use of mercury in batteries, electrical and electronic equipments, etc. [23-24]. The European Union is also engaged in action directly targeted at mitigating human exposure to mercury [25]. The European limits for allowable levels of mercury in fish are 0.5 ppm for fish in general, and 1.0 ppm for large predatory fish [26]. Furthermore, in some European states, additional controls have been adopted such as the prohibition of the use of dental amalgam in ordinary dental care, and also, in 2011, Parliamentary Assembly of the Council of Europe urges 47 member countries to take measures to prohibit or limit the use of mercury in dental amalgams [27].

In the same way, on 29 of September 2008, the United States Congress approved the Mercury Export Ban Act, which will ban the export of elemental mercury from the United States effectively in January 1, 2013 (USEPA 2010) [28]. The United States has established a series of measures to control mercury emissions. For that, United States Environmental Protection Agency (USEPA) has implanted, first of all, the management of mercury under the Clean Air Act, which regulates hazardous air pollutants by implanting a series of regulations and standards, including National Emission Standards for Hazardous Air Pollutants (NESHAP) Rules for iron and steel foundries, electric arc furnace steelmaking facilities and chlor-alkali plants. In October 2009, USEPA signed a consent decree committing the Agency to develop a final mercury emissions rule for coal- and oil-fired power plants by 2011 [28]. Under the Toxic Substances Control Act, USEPA is proposing new rules for the use of elemental mercury in certain products and is considering a series of management actions on mercury-containing products (switches/relays, button-cell batteries and measuring devices). On the other hand, the US Food and Drug Administration (FDA) attempts to minimize direct human exposure to mercury by adopting controls on mercury in fish, on cosmetic products and on dental amalgam. The FDA established a level limit for mercury content in fish, shellfish and crustaceans of 1 ppm methylmercury (US FDA 2005) [29]. The Occupational Safety and Health Administration (OSHA) established the Permissible Exposure Limit (PEL) for mercury vapor as 0.1 milligram per cubic meter of air as a time weighted average for 8 hours workday [7, 30].

Mercury is indeed a global concern and a world-wide common effort is now under way to reduce both, the supply and demand of mercury. The United Nations Environment Program (UNEP) established the Global Mercury Program with the purpose of delivering activities on mercury and supporting negotiations of international measures for its control [31, 32]. Thus a series of Governing Council decisions were established from 2001 including prevention, minimization and management of mercury waste and also its long-term storage and disposal.

6. Technologies for the Treatment of Mercury and Mercury-Containing Wastes

Different processes have been described for the recovery of mercury from waste including thermal, pyrometallurgical, hydrometallurgical and electrometallurgical technologies. In the same manner, a number of technologies have been developed to reduce the mercury mobility in environment, such as roasting/retorting, amalgamation, stabilization/solidification, vitrification and encapsulation. This chapter aims to review the most sustainable technologies for the treatment of mercury containing soil and waste.

6.1. Thermal Treatments

Thermal treatments, in general terms, involve the application of heat and reduced pressure to volatilize mercury from the contaminated medium. Roasting/retorting and incineration are the most commonly used for the treatment of mercury contaminated soils, sludges, sediments and wastes [33, 34].

Roasting/retorting operations separate the mercury from the rest of the waste stream and condense it for recovery or removal. In retorting techniques, mercury waste is sealed in a batch vessel and heated with the volatile gases released. Then, mercury vapor is condensed and liquid elemental mercury is collected. Roasting involves introducing air to the hot waste which oxidizes mercury compounds and helps its transport to a condenser. In the case of incineration, mercury waste is volatilized and converted to elemental mercury in high temperature regions of the furnace. As the flue gas cools, the elemental mercury is oxidized to ionic forms. Elemental mercury, mercuric chloride and mercuric oxide present in the vapor phase of flue gas, must be captured by adsorption onto porous solids or removed using a wet scrubber [35]. The Resource Conservation and Recovery Act (RCRA) of United States suggests that mercury containing waste can be treated by thermal method when mercury content is higher than 260 mg Hg kg^{-1} [9]. In general, the systems for carrying out the thermal desorption process requires a pretreatment unit of mercury containing waste and a mercury vapor recovery unit. Currently there are several commercial technologies, among them, thermal desorption system dry-based (MRT system of Sweden, AERC, SepraDyne and Mercury Recovery Technologies of United of States) and the wet-based (Nomura system of Japan) [36-40]. Wet based thermal desorption systems present several limitations, such as the production of mercury-containing sludge, the treatment required for the wastewater and the high cost. While dry-based thermal desorption systems are economically more efficient and generate lower amount of secondary waste.

The liquid elemental mercury collected from the condenser units can be reused or treated by amalgamation or solidification for disposal. The final treated residue needs additional treatments according to the final destination. If the material is reused in construction, total mercury content should be below limits established for this type of use, however, associated costs could be prohibitive. When the treated material is considered for landfill disposal, the total mercury reduction only has to be below the TCLP limit (0.20 mg lL^{-1}) [41]. Off-gasses also may require further treatment [35]. Undoubtedly, the most disadvantage of the use of the heat treatment technologies for the recovery of mercury is their high cost. A decreasing of cost is achieved working at reduced pressures to reduce the boiling temperature of mercury. When the process is carried out under vacuum, the problem of gas emissions and its additional cost treatment could be avoided.

Different factors may influence the heat treatment efficiency such as the nature of the waste, the particle size and the moisture content. When the waste to be treated contained a large amount of organic matter, an explosion hazard may occur in presence of oxygen. Relating to the particle size of the waste, large particles produce problems in the heat transfer and not securing the reduction of contamination of mercury. For small particles, the amount of particulate matter in the emissions gases can increase which implies a complication in its treatment and a higher economic cost. High moisture contents decrease the yield and increase the cost process.

Thermal desorption is an alternative process, especially for treatment and remediation of contaminated soil *in situ*. Kurchaski et al. (2005) [42] studied a low temperature thermal desorption process and the subsequent vapor capture, making sure that biological activity of the soil was not destroyed. Figure 2 shows a schematic diagram of a pilot installation for mercury volatilization and capture from the soil. The heating chamber was constructed of temperature resistant plates. A bank of fourteen infrared lamps (370 W) was installed as a heating source above the soil surface. Mercury vapors were absorbed in a scrubber made of stainless steel and vacuum pump. Although these experiments used a relatively expensive source of heat (infrared lamps), the technology was conceptually feasible.

Kunkel et al. (2006) [43] assessed the feasibility of using *in situ* thermal desorption process that, simultaneously applies heat and vacuum to the soil subsurface at temperatures up to 600 °C. There is no practical limit as to the treatment depths because the heaters were originally designed for use at depths reaching about 610 m. As the soil is heated the vapor pressure of the contaminant increases, allowing mass transfer to the gas phase. The typical vacuum applied is on the order of 0.02 bars. Two electric band heaters connected to temperature controllers and thermocouples placed within column thermowells were used to maintain the column temperature at the setpoint. The problem of *in situ* thermal desorption process is its high cost, the site accessibility for well installation and the potential need for control of water recharge for contaminated zones located below the water table.

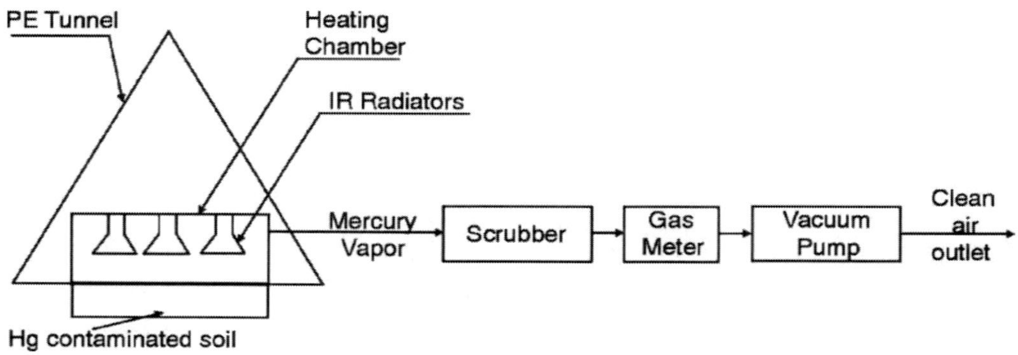

Figure 2. Scheme of Kucharsti et al. for the remediation on soil *in situ* [42].

To minimize the high cost of the thermal technologies, Navarro et al. (2009) [44] studied the potential of solar thermal desorption for removal of mercury from mining contaminated soils. They use solar energy as the primary energy source in a low and medium temperature solar furnace (Figure 3a).

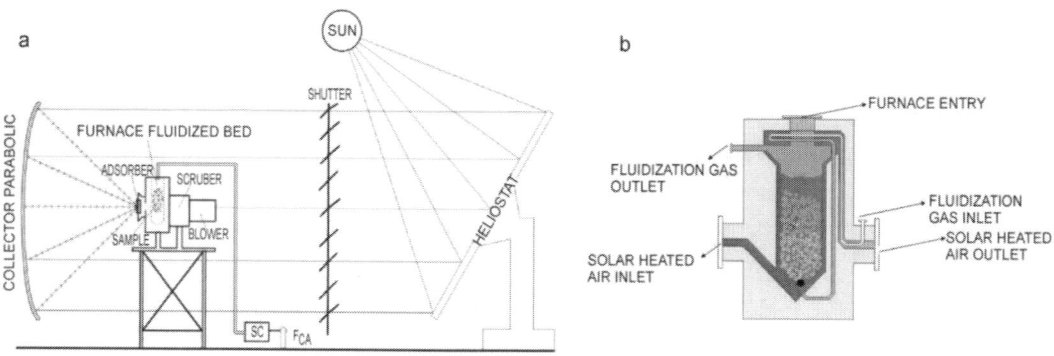

Figure 3. (a) Scheme of operation of the middle-temperature solar furnace (SC: system cooling, F_{CA}: activated-carbon filter). (b) Schematic drawing of the fluidized-bed module in the Solar Platform Almería, Spain [44].

A fluidized-bed reactor was used to test the feasibility of high-temperature solar desorption (Figure 3b). The system is composed of two main parts: the solar furnace facility and the solar fluidized-bed module connected to a mercury gas collection system with a final activated-carbon adsorption bed. All the equipment is insulated to minimize heat loss. The treatment consists of placing a soil sample in the fluidized-bed chamber, sealing the cover and connecting it to the gas recovery system. The heliostat controller automatically directs the solar radiation onto the concentrator dish, which in turn beams the concentrated solar rays onto a focal point located on the optical axis. The absorber module in the focal zone is gradually heated by concentrated solar radiation as the shutter is gradually opened to the reference temperature of 400 °C, and this temperature is maintained for approximately 30 min. During the treatment, a blower forces air through the absorber module, where it is heated, and part of the energy in this air is transferred into the fluidized-bed chamber through its steel walls, heating it and then preheating the air or gas that fluidizes the bed as it passes through the coil behind it. The technique was successfully applied to polluted samples and was demonstrated to be an alternative method for decontaminating hazardous waste sites where mercury is present because of its volatility [44].

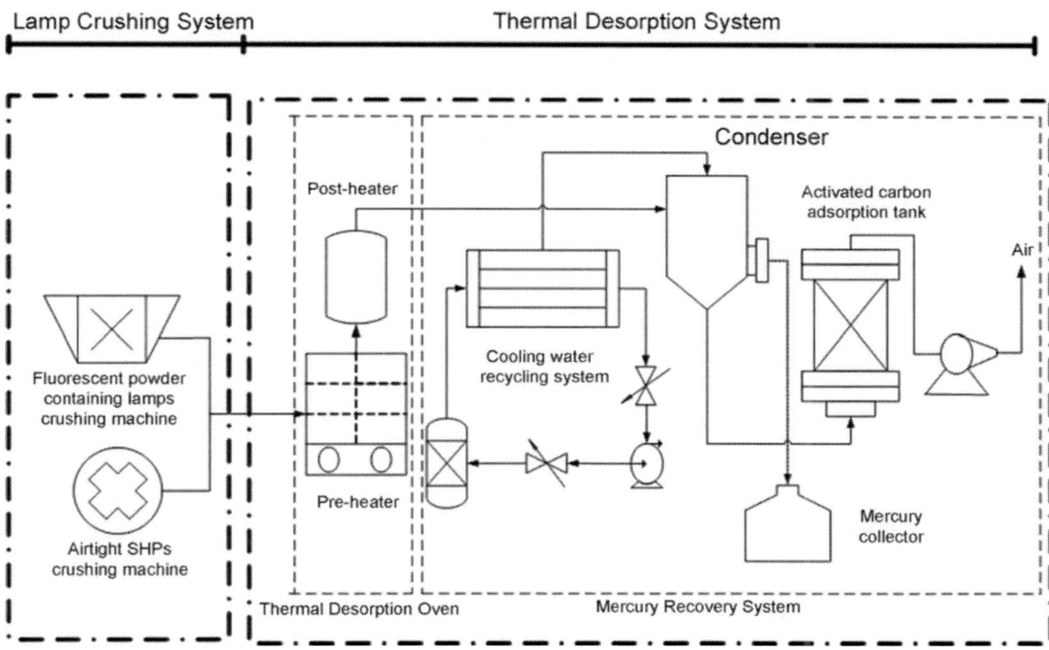

Figure. 4. Schematic diagram of full-scale thermal desorption process to recover the mercury from high-mercury-containing lamps [46].

Desorption thermal technologies are being also used for the recovery of mercury from lamps. Jang et al. (2005) [45] applied this technology at laboratory scale to pulverized spent fluorescent lamps using a muffle furnace. Above 400 °C, mercury was recovered almost completely. In 2009, Chang et al. [46] set up a full-scale thermal desorption process, monitoring the use of different pre-heating temperatures and desorption times, to treat and recover the mercury from super high pressure mercury lamps, fluorescent tube tailpipes, fluorescent tubes containing mercury-fluorescent powder, and cold cathode fluorescent lamps containing mercury-fluorescent powder. Figure 4 shows a schematic diagram of the full-scale thermal desorption process, which can be divided into a lamp crushing system and a thermal desorption system.

The pre-heating chamber could operate continuously at 700-800 ºC under vacuum conditions to vaporize the mercury from the lamps. The post-heating chamber operated at over 800 ºC to ensure that all the mercury from the mercury-containing lamps was vaporized. The mercury vapor gas flowed into the condenser to be condensed into liquid mercury at 25 ºC which was collected in a tank. Finally, uncondensed gas was then treated in an activated carbon adsorption tank. Vulcanized activated carbon was used as a sorbent to lower the mercury content enough to meet 0.3 mg m^{-3} in the exhaust gas [46]. In 2010, Chang et al., used a bench-scale thermal desorption process (Figure 5) to carry out thermal desorption batch experiments [47]. The system includes a 2.5 cm diameter high temperature tubular combustion-heated furnace and a mercury vapor analyzer.

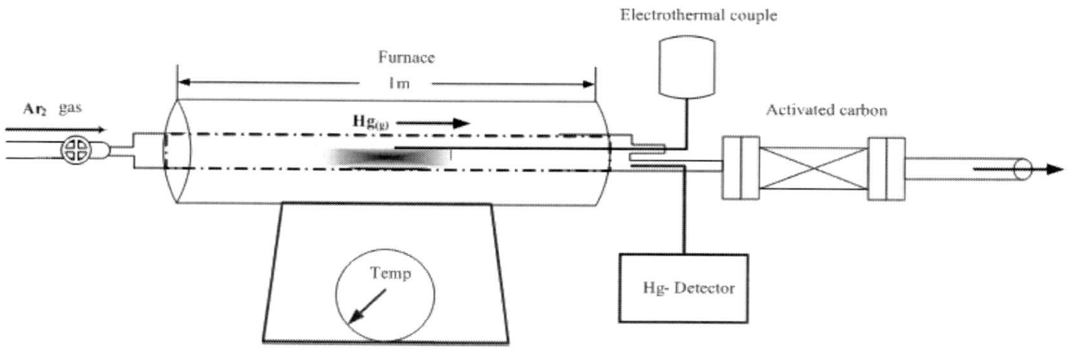

Figure 5. Bench scale thermal desorption setup [46, 47].

Table I. Experimental conditions of different thermal treatments

Authors	Waste	Time (Minmin)	T (ºC)	Treatment type	Efficiency (%)
Busto et al., 2011 [34]	Sludge sludge (chloralkaline)	60	>400	retorting	>90
Chang et al., 2009 [46]	lamps	>480	>350 *(vacuum)*	retorting	>81
Taube et al., 2008 [33]	soil	20	>470	roasting	99
Chang and Yen, 2006 [48]	soil	120	650 *(vacuum)*	incineration	98
Kucharski et al., 2005 [42]	soil *(in situ)*	10 days	167	roasting	60-70
Kunkel et al., 2006 [43]	soil *(in situ)*	695-78511-13 hours	244	roasting	99
Navarro et al., 2009 [44]	mine soil	30	≈<>400	solar thermal desorption	41.3-87

The authors showed that the thermal desorption efficiency was influenced by the complexity of the mercury compounds contained in the lamps and also temperatures from 550 to 850 °C are necessary for the mercury releases from amalgam and mercury/fluorescent powder.

Table I collects the experimental conditions of different thermal technologies applied to various wastes and their efficiency for mercury desorption.

6.2. Immobilization Technologies

The technologies developed with the view to achieve safe disposal of mercury, aim also its immobilization and thereby the reduction of its mobility towards the environment. Although most processes involve the use of two or three different technologies, they might be tentatively grouped as follows: amalgamation, stabilization/solidification, encapsulation and vitrification.

6.2.1. Amalgamation

Amalgamation is defined as the combination of mercury with metallic reagents such as copper, gold, zinc, aluminium, nickel, tin, gold, silver, etc., that results in a solid and non-volatile product, the amalgam. In fact, amalgams consist of semi-solid solutions obtained by dissolution of mercury in the solid metal. The process is reversible, so that mercury can be released from these alloys by heating. Amalgams, although solid, show a significant vapor pressure and solubility of mercury. This technique has been used for the extraction of precious metals from their ores (after amalgamation, the amalgam was subjected to a thermal treatment in order to volatilize mercury and thus, recover the precious metal). Although amalgamation is a relatively inexpensive and speedy process for the treatment of liquid mercury, disadvantages come from the difficulties to scale up and the needing of using dilute nitric acids to achieve high efficiency. The presence of water in the amalgam can lead to the formation of the hydroxide of the amalgamating metal which makes the process unsuitable due to the treatment of radioactive contaminated mercury. The method commonly used is unable to stabilize adequately instable forms of mercury in the hazardous waste material.

The process designed by Woodward (1991) [49] consists on putting liquid mercury in a disposable container and mixing it with a powdered amalgamating agent (copper, nickel, zinc) in ratio of 1:1 up to 1:3. This process yields to a permanent bonding of the mercury to the powder in a solid form. The container with the amalgam therein is safely and legally deposited in a landfill. Gorin et al. (1994) [50] proposed a copper amalgam. Elemental mercury is added to fine copper powder (325 mesh), so that the mixture contains 65 % mercury. The same authors proposed also a zinc amalgam, which contains 45 % mercury. Copper and zinc are previously cleaned. The amalgam with zinc does not fulfill the TLCP TCLP test requirements and thus it cannot be considered as an adequate treatment technique.

DeHg^{SM} (de-merk) [51, 52] is a batch metal amalgamation process that converts the mercury component in mixed waste from wide variety of waste matrices (shreddabble debris, nonshreddable debris, waste waters, soils, sludges, organics) to a nonhazardous final waste form at ambient temperature, suitable for land disposal. Depending on the degree of difficulty in obtaining final waste form stability, the DeHg process can be performed as either a single or a two step process. The process treats mercury mixed waste containing not only elemental mercury, but also ionic and complexed forms of mercury. The first stage of the process involves amalgamation of the elemental mercury component, while the second stage aims the stabilization of soluble mercury species. The process generates hydrogen gas as a byproduct, which is vented through control equipment.

Rehmat et al. (2001) [53] developed a stabilization method of liquid mercury in monolithic solid form for storage or disposal. The method consists in mixing powdered copper particles with liquid mercury at ambient temperature, forming a powdered copper/liquid mercury mixture and hardening the powdered copper/liquid mercury mixture to form an amalgam. The amount of mercury in the obtained amalgams is greater than 50 % by weight of the total amalgam. This amalgam is resistant to deformation, breakage, chemical corrosion, and environmental release of its mercury content. The mercury sequestered can be disposed of by burial or it can be separated from the copper and then recovered by distillation after storage.

In general term, the amalgamation gave rise to a solid material with a relatively quite high mercury vapor pressure and accordingly, this technique can only be used as a complementary treatment along with other immobilization processes.

6.2.2. Stabilization/Solidification (S/S) with Sulfur And and Selenium

To stabilize means to convert to a solid, hard, non-flowable mass or a solid, inert particulate material which resists common solvents and retains the materials contained therein when subjected to leaching conditions. Stabilization is a chemical immobilization of hazardous constituents, through chemical bonds to an immobile matrix, or chemical conversion to an immobile species, insoluble or non-toxic, thereby reducing vaporization or leaching to the environment [54]. Mostly stabilization is accompanied with a solidification process. Stabilization/solidification (S/S) is a mature technology for the treatment of different kind of wastes. In the case of mercury, S/S is principally achieved with sulfur or sulfur compounds. Selenium has been also used for this purpose.

Mercury reacts with powdered sulfur and/or liquid sulfur (polysulfide) to form mercuric sulfide.

$$Hg^{2+} + S^{2-} \rightarrow HgS_{(s)} \ (Ks_{black}=6.44 \times 10^{-53};\ Ks_{red}=2 \times 10^{-53}) \qquad [1]$$

Mercury sulfide is a stable or stabilized compound which as a result of its properties (nontoxic, chemically stable and even under corrosive conditions), allows permanent and safe storage of mercury. Mercuric sulfide is insoluble in water and in acidic solutions, but in alkaline solutions with excess of S^{2-} is solubilized as HgS_2^{2-}. This process is simple, effective, fast and inexpensive.

According to Gibbs free energy value (-11.4 kcal mol^{-1} for metacinnabar and -12.1 kcal mol^{-1} for cinnabar [1]) the direct reaction of elemental mercury and sulfur is favored. However, merely placing elemental mercury in bed of sulfur or in a volume of polysulfide does not yield the desired mercuric sulfide reaction. Sufficient heat must be available to overcome the energy of activation for the reaction to occur. Intensive mixing provides the necessary energy activation for the reaction to take place. To control the temperature reaction, it is necessary in some cases, to add water and/or several materials to the mixture in order to cool the reactants. The application of this method on soils and sludge spiked with elemental mercury yield low success, because the heat provided by mixing action or generated from the mercury-sulfur reaction quickly dissipates throughout the surrounding matrix materials and so it is difficult to overcome the energy of activation to initiate the mercury-sulfur reaction throughout the entire reaction mass [55]. Beside mercury in soils is not a continuous phase. Chemical stabilization of mercury in solid materials is difficult because of the dispersed nature of the contaminant within the solid matrix and the consequent problem of contacting a stabilizing agent with the widely dispersed contaminant. Also the use of inorganic sulfur compounds alone, often can release toxic hydrogen sulfide gas and can be prohibitively expensive for the treatment of mercury containing waste.

During the stabilization of soluble mercury by the formation of insoluble mercury sulfides, it is desirable to minimize the formation of mercury polysulfide complexes to prevent that mercury could be eluted or leached, by an adequate selection of the inorganic sulfur compound. Among other compounds used to minimize the mercury polysulfide formation are alkali metal sulfite, alkali metal bisulfite and alkali metal metabisulfite, etc. Similarly, the prevention and the minimization of the formation of hydrogen sulfide is obtained by limiting the amounts of hydrogen-containing sulfur compounds, present as impurities in the inorganic sulfur compounds, or by limiting the exposure of the stabilizing inorganic sulfur compounds to air and sunlight. The use of antioxidants (ascorbic acid, citric acid, phosphoric acid, butylated hydroxyanisole and butylated hydroxytoluene) will also reduce hydrogen sulfide formation.

Mercury selenide is a quite insoluble compound that is formed from elemental mercury and selenium. HgSe is as well known under its mineral name Tiemannite. Mercury selenide cannot be synthesized by mixing the elements at room temperature and gas phase reaction is normally required.

Litz et al. (2002) [56] describe a process which includes the following steps: a): combining the mercury-containing feed material with a polysulfide and reactive sulfur-containing compounds other than the polysulfide, to form a composite feed material. b): mixing the composite feed material to treat the reactive sulfur-containing compound with the mercury or mercury waste to form mercuric sulfide. The process provides an effective method for converting elemental and speciated mercury into an environmentally stable compound, such as HgS, utilizing mercury reactive material. To accomplish a high extent of reaction between the mercury and the mercury reactive material, the method involves mixing additives and any stirred vessels. The polysulfide acts as an activator of the reaction between the reactive sulfur-containing compound and the unstable forms of mercury. The polysulfide is preferably selected from the group consisting of calcium polysulfide, sodium polysulfide and other alkaline earth polysulfides and mixtures thereof. Such forms of polysulfide are readily soluble/suspendable in water. According to the authors, the final product fulfills the TCLP or contains no more than about 1000 ppm elemental mercury and/or environmentally unstable forms for speciated mercury.

The stabilization with sulfur has been also used for the treatment of spent mercury lamps, mercury thermometers, barometers, vacuum manometers and other devices containing glass-evolved mercury [57-60].

More recently new researches on S/S of mercury have been carried out by several authors [61-70]. Petit and Durecu (2007) [65] patented a process for the treatment of solid wastes polluted by mercury. The method consists of the oxidation of mercurial species and the sulfidation of the oxidized mercury species in an alkaline medium. For the stabilization of metallic mercury with sulfur, Rieviere-Huc et al. (2008) [66] designed a spherical and hermetically seal reactor which generates small quantities of HgS.

New researches have been carried out in the last two years to achieve the transformation of mercury in mercury sulfide. We point out the work of Fuhrmann et al. (2010) [67] to remove mercury from contaminated waste by applying a sulfur reagent to the waste. The stable mercury sulfide may then be removed from the waste leaving behind the treated waste itself *in situ* following removal of the contaminating mercury therefrom. The waste may be plain earth or soil, sediment, dredge spoils, sludge and other industrial waste of various forms contaminated by mercury dispersed therein. López et al. (2010) [68] achieved metacinnabar from direct reaction between liquid mercury and elemental sulfur at room temperature, by using the mechanical energy provided by a milling process. The method is simple, inexpensive and gave rise to a stable solid compound easily handled, transported and long-term stored, the HgS in form of black metacinnabar powder. Boyle et al. (2010) [69] generate mercury sulfide from elemental mercury which is injected into a reaction vessel containing vaporized sulfur. Then, elemental mercury reacts with a portion of the vaporized sulfur to form the mercury (II) sulfide. Elemental mercury is converted into a high purity mercury sulfide with the same physical and chemical characteristics of naturally occurring cinnabar. The product is a red pellet that is suitable for land disposal. Bonmann (2011) [70] patents a process for immobilizing mercury by producing crystalline mercury sulfide. The mercury sulfide was formed by reaction of elemental mercury and an additive material of elemental sulfur or a sulfur compound. This reaction takes place at temperature above the melting point of the additive material.

The stabilization of mercury as mercury selenide was investigated by several researchers. Jonsson (1999) [71] patented a method and arrangement for destruction of hazardous waste (strip lights, mercury lamps and other types of light sources mercury containing). Lamps in pieces were fed into an electric furnace at 1350-1400 °C, then the glass was melted down to a mass which spontaneously flows out, whereas the mercury was gassed off from the glass mass and was sucked into a chamber, which was connected with the electric furnace. The well balanced amount of selenium was dosed into the chamber, evaporated and mixed with the mercury gas. The mixed gas was sucked into a cooling device and then precipitated in solid form of mercury selenide that is a less mobile compound than elemental mercury.

Selenide process has also been patented for treatment of batteries [72]. In treating wastes, a rotary kiln was used to provide agitation of the material; selenium was added to the furnace under inert conditions and other components of the process are similar to those used for lamps. The process has not been applied to elemental mercury, although lamps do contain elemental mercury. This method presents potential risks due to the additional components of heat and selenium that it is a toxic metal. Platen (2006) [64] patented a method for disposing mercury using selenium or sulfur. The method consists in supplying mechanical energy to the treatment vessel containing the mercury and the selenium or sulfur. This vessel was sealed to prevent air from entering. This reactor can be stored with the material inside it and then it is not necessary any post treatment steps.

The formation of sulfides and selenides is an adequate technology for the stabilization of elemental mercury; however complementary processes as the encapsulation technologies contribute to a major isolation of the residue from the surrounding atmosphere. These processes will be described below.

6.2.3. Encapsulation

To encapsulate means to embed substances within a stable matrix. Molten and casting are part of the encapsulation process. Encapsulation is a physical immobilization of hazardous constituents by enveloping a waste in a non-porous, impermeable material. Micro and macroencapsulation can be defined depending on if the waste is fine-grained and well dispersed throughout the matrix so that, each particle is separately encapsulated, or if clumps of waste matrix or bulk waste are enclosed within the encapsulating material, respectively. In the case of mercury, the purpose of the encapsulation is to seal the mercury waste from the surrounding environment. Although encapsulation might be included in S/S technologies, the huge number of research development permit its classification as a separated group [52, 73].

Cement or concrete like products, are commonly used for waste encapsulation, due to their hardness, very long life and the capacity to be mould prior to setting. But in some cases, the success of the procedure is limited because the presence of contamination affects negatively the binding stage.

The methods used for encapsulation of mercury wastes can be classified in two categories depending on the use or not of pozzolanic materials in the process.

6.2.3.1. Encapsulation without Pozzolanic Materials

An encapsulation method described by Rechichi (2002) [74] involves adding hazardous waste material to a sellable composition of calcium carbonate and caustic magnesium oxide, forming slurry, and allowing the slurry to set to encapsulate the waste material.

Tests conducted on the encapsulated material indicated that virtually very small mercury content is leached out of the set composition which has a concrete-like appearance. Indeed, leach rates are less than the allowed leach rate of 5.00 μg $mlmL^{-1}$ mercury. The composition of calcium carbonate and caustic magnesium oxide can be formed by treating dolomite. Dolomite is a calcium magnesium carbonate found in nature. A true dolomite contents about 54 % calcium carbonate and 43 % magnesium carbonate. If the dolomite is heated between 500-800 °C, the magnesium carbonate releases carbon dioxide over the calcium carbonate, then the dolomite is converted to a settable dolomite comprising caustic magnesium oxide. Aluminium sulfate or magnesium sulfate is used to enhance the preferential decomposition of magnesium carbonate. Various additives (organic fillers and inorganic fillers) can be added to accelerate the formation of strong binding agents, and assist in the recrystallization of the composition to make it set. Other additives as sodium carbonate or carboxylic/polycarboxilic acids can be added to assist in the setting process by means of the liberation of carbon dioxide. Acidifying agents as citric, acetic and oxalic acid, among others, can be necessary for assist in the carbonization of MgO and $Mg(OH)_2$ to recrystallize the composition into a set material. Besides they can act as ligand to form complexes around the fillers helping to trap them in the set matrix.

Technologies based on the use of phosphate ceramics have been also described [75, 76]. The process, in general terms, consists of two reactions, including a chemical bondage of the elemental mercury as well as a microencapsulation within a matrix. In the first reaction, the elemental mercury is bond to the phosphate and various phosphate types can be formed, as follows:

$$Hg + H_3PO_4 \rightarrow Hg_3(PO_4)_2 / HgH(PO_4)$$

During the second reaction, the ceramic matrix is built up, within which the mercury compounds are microencapsulated:

$$MgO + KH_2PO_4 + 5H_2O \rightarrow MgKPO_4.6H_2O$$

Wang et al. (2000) [77] studied mercury encapsulation in chemically bonded phosphate ceramics, formed by the reaction of MgO and KH_2PO_4 in solution. They use K_2S for mercury stabilization. Once these components are thoroughly mixed, the slurry is allowed to set. Setting occurs in approximately 2 h and a ceramic waste form of high strength and low open porosity was formed. Samples were cured for 3 weeks for good curing. This process seems to be very efficient with waste streams such as ash waste streams that were generated by destruction combustibles, by destruction of organics with ferric oxide and ferric chloride or ferric phosphate and soils. Hg stabilization is not just a result of sulfidation, but it is also a consequence of the chemical immobilization of Hg as phosphate ceramics matrix.

6.2.3.2. Encapsulation with Pozzolanic Materials

Cement, cementitious material, sulfur polymer cement among others are used to achieve the encapsulation of mercury and mercury wastes due to their pozzolanic characteristics. Pozzolans are materials having hydraulic cementitious properties when mixed with free lime in presence of water. They contain principally oxides of silicon, aluminum and iron. Pozzolanic materials can be volcanic rocks, kiln dust, fly ash and industrial by-products, such as, granulated blast furnace slag and silica fume. Pozzolanic cements, including blends with Portland cement, can have performance advantages over straight Portland cements for the encapsulation of mercury wastes [54]. The fixation of metals using Portland cement and fly ash produces an impermeable and solid waste form at pH >10, that limits the solubility and the leachability of metals, but in the case of mercury, this metal does not form a low solubility hydroxide solid and accordingly, it is very difficult to stabilize mercury with cement based processes.

Zhang and Bishop (2002) [78] developed a process for the S/S of mercury-containing solid wastes using power activated carbon reactivated with sulfur (CS_2) and cement. After mercury stabilization by adsorption as mercuric sulfide on the power activated carbon with sulfur, the product was mixed with ordinary Portland cement for solidification. The process allowed the stabilization of 1000 mg kg^{-1} of Hg. Portland cement also was used by Zhuang et al. (2004) [79] for the stabilization of highly leachable mercury contaminants of brine purification sludge generated from the Hg-cell electrolysis process in chlorine production. Ferric lignin derivates were used along with the cement. These materials present the capability to immobilize the leachable Hg contaminants through the strong bonding strength with Hg-contaminants in brine purification sludge. The process was used for the stabilization of approximately 2000 t of brine purification sludge. Qian et al. (2003) [80] studied the solidification behavior of mercury salt in alkali activated slag (AAS) which consists principally of: 40.31 % CaO, 34.47 % SiO_2, 9.47 % Al_2O_3, 0.53 % Fe_2O_3 and 7.81 % MgO. Results indicated that the immobilization of Hg^{2+} ions in the AAS matrix is due to the formation of reaction products C-S-H and amorphous metal silicate precipitates. In this process, Hg^{2+} can react with soluble silicate to form insoluble amorphous metal silicate precipitates, then is competitively and gradually bound into the crystal lattice of C-S-H, replacing Ca^{2+} as mercury salt is added into the AAS.

Svensson and Allard (2008) [81] prepared cylindrical monoliths by mixing sand, cement, blast furnace slag, water, the sulfur source and mercury. The monoliths were long period cured under dry conditions to allow the development of pozzolanic reactions between cement and the blast furnace slag and also the formation of mercury sulfide. The leaching process showed the time increasing the leach rate decreasing.

Chan and Hall (2008) [82] gradually added a cementitious particulate filler material to the sludge obtained previously by amalgamation of mercury with copper powder and dilute nitric acid. The cementitious material is a mixture of ordinary Portland cement and blast furnace slag. The method provides a final waste form after allowing curing which is non toxic or less toxic than elemental mercury. It is also easy to handle and presents minimum health risks, good long term stability and contains minimum amounts of leachable substances and is thereby suitable for disposal. The aAuthors state that the method is particularly applicable to the safe disposal of elemental mercury which is contaminated with radioactive materials. Chen et al. (2009) [83] studied the effect of accelerated carbonation on the immobilization of mercury with Portland cement. Sodium carbonate was used as an internal carbon source and CO_2 as an external carbon source. The compressive strength of the solidified waste improved significantly by reducing porosity and consequently permeability of the samples. Other process described the use of zeolite along with Portland cement [84]. Thiol-functionalized zeolite, obtained by grafting the thiol group to the natural clinoptilolite zeolites, was first used to stabilize mercury solid wastes, after that Portland cement was used to solidify the sample. A final product with 1000 mg kg^{-1} of Hg, fulfills the TCLP leaching test.

6.2.3.3. Encapsulation with Thermoplastic Materials

The use of thermoplastic materials for encapsulation of mercury and mercury containing-waste has been increasing since the beginning of 2000, in such extension that these processes are known as Sulfur Polymer Stabilization Solidification processes (SPSS). The SPSS process is widely being used to convert mercury compounds into the highly insoluble HgS form and simultaneously encapsulate the waste. The process relies upon the use of a thermoplastic material which contains 95 % of elemental sulfur and 5 % of organic modifiers, such as dicyclopentadiene and oligomers of cyclopentadiene. This material is referred to in the literature as sulfur polymer cement (SPC), although it is not a cementitious material. SPC melts at approximately 112.7 ºC and sets rapidly upon cooling. It is relatively impermeable to water, compared to conventional concrete and has a high mechanical strength, approximately double that of conventional concrete. SPC is also well suited to harsh environments with high levels of mineral acids, corrosive electrolytes, or salt solutions [85]. This technology is applicable to mercury non-wastewaters, to mercury amalgams and to chemically fixed mercury (mercuric sulfide) [54]. SPC has a very long life and its strength greatly increase within the first few years after forming. Besides, other advantages of process are that it is carried out relatively at low temperature (120-140 ºC) and allows high concentration of mercury waste can be effectively treated. The absence of water in the final waste form offers less chemical breakdown, biodegradation, leaching and gas generation after disposal [86, 87]. For macroencapsulation, molten SPC is poured over and around the waste or debris and is then allowed to set into a monolithic waste form (Figure 6). For microencapsulation, Kalb et al (2002) [88] patented two-stage process for the treatment of liquid, elemental mercury. In the first one the stabilization of mercury is carried out by combining mercury containing waste with SPC under an inert atmosphere (Ar or N_2) at a temperature about 20-80 ºC.

Sodium sulfide, triisobutyl phosphine sulfide, calcium hydroxide, sodium hydroxide, calcium oxide or magnesium oxide or a combination of them are used as stabilizing additives. In the second stage involves the encapsulation by heating the mixture at temperature between 120-150 °C to form a molten product, which is then cast as a monolithic final waste form.

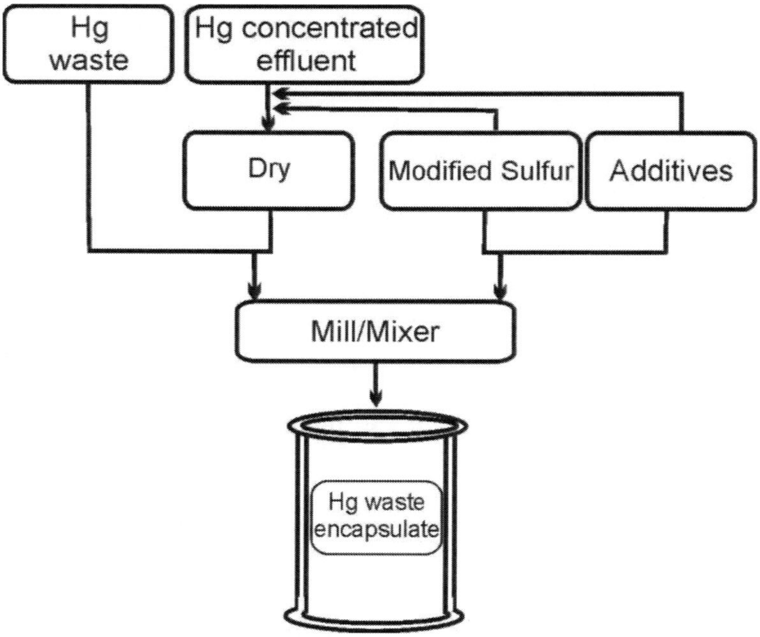

Figure 6. Scheme of the sulfur polymer stabilization/solidification process.

In 2002, Fuhrmann et al. [89] studied the formation of HgS by mixing elemental mercury with SPC and sodium sulfide which is the catalyzer. The molten mixture is poured into a mold to cool and solidify. The leachate mercury concentrations were low enough to meet EPA´s land disposal restrictions for mercury contaminated soils and sludge < 25 μg lL^{-1} [90]. The process was also demonstrated for mercury-containing soil. (Kalb, 2001) [85]. More recently, López et al. [91] patented also a two stage microencapsulation process for the stabilization of mercury using the polymeric sulfur cement. This process leads to a stable concrete-like sulfur matrix that allows the immobilization of mercury (Figure 7). Firstly, liquid mercury was transformed in metacinnabar by the energy supplied by a ball mill [68]. Then the metacinnabar was mixed with gravel, sand and a thermoplastic material (STX™ supplied by Starcrete™ Technologies Inc. Quebec, Canada) [92, 93]. The final product, in the form of a solid block containing up to 30 % Hg, exhibits excellent mechanical properties (compressive strength of 53-61 MPa and flexural strength of 7-10 MPa), low porosity (0.57 % P_{He}), very low total pore volume (0.63 x 10^{-2} cm^3 g^{-1}) and extremely low permeability (coefficient of water absorption by capillarity 0.07 g cm^{-2}). The TCLP test reveals a mercury concentration in leachates well below the 0.2 mg Ll^{-1} set out in USEPA Land Disposal Restrictions (LDRs) [90]. The monolithic end product exhibits very low mercury vapor emission and very low mercury leachability. Thus, the process developed stabilizes and solidifies liquid mercury by rendering inert materials and allowing the long-term storage of this waste [94, 95]. The process is now under scaling up to a demonstration plant which is been built in Almadén, Ciudad Real, Spain.

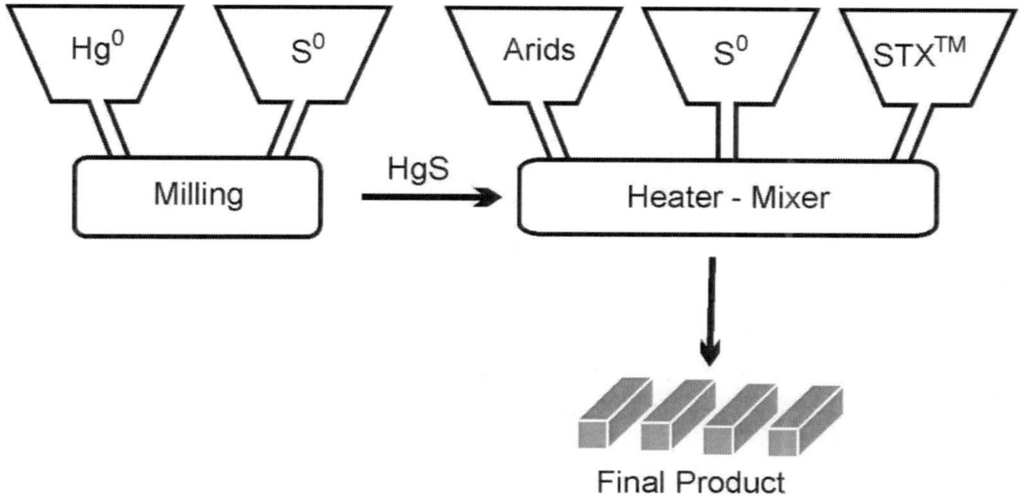

Figure 7. Scheme of the SPSS process used for the stabilization/solidification of mercury.

6.2.4. Vitrification

Vitrification is a high-temperature treatment to immobilize contaminants by its incorporation into the vitrified end product, which is chemically durable and leach resistant. This technology requires a large amount of energy to achieve vitrification temperatures and off-gases may require further treatment. It is used to treat mercury-contaminated soil and sediments *in situ* and *ex situ*. In vitrification process the electric current is passed through soil by an array of electrodes inserted vertically into the surface of the contaminated zone. As soil is non-conductive, a starter pattern of electrically conductive glass frit that contains graphite is placed on the soil in the paths of the electrodes. When power is fed to the electrodes, the graphite and glass frit conduct electricity through the soil, heating the surrounding area and directly melting the adjacent soil. Once molten, the soil becomes conductive. The melt grows out ward and down ward as power is gradually increased to a full constant operating level. The temperature of the contaminated soil could reach between 1600 and 2000 ºC. A single melt can treat a region up to 1000 tons. Vitrification is used to treat wastes to a depth of 6.1 m [9]. Two vitrification technologies have been used to treat mercury-contaminated soil and sediment: the Glass Furnace Technology (GFT) [96] and the Plasma Hearth Process (PHP) [97]. The patented GFT process developed by Minergy Corporation consists in mixing dried waste (moisture <10 %) with a flux material to control melting temperatures and improve the physical properties of the glass aggregate product. The waste is then fed to the glass furnace, used for melt minerals into glass. Into the furnace constructed of refractory brick temperature of 1600 ºC is reached. The sediments with mercury melts and metals are encapsulated within a glass matrix. The matrix flows out of the furnace as molten glass. Molten glass flows into a water-quenching tank, where the cool water is used to harden the molten glass into glass aggregate. Quenching with water improves the hardness of the glass, yielding it inert to the environment. The off-gas from the process enters the air quality control system, which consists of a wet scrubber, a fabric filter and a carbon filter. The wet scrubber eliminates sulfur dioxide and hydrogen chloride, the fabric filter removes particulate matter (dust) and the carbon filter captures mercury gas [9, 96].

The Plasma Hearth Process is a high-temperature thermal process for the treatment of mixed waste [97]. The process uses an electric arc to melt non-combustible material, inorganic material and volatilize and oxidize organic materials. The molten material is cooled into a stable glassy and/or crystalline waste form. Hazardous metals are incorporated in the molten phase. The organic fraction is partially combusted and pyrolyzed and then is conducted to the secondary combustion chamber where it is subjected to high temperature in the presence of excess air and a methane flame to complete combustion. The plasma hearth process chamber temperature is approximately 593-704 °C and plasma hearth process melt temperatures are in the range of approximately 1650-2200 ºC. The off-gas is quenched to approximately 204 ºC in an evaporative cooler. The off-gas then enters a bag house for to remove large particles, followed by a carbon filter for to remove the volatilized mercury and a high efficiency particulate air filter bank for to removes fine particulates. The advantage of the PHP is that is a single, high-temperature thermal process that minimizes the need for multiple treatment systems and for extensive sorting/segregating for large volumes of waste. The PHP has the potential to treat a wide range of wastes, while minimizing the need for sorting, reduce the final waste volumes, produce a leach resistant waste form and destroy organic contaminants [96, 97]. The problem of this process is the presence of chlorides, fluorides, sulfides and sulfates, which may interfere with the process, resulting in higher mobility of mercury in the vitrified product, but feeding additional slag-forming materials such as sand to the process, may compensate for the presence of chlorides, fluorides, sulfides and sulfates. During vitrification, the combustion of the organic content of the waste liberates heat, reducing the external energy requirements. This process may be advantageous in the treatment of wastes that contain a combination of mercury and organic contaminants or for the treatment of organo-mercury compounds. Mercury may be difficult to treat because of its high volatility and low solubility in glass (less than 0.1 %), but may be effectively treated at low concentrations [9].

7. Corollary

The global concern about the hazardousness of mercury, due to its adverse effects on human health and the environment, has promoted the huge number of researches on processes to immobilize the metal and the mercury-containing waste, to render inert materials, principally for the safe long-term disposal or storage. Nowadays new materials and/or new processes are being developed [54, 98] to achieve a sustainable framework in which industries and citizens can coexist side by side without forgetting the protection of the environment.

Acknowledgments

Dr. O. Rodriguez thanks to CSIC (Spain) for the grant JAEDoc_09_00121.

REFERENCES

[1] Lide, D.R. (1990-1991). *Handbook of Chemistry and Physics.* 71st edition. Boca Raton, Florida, CRC Press, Inc.

[2] Anthony, J.W., Bideaux, R.A., Bladh, K.W., Nichols, M.C. (2011). *Handbook of Mineralogy.* Mineralogical Society of America, Chantilly, VA 20151-1110, USA. Available from: http://www.handbookofmineralogy.org/.

[3] Clarkson, T.W., Marsh, D.O. (1982). Mercury toxicity in man, in: Prasad A.S. (Ed.), *Clinical, Biochemical, and Nutritional aspects of trace elements,* vol. 6, Alan R. Liss, Inc., pp. 549-568, New York.

[4] Risher, J.F. (2003). *Elemental mercury and inorganic mercury compounds: Human health aspects.* Agency for Toxic Substances and Disease Registry (ATSDR), Atlanta, Georgia, USA. Available from: http://www.who.int/ipcs/publications/cicad/en/cicad50.pdf

[5] Fitzgerald, W.F., Lamborg, C. H. (2005). Geochemistry of mercury in the environment. In: Lollar, B.S. (Ed*).), Environmental Geochemistry*, vol. 9. Elsevier, Amsterdam, pp. 107-148.

[6] Streets, D.G., Hao, J.M., Wu, Y., Jiang, J.K., Chan, M., Tian, H.Z., Feng, X.B., (2005). Anthropogenic mercury emissions in China. *Atmospheric Environment,* 39(40), 7789-7806.

[7] *Environment Canada and Health Canada. Risk Management Strategy for Mercury* (2010). Environment Canada. Available from: *http://www.ec.gc.ca/doc/mercure-mercury/1241/index_e.htm*

[8] *Tetra Tech, Inc.* (2010). Tetra Tech, Inc. Available from: www.tetratech.com/

[9] USEPA (2007). *Treatment technologies for mercury in soil, waste and water.* Washington, DC 204160. EPA-542-R-07-003. Available from: http://www.epa.gov/tio/download/remed/ 542r07003.pdf

[10] Fitzgerald, W.F., Engstrom, D.R., Mason, R.P., Nater, E.A. (1998). The case for atmospheric mercury contamination in remote areas. *Environmental Science and Technology*, 32(1), 1-6.

[11] Mason, R.P., Fitzgerald, W.F., Morel, F.M.M. (1994). The biogeochemical cycling of elemental mercury: anthropogenic influences. *Geochimica et Cosmochimica Acta*, 58(15), 3191-3198.

[12] Wagner-Döbler, I. (2003). Pilot plant for bioremediation of mercury-containing industrial wastewater. *Applied Microbiology Biotechnology*, 62(2-3), 124-133.

[13] USEPA. *Mercuy. Health Effects*. (2012). Available from: http://www.epa.gov/mercury/ effects.htm;

[14] Olson. D.A. *Mercury Toxicity* . (2011). Available from: http://emedicine.medscape. com/article/1175560-overview

[15] *Elemental mercury and inorganic mercury compounds: Human health aspects.* Concise International Chemical Assessment Document 50. World Health Organization 2003. ISBN 9241530502. Available from: *http://www.who.int/ipcs/*publica*tions/ cicad/en/cicad50.pdf*

[16] *Agency for to*xic substa*nces ¬ Disease* registry. *Toxicological profile for mercury.* 2010. Available from: http://www.atsdr.cdc.gov/toxprofiles/tp46.html

[17] USEPA. *Intefrated Risk Information System. Mercury elemental.* 2011.Availble from: http://www.epa.gov/iris/subst/0370.htm

[18] USEPA. *Intefrated Risk Information System. Mercuric chloride.* 2011. Available from: http://www.epa.gov/iris/subst/0692.htm

[19] Ministry of the Environment. *National Institute for Minamata Disease*. Available from: http://www.nimd.go.jp/english/

[20] USEPA. *Intefrated Risk Information System. Methylmercury*. 2011. Available from: http://www.epa.gov/iris/subst/0073.htm

[21] Directive 2000/60/EC of the European Parliament and of the Council of 23 October 2000 establishing a framework for Community action in the field of water policy. Available from: http://eur-lex.europa.eu/LexUriServ/LexUriServ.do?uri=CELEX:32000L0060:EN:NOT

[22] Regulation (EC) No 1102/2008 of the European parliament and of the Council of 22 October 2008 on the banning of exports of metallic mercury and certain mercury compounds and mixtures and the safe storage of metallic mercury. Available from: http://eur-lex. *europa.eu/LexUriServ/LexUriServ.do?uri=OJ:L:2008:304:0075: 0079: EN:PDF*

[23] Directive 2002/95/EC of the European Parliament and of the Council of 27 January 2003 on the restriction of the use of certain hazardous substances in electrical and electronic equipment. Available from: http://eur-lex.europa.eu/LexUriServ/LexUriServ.do?uri=OJ:L:2003:037:0019:0023:en:PDF

[24] Directive 2006/66/EC of the European Parliament and of the Council of 6 September 2006 on batteries and accumulators and waste batteries and accumulators and repealing. Directive 91/157/EEC (Text with EEA relevance). Available from: http://eur-lex.europa.eu/LexUriServ/LexUriServ.do?uri=CELEX:32006L0066:en:NOT

[25] Communication from the Commission to the Council and the European Parliament - Community Strategy Concerning Mercury {SEC(2005) 101}. Available from: http://eur-lex.europa.eu/LexUriServ/LexUriServ.do?uri=CELEX:52005DC0020:EN: NOT

[26] Report of the twenty-fourth session of the Codex Committee on food additives and contaminants. Codex Circular Letter CL 1992/8-FAC. ALINORM 93/12. Available from: www.codexalimentarius.net/download/report/18/al93_12e.pdf

[27] *Health hazards of heavy metals and other metals*. Social, Health and Family Affairs Committee. Parliamentary Assembly. Council of Europe. Doc 12613 – 12 May 2011. Available from: *http://www.assembly.coe.*int/Docum*ents/Working Docs/Doc11/*EDOC 12613.pdf

[28] *USEPA (2010). Mercury*. Available from*: http://www.epa.gov/mercury/*

[29] US FDA (U.S. Food and Drug Administration) (2005). *Guidance for industry and* FDA *Staff. Class II special controls guidance document: Dental amalgam, mercury, and amalgam alloy*. Available from: www.fda.gov/downloads/MedicalDevices/DeviceRegulationandGuidance/GuidanceDocuments/ucm073330.pdf

[30] OSHA (Occupational Safety and Health Administration). *Occupational safety and health guideline for mercury vapor exposure limits.* Available from: http://www.osha.gov/SLTC/healthguidelines/mercuryvapor/recognition.html

[31] UNEP (United Nations Environment Programme) (2002). *Global Mercury Assessment.* Geneva, Switzerland. UNEP Chemicals. Available from: www.chem.unep.ch/mercury/Report/Final report/final-assessment-report-25nov02.pdf

[32] UNEP (United Nations Environment Programme). Global Mercury Partnership (2008). *Mercury fate and transport in the global atmosphere: Measurements, models and policy implications*. Available from: www.chem.unep.ch/mercury/Sector-Specific-Information/Full_Report.pdf

[33] Taube, F., Pommer, L., Larson, T., Shchukarev, A., Nordin, A. (2008). Soil remediation - mercury speciation in soil and vapor phase during thermal treatment. *Water Air Soil Pollution*, 193(1-4), 155-163.

[34] Busto, Y., Cabrera, X., Tack, F.M.G., Verloo, M.G. (2011). Potential of thermal treatment for decontamination of mercury containing wastes from chlor-alkali industry. *Journal of Hazardous Materials*, 186(1), 114-118.

[35] USEPA (2001). Proceedings and summary report. *Workshop on mercury in products, processes, waste and the environment: Eliminating, reducing and managing risks from non-combustion sources.* Baltimore, MD. EPA/625/R-00/014. Available from: http://www.clu-in.org/download/contaminantfocus/mercury/workshop-on-mercury-in-products.pdf

[36] *Mercury Recovery Technology*. Available from*: www.mrtsystem.com*

[37] *AERC Recycling solutions.* Available from: http://www.aercrecycling.com/information/fluor_process.php

[38] Morris, M.I., Osborne-Lee, I.W., Hulet, G.A. (2002). *Demonstration of new technologies required for the treatment of mixed waste contaminated with ≥ 260ppm mercury.* ORNL/TM-2001/147. U.S. Department of Energy. Office of Science and Technology. Available from: http://www.ornl.gov/~webworks/cppr/y2002/rpt/111527.pdf

[39] DOE/EM-0633. 2002. *The Sepradyne Tm – Raduce system for recovery of mercury from mixed waste.* TRU and mixed waste focus area. US Department of Energy. Office of Environmental Management. Office of Science and Technology. Summary Report. Brookhaven National Laboratory. Upton , New York. Available from*: http://clu-*in.org/download/contaminantfocus/mercury/DOE_SepraDyne.pdf

[40] *Mercury recovery Services.* Available from: http://hgremoval.com/small-volume/index.html

[41] *Frequent Questions about Regulations that Affect the Management and Disposal of Mercury-Containing Light Bulbs (Lamps).* 2011. Regulatory Overview. How are mercury-containing bulbs (called "lamps" in the regulations) regulated?. Federal Laws. Availble from: http://www.epa.gov/osw/hazard/wastetypes/universal/lamps/faqs.htm

[42] Kucharski, R., Zielonka, U., Sas-Nowosielska, A., Kuperberg, J.M., Worsztynowicz, A., Szdzuj, J. (2005). A method of mercury removal from topsoil using low-thermal application. *Environmental Monitoring and Assessment,* 104(1-3), 341-51.

[43] Kunkel, A.M., Seibert, J.J., Elliott, L.J., Kelley, R., Katz, L.E., Pope, G.A. (2006). Remediation of elemental mercury using *in situ* thermal desorption (ISTD). *Environmental Science and Technology*, 40(7), 2384-2389.

[44] Navarro, A., Cañadas, I., Martinez, D., Rodríguez, J., Mendoza, J.L. (2009). Application of solar desorption to remediation of mercury-contaminated soils. *Solar Energy*, 83(8), 1405-1414.

[45] Jang, M., Hong, S.M. Park, J.K. (2005). Characterization and recovery of mercury from spent fluorescent lamps. *Waste Management*, 25, 5-14.

[46] Chang, T.C., You, S.J., Yu, B.S., Chem, C.M., Chiu, Y.C. (2009). Treating high mercury - containing lamps using full-scale thermal desorption technology. *Journal of Hazardous Materials*, 162, 967-972.

[47] Chang, T.C., Chen, C.M., Lee, Y.F., You, S.J. (2010). Mercury recovery from cold cathode fluorescent lamps using thermal desorption technology. *Waste Management and Research*, 28(5), 455-460.

[48] Chang, T.C., Yen, J.H. (2006). On-site mercury-contaminated soils remediation by using thermal desorption technology. *Journal of Hazardous Materials,* 128, 208-217.

[49] Woodward, J.C. (1991*). Process for treating mercury in preparation for disposal.* Patent number US 5,034,054.

[50] Gorin, A.H., Leckey, J.H., Nulf, L.W. (1994). *Final disposal options for mercury/uranium mixed wastes from the Oak Ridge Reservation.* Y/DZ-1106. Managed by Lockneed Martin Energy Systems, Inc. For the United States Department of Energy. Available from:
http://www.osti.gov/bridge/servlets/purl/290905-stQxjA/webviewable/290905.pdf

[51] DOE 1999a. U.S. Department of Energy. *Demonstration of NFS DeHg process for stabilizing mercury (<260 ppm) contaminated mixed waste*. Mixed waste focus area, innovative technology summary report, DOE/EM-0468, prepared for Office of Environmental Management, Office of Science and Technology. Available from:
https://www.dndkm.org/DOEKMDocuments/ITSR/TRUMixedWaste/Demonstration%20of%20NFS%20DeHg%20Process%20for%20Stabilizing%20Mercury%20(260%20ppm)%20Contaminated%20Mixed%20Waste.pdf

[52] DOE 1999b. U.S. Department of Energy. *Mercury contamination - amalgamate (contract with NFS and ADA) demonstration of DeHg*^{SM} *process*. Mixed waste focus area, innovative technology summary report, DOE/EM-0471, prepared for Office of Environmental Management, Office of Science and Technology. Available from: https://www.dndkm.org/DOEKMDocuments/ITSR/TRUMixedWaste/Mercury%20Contamination-%20Amalgamate%20-Demonstration%20of%20DeHg%20Process.pdf

[53] Rehmat, A.G., Goyal, A., Laird, D.H. (2001). *Method for stabilization of liquid mercury*. Patent number US 6,312,499 B1.

[54] USEPA Contract 68-W4-0005, WA No. R11032 (1998) *Technologies for immobilizing high mercury subcategory wastes.* Science Applications International Corporation 11251 Roger Bacon Drive Reston, VA 20190. Available from:
http://www.epa.gov/osw/hazard/tsd/ldr/mercury/imoblzn2.pdf

[55] Broderick T.E., Roth R.L, Carlson A.L. (2005). *Method for fixating sludges and soils contaminated with mercury and other heavy metals*. Patent number US 6.911.570 B2.

[56] Litz, J.E., Broderick, T., Stewart, R.M. (2002). *Method and apparatus for stabilizing liquid elemental mercury*. Patent number US 6.403.044 B1.

[57] Mosin, A.A., Borbat, V.F., Mukhin, V.A. (2003). *Method for detoxification of mercury - containing waste*. Patent number RU 2209695 (C1).

[58] Borbat, V.F., Mukhin, V.A., Kanushin, I.F. (2004). *Method of decontamination of mercury - containing wastes.* Patent number RU 2228227 (C1).

[59] Plunge, R.A., Plunge, A., Griskonid, E. (2003*). Method for utilization of elemental mercury*. Patent number LT 2003030 (A).

[60] Xiong, Z., He, F., Zhao, D., Barnett, M.O. (2009). Immobilization of mercury in sediment using stabilized iron sulfide nanoparticles. *Water Research,* 43(20), 5171-5179.

[61] Svensson, M., Allard, B., Düker, A. (2006). Formation of HgS - mixing HgO or elemental Hg with S, FeS or FeS_2. *Science of the Total Environment,* 368(1), 418-423.

[62] Svensson, M., Düker, A., Allard, B. (2006). Formation of cinnabar - estimation of favourable conditions in a proposed Swedish repository. *Journal of Hazardous Materials*, B136, 830-836.

[63] Piao, H., Bishop, P.L. (2006). Stabilization of mercury-containing wastes using sulfide. *Environmental Pollution*, 139(3), 498-506.

[64] Platen, J.M.P. (2006). *Disposing of mercury using selenium or sulfur, involves supplying mechanical energy to sealed treatment vessel.* Patent number SE 527661 (C2).

[65] Petit, M.T., Durecu, S. (2007). *Treatment of solid wastes e.g. fiber-cement polluted by mercury, useful to fix mercurial species e.g. elemental mercury as sulfur amalgam in solid wastes, comprises oxidization and sulfidation of mercurial species in alkaline medium.* Patent number FR 2900074 (A1).

[66] Rieviere-Huc, C., Huc, V., Bosse, E. (2008). *Method for stabilization of metallic mercury using sulphur.* Patent number US 2008019900 (A1).

[67] Fuhrmann, M., Heiser, J., Kalb, P. (2010). *Mercury contamination extraction.* Patent number US7692058 B2.

[68] López, F.A., López-Delgado, A., Padilla, I., Tayibi, H., Alguacil, F.J. (2010). Formation of metacinnabar by milling of liquid mercury and elemental sulfur for long term mercury storage. *Science of Total Environmental*, 408(20), 4341-4345.

[69] Boyle, J.M., Lawrence, B.J., Schreffler, S.A. (2010). *Method and apparatus for generating mercury (II) sulphide from elemental mercury.* Patent number US 7691361.

[70] Bonmann, C. (2011). *Process and device for inmobilizing mercury by producing crystalline mercury sulphide for subsequent disposal of mercury sulphide*. Patent number US7914757 (B1).

[71] Jonsson, T. (1999). *Method and arrangement for destruction of hazardous waste*. Patent number WO/1999/004913 (A1).

[72] USEPA/600/R-03/048 (2002*). Preliminary analysis of alternatives for the long term management of excess mercury.* National Risk Management Research Laboratory. Office of Research and Development USEPA. Cincinnati, Ohio.

[73] Randall, P., Chattopadhyay, S. (2004). Advances in encapsulation technologies for the management of mercury-contaminated hazardous wastes. *Journal of Hazardous Materials*, B114, 211-223.

[74] Rechichi, D. (2002). *Encapsulation of hazardous waste materials.* Patent number US 6399848 B1.

[75] *Extended summary on pretreatment technologies for metallic mercury (2010).* BiPRO (Beratungsgesellschaft für integrierte Problemlösungen). 07.0307/2009/530302. European Commission, Brussels.

[76] Liu, Z., Qian, G., Zhou, J., Li, C., Xu, Y., Qin, Z. (2008). Improvement of ground granulated blast furnace slag on stabilization/solidification of simulated mercury doped wastes in chemically bonded phosphate ceramics. *Journal of Hazardous Materials*, 157, 146-153.

[77] Wang, A.S., Sing, D., Jeong, S.Y. (2000). *Mercury stabilization in chemically bonded phosphate ceramics*. Environmental Protection Agency's Workshop on Mercury Products, Processes, Waste, and the Environment: Eliminating, Reducing and Managing Risks, Baltimore, MD.

[78] Zhang, J., Bishop, L. (2002). Stabilization/solidification (S/S) of mercury containing wastes using reactivated carbon and Portland cement. *Journal of Hazardous Materials*, B92, 199-212.

[79] Zhuang, J.M., Lo, T., Walsh, T., Lam, T. (2004). Stabilization of high mercury contaminated brine purification sludge. *Journal of Hazardous Materials*, 113(1-3), 157-164.

[80] Qian, G., Sun, D.D., Tay, J.H. (2003). Characterization of mercury - and zinc -doped alkali-activated slag matrix. Part I. Mercury. *Cement and Concrete Research*, 33, 1251-1256.

[81] Svensson, M., Allard, B. (2008). Leaching of mercury containing cement monoliths aged for one year. *Waste Management*, 28, 597-603.

[82] Chan H.B., Hall, R. (2008). *Treatment of elemental mercury*. Patent number: 20080234529.

[83] Chen, Q., Ke, Y., Zhang, L., Tyrer, M., Hills, C.D., Xue, G. (2009). Application of accelerated carbonation with a combination of Na_2CO_3 and CO_2 in cement-based solidification/stabilization of heavy metal-bearing sediment. *Journal of Hazardous Materials*, 166, 421-427.

[84] Zhang, X.Y., Wang, Q.C., Zhang S.Q., Sun, X.J., Zhang, Z.S. (2009). Stabilization/Solidification (S/S) of mercury contaminated hazardous wastes using thiol-functionalized zeolite and Portland cement. *Journal of Hazardous Materials*, 168(2-3), 1575-1580.

[85] Kalb, P.D., Adams, J.W., Milian, L.W. (2001). *Sulfur Polymer Stabilization/Solidification (SPSS) treatment of mixed waste mercury recovered from environmental restoration activities at BNL.* BNL-52614 Formal Report. Brookhaven National Laboratory. Envorinmental Sciences Department. Under contract nº DE-AC02-98CH10886. USA.

[86] Hamby, D.M. (1996). Site remediation techniques supporting environmental restoration activities-a review. *The Science of the Total Environment*, 191, 203-224.

[87] Bowerman, B., Adams, J., Kalb, P., Wan, R-Y., LeVier, M. (2003*). Using the sulfur polymer stabilization/solidification process to treat residual mercury wastes from gold mining operations.* BNL-71499-2003-CP. U.S. Department of Energy.

[88] Kalb, P.D., Melamed, D., Patel, B.R., Fuhrmann, M. (2002). *Treatment of mercury containing waste*. Patent number US6399849.

[89] Fuhrmann, M., Melaned, D., Kalb, P.D., Adams, J.W., Milian, L.W. (2002). Sulfur polymer solidification/stabilization of elemental mercury waste. *Waste Management*, 22, 32-333.

[90] USEPA (2002). Code of Federal Regulations. Title 40, Part 268. *Land Disposal Restrictions.*

[91] López, F.A., López-Delgado, A., Alguacil, F.J., Alonso, M. (2009). *Procedimiento de estabilización de mercurio líquido mediante cemento polimérico de azufre, vía sulfuro de mercurio*. Patent number ES 2355001 A1.

[92] Vroom, H. (1977). *Sulfur cements, process for making same and sulfur concretes made therefrom*. Patent US 4058500.

[93] *STARTcrete™* (2000). Technologies Inc. Laboratory Procedure for Producing STARcretes™ Test Specimens. Tehnical Report.

[94] López-Delgado, A., López, F.A., Alguacil, F.J., Padilla, I., Guerrero, A. (2012). A microencapsulation process of liquid mercury by sulfur polymer stabilization/solidification technology. Part I: characterization of materials. *Revista de Metalurgia* 48 (1), 1-13 .

[95] López-Delgado, A., Guerrero, A., López, F.A., Pérez, C., Alguacil, F.A. (2012). A microencapsulation process of liquid mercury by sulfur polymer stabilization/solidification technology. Part II: durability of materials. *Revista de Metalurgia* 48 (1), 14-23.

[96] Glass furnace technology. Available from: http://www.minergy.com/technology/glass-furnace-technology/

[97] US Department of Energy, (1998). *Plasma Hearth Process*. Mixed Waste Focus Area. OST Reference #26. Office of Environmental Management. Office of Science and Technology. Available from: http://costperformance.org/pdf/itsr26.pdf

[98] *Xtaltite® Waste Forms*. Toxic Waste Disposal – Nuclear Waste Disposal. (2003). Greg Eaton and Associates Pty. Ltd. P.O. Box 8032, Rhyll, Victoria, Australia, 3923. Available from: *http://www.eatonco.net/xtal.htm*

In: Mercury: Sources, Applications and Health Impacts
Editors: Ki-Hyun Kim and Richard J. C. Brown
ISBN: 978-1-62257-721-7

Chapter 7

MERCURY TOXICITY ALTERS THE ANTIOXIDANT SYSTEM OF GROWING CUCUMBER SEEDLINGS

***Denise Cargnelutti*[1], *Fernando Teixeira Nicoloso*[2], *Luciane Almeri Tabaldi*[2], *Valderi Luiz Dressler*[3], *Vera Maria Morsch*[3] and *Maria Rosa Chitolina Schetinger*[3]**

[1]Universidade Federal da Fronteira Sul, Campus de Erechim, Erechim, RS, Brazil
[2]Departamento de Biologia, Centro de Ciências Naturais e Exatas, Universidade Federal de Santa Maria, Santa Maria, RS, Brazil
[3]Departamento de Química, Centro de Ciências Naturais e Exatas, Universidade Federal de Santa Maria, Santa Maria, RS, Brazil

ABSTRACT

Mercury (Hg) is a heavy metal with pronounced toxic effects for plants and animals. In this study, the toxicity of Hg was evaluated on cucumber (*Cucumis sativus* L.) seedlings treated for 10 and 15 days in agar-based medium containing $HgCl_2$ at 0-500 μM. At both growth stages, the seedlings were analyzed for changes in electrolyte leakage percentage (ELP), hydrogen peroxide, carotenoids, superoxide dismutase (SOD), delta-aminolevulic acid dehydratase (ALA-D), ascorbic acid (ASA) and non-protein thiols (NPSH). Results showed that the increase of ELP and H_2O_2 production in seedlings might be related to decreased superoxide dismutase (SOD) activity with a consequent increase in membrane and protein damage (such as ALA-D). Mercury stress increased the levels of ascorbic acid (ASA), non-protein thiols (NPSH), and carotenoids in seedlings at the initial period of exposure. These antioxidant systems might play an important role in making Hg harmless, in turn making the 10-day-old seedlings tolerant at low concentrations. However, the antioxidant systems were not able to protect from the toxicity caused by higher levels of Hg (250 and 500 μM) and increased exposure times, resulting in the negative effects observed in the growth of seedlings. Moreover, at 50 μM $HgCl_2$, a compensatory effect was observed for the antioxidant defense system, which could be attributed to the hormetic dose response. Therefore, Hg toxicity triggered alterations in the antioxidant system which reduced oxidative damage in cucumber seedlings.

Keywords: Mercury, Cucumber, Superoxide dismutase, Non-protein thiol groups, δ-aminolevulinic acid dehydratase

1. Introduction

Heavy metal contamination is one the most serious environmental problems for plant productivity and it is also a threat to human health. Factors such as mining or industrial activities, automotive emission, and use of metal-enriched materials such as chemicals fertilizers, farm manures, sewage sludge, and wastewater irrigation can contribute to this contamination (Webber, 1981; Freedman and Hutchinson, 1981) of natural ecosystems (Alumaa et al., 2002; Segura-Muñoz et al., 2006).

Mercury (Hg) is regarded as a non-essential element, with no known physiological function for organisms. One of the characteristic effects of metal poisoning, observable at an early stage, is a reduction of both seedlings survival (Crane et al., 2012), cell proliferation and growth (Schützendübel et al., 2001; Israr and Sahi, 2006; Zhou et al., 2007). In addition, Hg^{2+} alters the transpiration (Zhang and Tyerman, 1999), mineral nutrition (Boening, 2000), chlorophyll metabolism (Morsch et al., 2002; Cargnelutti et al., 2006) and photosynthetic integrity (Israr et al., 2006), thus these parameters may be used as indicators of its phytotoxicity.

However, this metal did not trigger a clear increase in lipid peroxidation in epidermal cells, even though the cell viability was drastically reduced. Generally, the strategy adopted by plants to avoid the build-up of excess metal levels is achieved by the use of various mechanisms such as the reduction of metal uptake into the cytosol by a number of ligands or efflux from the cytosol either into the apoplast or into the vacuole. It is also possible that more than one mechanism is involved in the reduction of the toxicity of a particular metal, such as phytochelatin (PC) production (Hall, 2002). Mercury possesses high affinity to sulfhydryl groups of glutathione (GSH), this peptide being an important non-enzymatic antioxidant. Also, GSH participates in the formation of PC, which might play an important side role in Hg detoxification (Rellán-Álvarez et al., 2006). Mercury has been demonstrated to stimulate the formation of reactive oxygen species (ROS) (Cho and Park, 2000), including superoxide radicals ($O_2^{\bullet-}$), hydrogen peroxide (H_2O_2) and hydroxyl radicals ($^{\bullet}OH$), all of which attack proteins (Romero-Puertas et al., 2002; Noriega et al., 2006), lipids (Sandalio et al., 2001) and nucleic acids (Gichner et al., 2006), leading to oxidative stress (Lomonte et al., 2010). Among the proteins that can suffer oxidation and damage is the enzyme δ-aminolevulinic acid dehydratase (ALA-D). ALA-D catalyzes the asymmetric condensation of 2 molecules of δ-aminolevulinic acid (ALA) to porphobilinogen (Gibson et al., 1955) and is sensitive to heavy metals because of its sulfhydrylic nature (Rocha et al., 1995, Morsch et al., 2002). The synthesis of porphobilinogen promotes the formation of porphyrins, hemes, and chlorophylls, which are essential for adequate aerobic metabolism and for photosynthesis (Jaffe et al., 2000).

There is substantial evidence indicating that Cu and Cd toxicity in higher plants is mediated by oxidative stress (Vitória et al., 2001; Fediuc and Erdei, 2002). However, there are few reports on the mechanisms of Hg-induced oxidative stress and tolerance in higher plants. Ortega-Villasante et al. (2005) reported that Hg causes serious oxidative damage even at very short exposure times. Mercury has been demonstrated to stimulate the formation of reactive oxygen species (ROS) (Cho and Park, 2000), including superoxide radicals ($O_2^{\bullet -}$), hydrogen peroxide (H_2O_2) and hydroxyl radicals ($^{\bullet}OH$), all of which attack proteins (Romero-Puertas et al., 2002; Noriega et al., 2006), lipids (Sandalio et al., 2001) and nucleic acids (Gichner et al., 2006), leading to oxidative stress.

In order to combat metal toxicity, plant cells have antioxidants such as carotenoids (Kenneth et al., 2000; Polyakov et al., 2001), GSH (Foyer et al., 2001, Foyer, 2002; Foyer and Noctor, 2005) and ascorbate (Dipierro et al., 2005), and also antioxidant enzymes such as superoxide dismutase, ascorbate peroxidase, catalase and glutathione reductase, which participate in scavenging ROS and hence protect the cells from injury (Matés, 2000, Alscher et al., 2002).

Similar for other plant species (Sobrino-Plata et al., 2009), *Cucumis sativus* is known to accumulate toxic metals under laboratory conditions and has been selected as one of the test plant species due to its sensitivity to a wide range of contaminants (Cargnelutti et al., 2006; Pereira et al., 2006). Thus, the objective of the present study was to contribute to the understanding of the toxicology of mercury. In order to obtain these results, cucumber seedlings were used to evaluate the effect of this metal on the antioxidant system and its relation to ALA-D activity (an enzyme involved in the metabolism of chlorophyll).

2. Material and Methods

2.1. Plant Material and Growth Conditions

Seeds of cucumber (*Cucumis sativus* L.) obtained from Feltrin Ltd. (Santa Maria/RS, Brazil) were germinated in glass recipients (100 mL) containing 20 mL of 10% of Murashige and Skoog (1962) medium, supplemented with 0.6% agar and various $HgCl_2$ levels. Seedlings were exposed to 0.5, 50, 250 and 500 μM of $HgCl_2$. These concentrations had been chosen due the highest mercury (Hg) concentrations found in soil, ranging from 15 to 300 $\mu g\ g^{-1}$ dry weight (55 – 1104.94 μM Hg) (Cavallini et al., 1999). Moreover, in a previous study (Cargnelutti et al., 2006), it was shown that Hg ranging from 50 to 500 μM was toxic to the cucumber seedlings.

The medium pH was adjusted to 5.8. Each experimental unit consisted of 6 seeds, totalizing 15 replicates per treatment. After the radicle broke through, the seedlings were maintained in a growth chamber with controlled temperature (25±1°C) and photoperiod (16 h light; light intensity of 35 $\mu mol\ m^{-2}\ s^{-1}$ at plant level).

Although 10 to 15 days is a rather long exposure period to such high concentrations of Hg, it was selected to verify if there would be alterations in the biochemical parameters evaluated both at a small interval of time and at a prolonged time of exposure, which could be closer to environmental conditions.

2.2. Determination of Electrolyte Leakage Percentage

ELP was used to assess membrane permeability and it was measured using an electrical conductivity meter (Digimed, CD21, Brazil). The procedure used was based on the method of Zhu et al. (2004), with some modifications. Plant samples were separated into 5 g segments and placed in individual stoppered vials containing 50 mL of distilled water after washes with distilled water to remove surface contamination. These samples were incubated at room temperature (25°C) on a shaker (Nova Técnica, NT-145, Brazil) (100 rpm) for 24 h. Electrical conductivity of the bathing solution (EC1) was read after incubation. Samples were then placed in a thermostatic water bath at 95°C for 15 min and the second reading (EC2) was determined after cooling the bathing solutions to room temperature. ELP was calculated as EC1/EC2 and expressed in percentage (%).

2.3. Determination of Hydrogen Peroxide

The H_2O_2 contents of both control and treated seedlings were determined according to Loreto and Velikova (2001). Approximately 100 mg of seedlings were homogenized at 4°C in 2 mL of 0.1% (w/v) trichloroacetic acid (TCA). The homogenate was centrifuged at 12000 x g for 15 min and 0.5 mL of 10 mM potassium phosphate buffer pH 7.0 and 1 mL of 1 M KI. The H_2O_2 content of the supernatant was evaluated by comparing its absorbance at 390 nm with a standard calibration curve, using a molar extinction coefficient of 0.28 μM^{-1} cm^{-1}. The H_2O_2 content was expressed as $\mu mol\ g^{-1}$ fresh weight.

2.4. Superoxide Dismutase (E.C 1.15.1.1)

The activity of superoxide dismutase was assayed according to Mc Cord and Fridovich (1969). About 200 mg fresh tissues were homogenized in 5 mLl of 100 mM K-phosphate buffer (pH 7.8) containing 0.1 mM EDTA, 0.1% (v/v) Triton X-100 and 2% (w/v) polyvinyl pyrrolidone (PVP). The extract was filtered and centrifuged at 22000 x g for 10 min at 4°C, and the supernatant was utilized for assays. The assay mixture consisted of a total volume of 1 mL, containing buffer glycine (pH 10.5), 60 mM epinephrine and enzyme. Epinephrine was the last component to be added. The adrenochrome formation in the next 4 min was recorded at 480 nm with a UV- Vis spectrophotometer (Hitachi, U-2001 UV/Vis Spectrophotometer, USA). This method is based on the ability of SOD to inhibit the autoxidation of epinephrine at an alkaline pH. Since the oxidation of epinephrine leads to the production of a pink adrenochrome, the rate of increase of absorbance at 480 nm, which represents the rate of autoxidation of epinephrine, can be conveniently followed. SOD has been found to inhibit this radical-mediated process.

2.5. Non-Protein Thiol Content

Non-protein thiol (NPSH) content in seedlings was measured spectrophotometrically with Ellman's reagent (Ellman, 1959). Reaction was read at 412 nm after the addition of 10 mM 5-5-dithio-bis (2-nitrobenzoic acid) (DTNB) (0.05 ml). Treated seedlings were homogenized in 10 mM Tris/HCl, pH 7.4, centrifuged at 4000 x g at 4°C for 10 min, and supernatants were used for total thiol group determination. NPSH were determined in the fraction obtained after mixing 1 volume of supernatant with 1 volume of 4% TCA (to precipitate proteins) followed by centrifugation and neutralization (to pH 7.5) with 1 M Tris as described by Jacques-Silva et al. (2001). A standard curve using cysteine was used to calculate the content of thiol groups in samples, and was expressed as μmol NPSH g^{-1} fresh weight.

2.6. Ascorbic Acid Content

Ascorbic acid (ASA) determination was performed as described by Jacques-Silva et al. (2001). Briefly, seedlings were homogenized in 10 mM Tris/HCl, pH 7.4, centrifuged at 4000 x g for 10 min and protein was removed by dilution with 1 volume of 4% TCA followed by centrifugation. An aliquot of the sample was incubated at 37°C in a medium containing 4.5 mg mL^{-1} dinitrophenylhydrazine, 0.6 mg mL^{-1} thiourea, 0.075 mg mL^{-1} $CuSO_4$, and 0.675 mol L^{-1} H_2SO_4 (final volume 1 mL). After 3 h, 1 ml of 65% H_2SO_4 was added and samples were read at 520 nm and were expressed as μg ASA g^{-1} fresh weight. A standard curve was constructed using ASA.

2.7. Carotenoids Determination

Cotyledons were weighed and used for carotenoid determination. Carotenoids were extracted following the method of Hiscox and Israelstam (1979). Carotenoid content was calculated using the formulae (Lichtenthaler, 1987) given below: carotenoid (mg g^{-1} fresh weight) = ((100 x A_{470}) – (2.27 x chl a) – (81.4 x chl b) / 227) x 50/500. Hundred milligrams chopped fresh cotyledon sample was incubated at 65°C in dimethylsulfoxide (DMSO) until the pigments were completely bleached. Absorbance of the solution was then measured at 470 nm with a spectrophotometer (Celm E-205D, Brazil). Carotenoid content were expressed as mg g^{-1} fresh weight.

2.8. Estimation of Delta-Aminolevulinic Acid Dehydratase (ALA-D; E.C. 4.2.1.24) Activity

Cucumber cotyledons were homogenized in 10 mM Tris-HCl buffer, pH 9.0, at a proportion of 1:1 (w/v). The homogenate was centrifuged at 12000 x g at 4°C for 10 min to yield a supernatant (S1) that was used for the enzyme assay. The supernatant was pre-treated with 0.1% Triton X-100 and 0.5 mM dithiothreitol (DTT). ALA-D activity was assayed as described by Morsch et al. (2002) by measuring the rate of porphobilinogen (PBG) formation. The incubation medium for the assays contained 100 mM Tris-HCl buffer, pH 9.0. For the enzyme assay, the final concentration of ALA was 3.6 mM. Incubation was started by adding 100 μL of the tissue preparation to a final volume of 400 μL. The product of the reaction was determined with the Ehrlich reagent at 555 nm using a molar absorption coefficient of 6.1×10^4 M^{-1} cm^{-1} (Sassa, 1982) for the Ehrlich-porphobilinogen salt. ALA-D activity was expressed as nmol PBG mg^{-1} protein h^{-1}.

2.9. Protein Determination

In all the enzyme preparations, protein was determined by the method of Bradford (1976) using bovine serum albumin as standard.

2.10. Statistical Analysis

Data were submitted to variance analyses (one-way ANOVA) and treatments means were compared by Tukey range at 5% of error probability using the SOC statistic package (Software Científico: NTIA/EMBRAPA). Treatments were presented as mean ± S.D. of three replicates.

3. Results

3.1. Electrolyte Leakage Percentage and Hydrogen Peroxide Content

Electrolyte leakage percentage (ELP) represents cell membrane injury. Fig. 1A shows that increased ELP only occurred at a prolonged period of Hg exposure (15 days), where at 500 µM $HgCl_2$ there was a significant increase of 172.7%, compared to the control. Compared with 15-day-old seedlings, 10-day-old seedlings showed higher ELP levels for all Hg treatments.

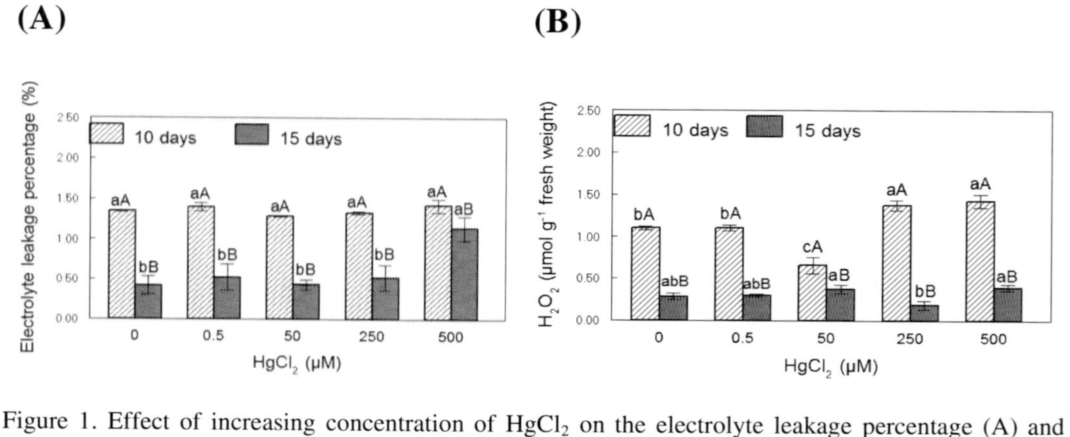

Figure 1. Effect of increasing concentration of $HgCl_2$ on the electrolyte leakage percentage (A) and hydrogen peroxide content (B) at 10- and 15-days-old cucumber seedlings. Data represent the mean ± S.D. of three different experiments. Different lower case and capital letters indicated significant differences between treatments and time among the same treatment, respectively ($p<0.05$).

As shown in Fig. 1B, the exposure of cucumber seedlings to 50 μM $HgCl_2$ for 10 days decreased the levels of H_2O_2 content by about 60%, when compared to control. At the higher concentrations (250 and 500μM $HgCl_2$), a significant increase of H_2O_2 content was observed. On the other hand, the concentration of H_2O_2 did not show any change in the 15-day-old seedlings. Compared with 15-day-old seedlings, 10-day-old seedlings showed higher H_2O_2 levels for all Hg treatments.

3.2. Superoxide Dismutase Activity

Among the various enzymes involved in the scavenging of reactive oxygen species (ROS), superoxide dismutase (SOD) can be considered a key enzyme. SOD activity varied as a function of both exposure time and Hg concentration (Fig. 2A). For 10-day-old seedlings, SOD activity decreased at 0.5, 250 and 500 μM $HgCl_2$.Similarly, 15-day-old seedlings showed the highest level of SOD activity at 50 μM $HgCl_2$ (Fig. 2A) and the lowest level at 250 μM $HgCl_2$ (Fig. 2A). Moreover, SOD levels for 10- and 15-day-old seedlings were positively correlated (0.77) (Table 1).

Table 1. Correlation coefficient between Hg-exposure times (10 and 15 days) for various studied parameters of cucumber seedlings

Parameters	Pearson's coefficient
ELP content	- 0.15†
SOD activity	+ 0.77**
ALA-D activity	+ 0.24†
cCarotenoids content	+ 0.71*
H_2O_2 content	- 0.27†
ASA content	+ 0.95**
NPSH content	+ 0.92**

†Statistically non-significant.
*Statistically significant at $p<0.005$.
**Statistically significant at $p=0.001$.

3.3. Concentration of Antioxidants Compounds

Non-protein thiol (NPSH) content also varied as a function of both exposure time and Hg concentration (Fig. 2B). For 10-day-old seedlings, NPSH content increased for all Hg treatments. On the other hand, at the highest exposure time (15 days), NPSH content increased by about 20% and 232%, respectively, at 0.5 and 250 μM $HgCl_2$, and decreased by about 80% and 75%, respectively, at 50 and 500 μM $HgCl_2$. Moreover, NPSH contents for 10- and 15-day-old seedlings were positively correlated (0.92) (Table 1).

The effects of Hg on ascorbic acid (ASA) concentration are shown in Fig. 2C. Regardless of the time of Hg exposure, the ASA content increased as a function of Hg concentration, except for 0.5 μM at 15 days. The maximum accumulation of ASA was 232.6 μg ASA g^{-1} fresh weight in 10-day-old seedlings treated with 500 μM $HgCl_2$. Compared with 15-day-old seedlings, 10-day-old seedlings showed reduced ASA levels at Hg levels ranging from nil to

50 μM. In addition, ASA content for 10- and 15-day-old seedlings were positively correlated (0.95) (Table 1).

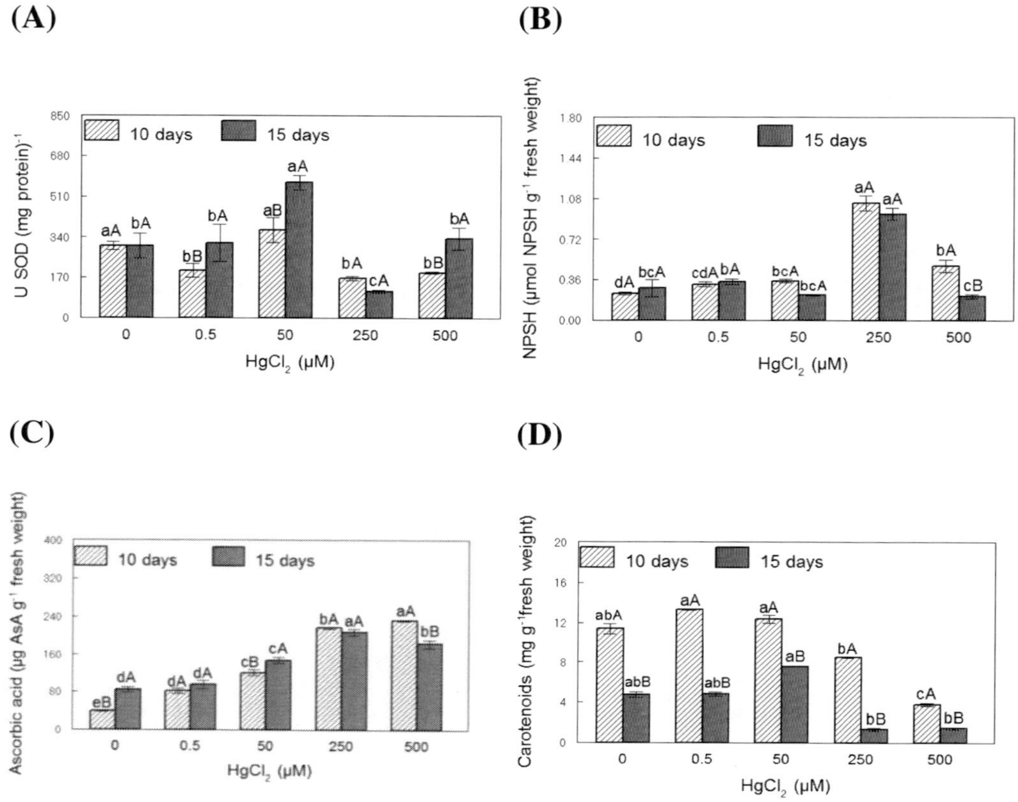

Figure 2. Effect of increasing concentration of $HgCl_2$ on the superoxide dismutase activity (A), and non-protein thiols (B), ascorbic acid (C) carotenoid content (D) of 10- and 15-days- old cucumber seedlings. Data represent the mean ± S.D. of three different experiments. Different lower case and capital letters indicated significant differences between treatments and time among the same treatment, respectively ($p<0.05$).

As shown in Fig. 2D, carotenoid concentration decreased upon addition of $HgCl_2$ levels ranging from 250 to 500 μM for 10-day-old seedlings. At 250 and 500 μM $HgCl_2$, the carotenoid concentration decreased by 26% and 67% for 10-day-old seedlings, respectively, in comparison withto the control. In addition, carotenoid content for 10- and 15-day-old seedlings were positively correlated (0.71) (Table 1).

3.4. Cotyledons δ-ALA-D Activity

Hg-exposure induced a significant reduction of ALA-D activity (Fig. 3).At the highest concentration of Hg (500 μM $HgCl_2$), ALA-D activity at 10 and 15 days was decreased by 99% and 95%, respectively, when compared to the control.

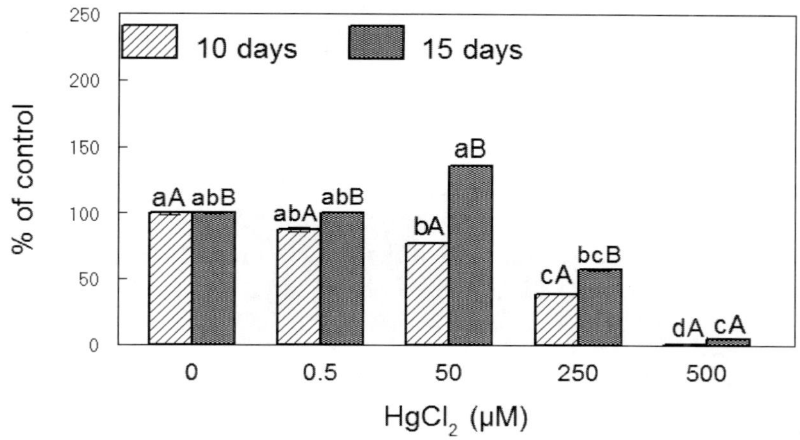

Figure 3. Effect of increasing concentration of $HgCl_2$ on delta-aminolevulinic acid dehydratase activity of 10- and 15-days- old cucumber seedlings. Data represent the mean ± S.D. of three different experiments. The control specific activity (without mercury) that represents 100% was 11.98± 1.17 and 3.37±0.1 nmol PBG min^{-1} mg^{-1} protein, for 10 and 15 days, respectively. Different lower case and capital letters indicated significant differences between treatments and time among the same treatment, respectively ($p<0.05$).

4. DISCUSSION

In our previous study (Cargnelutti et al., 2006), it was reported that for 10- and 15-day-old cucumber seedlings, exposed to 0.5, 50, 250 and 500 μM $HgCl_2$, the Hg content in the cotyledons was, on average, 3.7, 6.8, 1992 and 4854-fold higher than the control, respectively (Table 2). Moreover, there was higher Hg accumulation in the root system, which was 8, 2003, 18423 and 40817-fold higher than the control. These results are similar with that reported by other researchers (Castro et al., 2009; Sierra et al., 2009; Lomonte et al., 2010; Gao et al., 2010; Pérez-Sanz et al., 2012). The high Hg accumulation in the roots has been reported as a tolerance mechanism for different plant species (Castro et al., 2009). In general, Hg binds to cell walls and membranes rather than in the soluble fractions which features general strategy to cope with environmental Hg exposure (Castro et al., 2009; Válega et al., 2009). Nevertheless, phytochelatins also chelate Hg under environmental exposure, contributing for tolerance (Válega et al., 2009). On the other hand, Wu et al. (2011) in an study about accumulation, subcellular distribution and toxicity of inorganic Hg and methylmercury in marine phytoplankton found that a higher percentage of Hg(II) was bounded to the cellular debris fraction in *Thalassiosira pseudonana*, implying that the cellular debris may play an important role in Hg(II) detoxification.

In addition, the roots and shoot length of 10- and 15-day-old seedlings decreased for all Hg treatments (Cargnelutti et al., 2006). Such data reported by Cargnelutti et al. (2006) corroborate the data of the present study, which confirm the intoxication of cucumber seedlings by Hg. In the present study, 15-day-old cucumber seedlings only showed an increase in the electrolyte leakage percentage (ELP) level at 500 μM $HgCl_2$ (Fig. 1A), which strongly suggests the disruption of membrane integrity only by Hg at this level. In contrast, Mishra et al. (2006) reported that there was a significant correlation between Cd supply and

ELP in the roots of *Bacopa monnieri*. Moreover, ELP and H_2O_2 contents were greater in the 10-day-old seedlings than in the 15-day-old seedlings, which could be due to the reduced metabolism with increased exposure time. The germination process is characterized by an accelerated metabolism, which culminates with an increasing production of ROS, such as H_2O_2. These changes in ROS levels might induce the oxidation of antioxidant compounds and other molecules such as ascorbic acid (ASA) and enzymes (such as ALA-D) (Fig. 2C).

Table 2. Mercury content of cucumber seedling growth under increasing concentrations of $HgCl_2$ for 10 or 15 days

Hg treatment (μM $HgCl_2$)	Hg content (μg g^{-1} dry wt.)	
	Cotyledons	Root
Day-10		
0	0.67±0.17	0.60±0.11
0.5	3.40±1.47	6.13±0.74
50	552.33±43.5*	1284.33±61.5*
250	1800.33±50.5*	12498±78*
500	4734.33±63.5*	33377±55*
Day-15		
0	1.4±0.27	0.79±0.05
0.5	3.38±0.13	4.43±0.1
50	759±22*	1474.33±21.5*
250	1816.33±44.5*	12654±45*
500	3698±60*	20545±42*

Data represent mean values ± S.D. based on independent determination. * Different from control to $p < 0.05$. (Adapted from Cargnelutti and Coworkers, 2006).

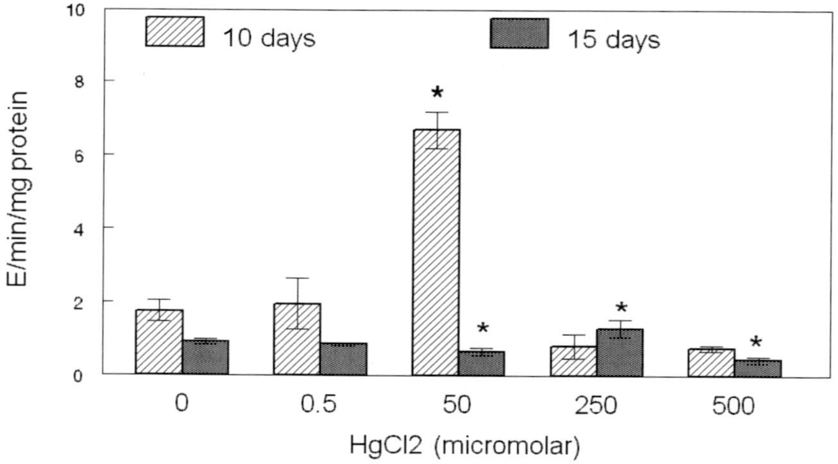

Figure 4. Effect of increasing concentration of $HgCl_2$ on catalase activity of 10- and 15-day old cucumber seedlings. Data represent the mean ± S.D. of three different experiments. *Different from control to $p<0.05$. (Adapted from Cargnelutti and Coworkers, 2006).

The present investigation clearly indicates that Hg-exposure resulted in increased H_2O_2 content in seedlings (Fig. 1B), which could be due to a decrease in catalase (CAT) activity, as was found in a previous study from the same authors (Fig. 4) (Cargnelutti et al., 2006) and/or in ascorbate peroxidase (APX) activity, as was found by Rellán-Álvarez et al. (2005). CAT and APX are key enzymes involved in the removal of toxic peroxide (Lin and Kao, 2000). According to Aebi (1984), high H_2O_2 levels inhibit catalase activity by oxidizing their thiol groups, leading to an increase in the H_2O_2 content. Rellán-Álvarez et al. (2005) reported that APX activity decreased in the presence of 30 μM Hg and the decrease was also related to a reduction of the cytosolic APX protein. However, in the present study, the decreased H_2O_2 accumulation observed at levels of 50 and 250 μM $HgCl_2$, for 10 and 15 days, respectively, could be due to an increase in CAT activity as had been observed in the previous work (Cargnelutti et al., 2006). In addition, the higher ELP levels in the 10-day-old seedlings may be related to H_2O_2 concentration, which could have caused greater oxidation of membrane lipids, thus triggering electrolytes leakage. With exception of the data for the activities of SOD and ALA-D, the pattern for the other parameters analyzed in 10- and 15- day-old seedlings in relation to Hg levels was either activation or inhibition, respectively (Scheme 1).

Parameter	Time	
	10 days	15 days
Electrolyte leakage percentage	↑	↓
Hydrogen peroxide	↑	↓
Superoxide dismutase activity	↓	↑
Catalase activity	↑	↓
Non-protein thiols groups	↑	↓
Ascorbic acid	↑	↓
Carotenoids	↑	↓
Delta-aminolevulinic acid dehydratase	-	-

↑: Increase; ↓: Decrease.

Scheme 1. Plant response to Hg stress.

Superoxide dismutase (SOD) scavenges $O_2^{\bullet -}$ radicals to protect from cellular oxidative damage and is usually considered the first line of defense against oxidative stress. In the present study, a biphasic effect was observed in SOD activity at 10 and 15 days. This result also could be attributed to the hormetic dose response (Calabrese, 1999; Calabrese et al., 2005). Low concentrations and short exposure times may produce an effect similar to high concentrations at any period of time. This trend was compatible to SOD activity reported in grape leaves (Ma, 1998), seedlings of tomato (Cho and Park, 2000), and in seedlings of *Sesbania drummondii* (Israr et al., 2006) cultivated under Hg treatments. Moreover, with an increase of both exposure time and Hg concentration, there may be an increase in the production of ROS, causing greater damage to cells. Low levels of SOD may be related with the increase of H_2O_2 levels, because H_2O_2 may inactivate enzymes by oxidizing their thiol groups, as for instance SOD (Charles and Halliwell, 1980; Bowler et al., 1994). Moreover, Lomonte et al. (2010) reported that high concentration of Hg might directly inhibit chloroplastic SOD catalysis, as most of the SOD in plant photosynthetic tissue is present in the chloroplasts.

For 10-day-old cucumber seedlings, NPSH content was increased for all Hg treatments. In contrast, at 15 days NPSH content increased only at 0.5 and 250 μM $HgCl_2$. Ortega-Villasante et al. (2005) also found increased NPSH content in *Medicago sativa* at a short exposure period. Such results could be due to the reduction of the availability of the metals by a cytoplasmatic detoxification mechanism. Moreover, our previous studies have showed that Hg is able to increase NPSH level in *Pfaffia glomerata* plantlets (Calgaroto et al., 2010). Moreover, Lomonte et al (2010b) reported that some plant species grown in potting mix spiked with mercury(II) showed higher levels of sulphur in their tissues. The same author suggested that Hg stress activated the defense mechanisms such as the enhanced production of sulphur compound. In agreement with Patra et al. (2004), Hg possesses a high affinity for NPSH and the Hg-induced increase in GSH levels might reflect the defensive mechanism against oxidative stress triggered by Hg (Zhou et al., 2007). For 15-day-old seedlings, the NPSH content was decreased at 50 and 500 μM $HgCl_2$ (Fig. 2B), which may be related to the increased H_2O_2 levels that would be involved in NPSH and ASA oxidation. The increase in antioxidant levels decreases the amount of oxidized biomolecules, but these detoxicants might be not be sufficient to protect cells against injury (Ranieri et al., 1993). In the present investigation, ASA levels increased for all Hg-treatments at 10 and 15 days. Similarly, Sinha et al. (1996) reported an increase in ASA content in *Baccopa monieri* plants treated with Hg, showing a significant increase in ASA content during the initial period of metal exposure. ASA is an important component of the plant antioxidant defense system, serves as a reductant for the removal of H_2O_2 (Noctor and Foyer, 1998) and was demonstrated to detoxify Hg in *Chlorella vulgaris*, thus protecting the integrity of –SH groups (Rai, 1979). Moreover, the positive correlation observed between the exposure times (10 and 15 days) for SOD activity, ASA and NPSH contents indicates that Hg caused a stress pattern that was maintained throughout the entire period of Hg-exposure.

Contrary to the ASA levels, an increase in the carotenoid content was found at levels of $HgCl_2$ up to 50 μM, corroborating with data reported for *Pistia stratiotes* under chromium-treatments (Sinha et al., 2005). These findings demonstrate that the antioxidant power of carotenoids might protect chlorophyll against the attack of free radicals. In present investigation, above 250 μM Hg level, a significant decline in the carotenoid content was observed for 10- and 15- day-old seedlings (Fig. 2D). Several studies have shown that Hg in the substrate decreased the levels of photosynthetic pigments, chlorophylls and carotenoids at a prolonged duration of exposure (Sobrino-Plata et al., 2009). Mercury also strongly inhibits the photosynthetic electron transport chain, being that photosystem II is the most sensitive target (Bernier et al., 1993; Bernier and Carpentier, 1995). Contrarily, Calgaroto et al. (2010) reported no alteration in the concentration of carotenoids and chlorophylls in *Pfaffia glomerata* plantlets exposed at Hg. The same authors suggest that the increase in proline content might account for the stability of photosynthetic pigments in *P. glomerata*. The reduced synthesis of porphobilinogen, the committed precursor of chlorophyll, in chill- and heat-stressed wheat seedlings, demonstrated that the inhibition of chlorophyll biosynthesis is partly due to the impairment of ALA biosynthesis (Tewari and Tripathy, 1998). The results of the present investigation showed that the ALA-D activity was strongly inhibited in the presence of higher Hg levels (Fig. 3). However, at 15 days, the ALA-D activity showed a tendency to increase at 50 μM $HgCl_2$, which might be a consequence of either the accelerated metabolism for the synthesis of chlorophyll or the hormetic dose response (Calabrese, 1999; Calabrese et al., 2005). The decrease in ALA-D activity (Fig. 3) observed at concentrations higher than 50 and 250 μM $HgCl_2$ at 10 and 15 days of exposure, respectively, may be due to both an increased H_2O_2 level (Fig 1B) and decreased NPSH content (Fig. 2B). Similarly,

Morsch et al. (2002) and Calgaroto et al. (2010) showed that Hg was a potent inhibitor of ALA-D activity in radishes and *P. glomerata* plantlets, respectively. At low Hg concentrations, the pool of enzymatic activity was not inhibited. Indeed, Cargnelutti et al. (2006) reported that the chlorophyll level was reduced at a level of Hg above 50 μM. Due to sensibility of ALA-D to Hg treatment, it can be suggested that the activity of this enzyme could be used as a sensitive environmental marker for the presence of mercury in plants.

The different levels of enzymatic activity are not exclusively dependent on enzyme synthesis, but on various other factors as well. Thus, in futures studies it will be necessary to analyze the effects of Hg on enzyme expression in order to obtain further understanding of the oxidative stress caused by Hg. Moreover, as reflected in some of the results, a similar pattern was obtained in seedlings grown for 10 and 15 days. This could mean that the physiological response reaches a plateau at ten days (Ortega-Villasante et al., 2005). Thus, in future studies the exposure time should be shortened in order to obtain the effects of Hg in the early stages of exposure.

In conclusion, the increase of ELP and H_2O_2 production in cucumber seedlings might be related to decreased superoxide dismutase activity with a consequent increase in membrane and protein damage (such as ALA-D). Mercury stress increased the levels of ASA, NPSH, and carotenoids in seedlings of *C. sativus* at the initial period of exposure. These antioxidant systems might play an important role in making Hg harmless, in turn making the seedlings tolerant at low concentrations and short exposure times. However, the antioxidant systems are not able to protect from the toxicity caused by higher levels of Hg and increased exposure times, resulting in the negative effects observed in the growth of cucumber seedlings. Considering the results as a whole, we hope to contribute to a better understanding of the oxidative stress conditions generated by mercury in plants (Scheme 2).

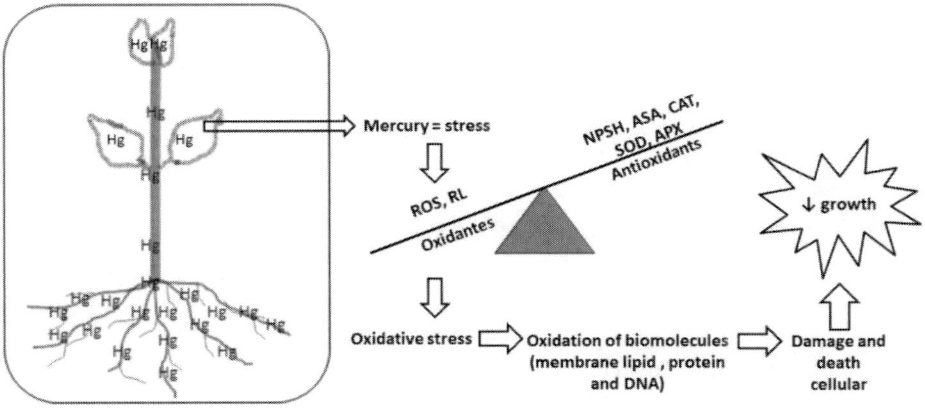

ROS: Reactive Oxygen Species; RL: Free Radical; NPSH: Non-protein thiols; ASA: Ascorbic Acid; CAT: Catalase; APX: Ascorbate Peroxidase; SOD: Superoxide Dismutase; ↓: reduction.

Scheme 2.

Due the high toxicity of Hg on the physiological processes of plants many works have focused in the study of the interactions between Hg and other ions and substances such as carbon monoxide (Meng et al., 2011), zinc (Calgaroto et al., 2011), salicylic acid (Zhou et al., 2009) and compounds of organoselenium (Cargnelutti et al., data non-published). These studies showed that in spite of the high toxicity of Hg to many plant species, the presence of these elements had the potential to improve the plant performance against Hg stress. Therefore, the screening of plant species for Hg tolerance (Shiyab et al., 2009; Dirilgen, 2011; Pérez-Sanz et al., 2012) as well the experimental systems concerning the interaction between Hg and other compounds that can alleviate the Hg toxicity should be the target of future investigations.

Acknowledgments

The authors wish to thank to Coordenação e Aperfeiçoamento de Pessoal de Nível Superior (CAPES), Conselho Nacional de Desenvolvimento Científico e Tecnológico (CNPq), and Fundação de Amparo à Pesquisa de Estado do Rio Grande do Sul (FAPERGS). We also thank the comments of three anonymous reviewers which allowed substantial improvement of the manuscript.

References

Aebi, H., 1984. Catalase in vitro. *Methods Enzymology* 105, 121-126.

Alumaa, P., Kirso, U., Petersell, V., Steinnes, E., 2002. Sorption of toxic heavy metals to soil. International *Journal of Hygiene and Environmental Health* 204, 375–376.

Bernier, M., Carpentier, R., 1995. The action of mercury on the binding of the extrinsic polypeptides associated with the water oxidizing complex of photosystem II. *Federation of European Biochemical Societies, Letters* 360, 251–254.

Bernier, M., Popovic, R., Carpentier, R., 1993. Mercury inhibition at the donor side of photosystem II is reversed by chloride. *Federation of European Biochemical Societies, Letters* 321, 19–23.

Boening, D.W., 2000. Ecological Effects, Transport, and Fate of Mercury: a general review. *Chemosphere* 40, 1335-1351.

Bowler, C., Camp, V.W., Montagu, V., Inzé, D., 1994. Superoxide dismutase in plants. *CRC Critical Reviews in Plant Science* 13, 199–218.

Bradford, M.M., 1976. A rapid and sensitive method for the quantitation of microgram quantity of protein utilizing the principle of protein-dye binding. *Analytical Biochemistry* 72, 248-254.

Calabrese, E.J., 1999. Evidence that Hormesis Represents an ''Overcom-pensation'' Response to a Disruption in Homeostasis. *Ecotoxicology and Environmental Safety* 42, 135–137.

Calabrese, E.J., Blain, R., 2005. The occurrence of hermetic dose responses in the toxicological literature, the hormesis database: an overview. *Toxicology Applied Pharmacology, Reviewer* 202, 289–301

Calgaroto, N.S., Castro, G.Y., Cargnelutti, D., Pereira, L.B.P., Gonçalves, J.F., Rossato, L.V., Antes, F.G., Dressler, V.L., Flores, E.M.M., Schetinger, M.R.C., Nicoloso, F.T., 2010. Antioxidant system activation by mercury in *Pfaffia glomerata* plantlets. *Biometals*, 23, 295–305.

Calgaroto, N.S., Cargnelutti, D., Rossato, L.V., Farias, J.G., Nunes, S.T., Tabaldi, L.A., Antes, F.G., Flores, E.M.M., Schetinger, M.R.C., Nicoloso, F.T., 2011. Zinc alleviates mercury-induced oxidative stress in *Pfaffia glomerata* (Spreng.) Perdersen. *Biometals*, 24, 959-971.

Castro, R., Pereira, S., Lima, A., Corticeiro, S., Válega, M., Pereira, E., Duarte, A., Figueira. E., 2009. Accumulation, distribution and cellular partitioning of mercury in several halophytes of a contaminated salt marsh. *Chemosphere*, 76, 1348-1355.

Cargnelutti, D., Tabaldi, L.A., Spanevello, R.M., Jucoski, G.O., Battisti, V., Redin, M., Linares, C.E.B., Dressler, Flores, E.M.M., Nicoloso, F.T., Morsch, V.M., Schetinger, M.R.C., 2006. Mercury toxicity induces oxidative stress in growing cucumber seedlings. *Chemosphere* 65, 999–1006.

Cavallini, A., Natali, L., Durante, M., Maserti, B., 1999. Mercury uptake, distribution and DNA affinity in durum wheat (*Triticum durum* Desf.) plants. *Science of the Total Environment* 243, 119-127.

Charles, S.A., Halliwell, B., 1980. Effects of hydrogen peroxide on spinach (*Spinacia oleracea*) chloroplast fructose biphosphatase. *Biochemical Journal* 189, 373–376.

Cho, U.-H., Park, J.O., 2000. Mercury-induced oxidative stress in tomato seedlings. *Plant Science* 156, 1-9.

Crane, S., Barkay, T., Dighton, J., 2012. The effect of mercury on the establishment of *Pinus rigida* seedlings and the development of their ectomycorrhizal communities. *Fungal Ecology*, 5, 245-251.

Dipierro, N., Mondelli, D., Paciolla, C., Brunetti, G., Dipierro, S., 2005. Changes in the ascorbate system in the response of pumpkin (*Cucurbita pepo* L.) roots to aluminium stress. *Journal of Plant Physiology* 162, 529–536.

Dirilgen, N., 2011. Mercury and lead: Assessing the toxic effects on growth and metal accumulation by *Lemna minor. Ecotoxicology and Environmental Safety* 74, 48–54.

Ellman, G.L., 1959. Tissue sulfhydryl groups. *Archives of Biochemistry and Biophysics* 82, 70–77.

Fediuc, E., Erdei, L., 2002. Physiological and biochemical aspects of cadmium toxicity and protective mechanisms induced in *Phragmites australis* and *Typha latifolia. Pant Physiology* 159, 265–271.

Foyer, C., Noctor, G., 2005. Redox homeostasis and antioxidant signaling: A metabolic interface between stress perception and physiology responses. Current Perspective Assay. *The Plant Cell* 17, 1866–1875.

Foyer, C., Theodoulou, F.L., Delrot, S., 2001. The functions of inter-and intracellular glutathione transport systems in plants. *Trends in Plant Science Review* 6, 486–492.

Foyer, C.H., 2002. The contribution of photosynthetic oxygen metabolism to oxidative stress in plants. In *Oxidative stress in Plants* (Inzé, D. and Van Montagu, eds), pp. 33–68, Taylor and Francis.

Freedman, B., Hutchinson, T.C., 1981. Souces of metal and elemental contaminations of terrestrial environments, in: Lepp, N.W., (Ed.), *Effect of heavy metal pollution on plants: Metals in environment, Vol. II*. London and New Jersey: Applied Science Publications, pp. 35–94.

Gao, S., Ou-yang, C., Tang, L., Zhu, J-Q., Xu, Y., Wang, S-H., Chen, F., 2010. Growth and antioxidant responses in *Jatropha curcas* seedling exposed to mercury toxicity. *Journal of Hazardous Materials*, 182, 591-597.

Gichner, T., Patková, Z., Száková, J., Demnerová, J., 2006. Toxicity and DNA damage in tobacco and potato plants growing on soil polluted with heavy metals *Ecotoxicology and Environmental Safety* 65, 420–426.

Hall, J.L., 2002. Cellular mechanisms for heavy metal detoxification and tolerance. *Journal of Experimental Botany* 53, 1–11.

Hiscox, J.D., Israelstam, G.F., 1979. A method for the extraction of chlorophyll from leaf tissue without maceration. *Canadian Journal of Botany* 57, 1132-1334.

Hung, S-S, Yu, C-W., Lin, C.H., 2005. Hydrogen peroxide function as a stress signal in plants. *Botanical Bulletin of Academia Sinica* 46, 1–10.

Israr, M., Sahi, S.V., 2006. Antioxidative responses to mercury in the cell cultures of *Sesbania drummondii. Plant Physiology and Biochemistry* 44, 590–595

Israr, M., Sahi, S., Datta, R., Sarkar, D., 2006. Bioaccumulation and physiological effects of mercury in *Sesbania drummondii. Chemosphere* 65, 591-598.

Jacques-Silva, M.C., Nogueira, C.W., Broch, L.C., Flores, E.M.M., Rocha, J.B.T., 2001. Diphenyl diselenide and ascorbic acid changes deposition of selenium and ascorbic acid in liver and brain of mice. *Pharmacology and Toxicology* 88, 119–125.

Kenneth, E., Pallet, K.E., Young, J., 2000. In: Alscher, R.G., Hess, J.L. (Eds.), *Carotenoids antioxidants in higher plants*. CRC Press, pp. 60–81.

Lin, C.C., Kao, C.H., 2000. Effect of NaCl stress on H2O2 metabolism in rice leaves. *Plant Growth Regulation* 30, 151–155.

Lichtenthaler, H.K., 1987. Chlorophylls and carotenoids: pigments of photosynthetic biomemranes. *Methods Enzymology* 148, 350–82.

Lomonte, C., Sgherri, C., Baker, A.J.M., Kolev, S.D., Navari-Izzo, F., 2010a. Antioxidative response of *Atriplex codonocarpa* to mercury. *Environmental and Experimental Botany*, 69, 9-16.

Lomonte, C., Doronila, A.I., Gregory, D., Baker, A.J.M., Kolev, S.D., 2010b. Phytotoxicity of biosolids and screening of selected plant species with potential for mercury phytoextraction. *Journal of Hazardous Materials*, 173, 494-501.

Loreto, F., Velikova, V., 2001. Isoprene produced by leaves protects the photosynthetic apparatus against ozone damage, quences ozone products, and reduces lipid peroxidation of cellular membranes. *Plant Physiology* 127, 1781–7.

Ma, C., 1998. *Hg harm on cell membrane of rape leaf and cell endogenous protection effect.* Yingyong Shengtai Xuebao 9, 323-326.

Matés, J.M., 2000. Effects of antioxidant enzymes in the molecular control of reactive oxygen species toxicology. *Toxicology* 153, 83–104.

Meng, D.K., Chen, J., Yang. Z.M., 2011. Enhancement of tolerance of Indian mustard (*Brassica juncea*) to mercury by carbon monoxide. *Journal of Hazardous Materials,* 186, 1823-1829.

Mc Cord, J.M., Fridovich, I., 1969. Superoxide dismutase: an enzymic function for erythrocuprein (hemocuprein). *Journal of Biological Chemistry* 244, 6049–6055.

Mishra, S., Srivastava, S., Tripathia, R.D., Govindarajan, R., Kuriakose, S.V., Prasad, M.N.V., 2006. Phytochelatin synthesis and response of antioxidants during cadmium stress in *Bacopa monnieri* L. *Plant Physiology and Biochemistry* 44, 25–37.

Morsch, V.M., Schetinger, M.R.C., Martins, A.F., Rocha, J.B.T., 2002. Effects of cadmium, lead, mercury and zinc on δ-aminolevulinic acid dehydratase activity from radish leaves. *Biologia Plantarum* 45 (1), 85–89.

Murashige, T., Skoog, F., 1962. A revised medium for rapid growth and bioassays with tabbaco tissue culture. *Plant Physiology* 15, 473-497.

Noctor, G., Foyer, C.H., 1998. Ascorbate and gluthatione: keeping active oxygen under control. *Annual Review of Plant Physiolo*gy (Plant and Molecular Biology) 49, 249–279.

Noriega, G.O., Balestrasse, K.B., Batlle, A., Tomaro, M.L., 2006. Cadmium induced oxidative stress in soybean plants also by the accumulation of δ-aminolevulinic acid. *Biometals*. DOI 10.1007/s10534-006-9077-0

Ortega-Villasante, C., Rellán-Álvarez, R., Del Campo, F.F., Carpena-Ruiz, R.O., Hernández, L.E., 2005. Cellular damage induced by cadmium and mercury in *Medicago sativa*. *Journal of Experimental Botany* 56, 2239–2251.

Patra, M., Bhowmik, N., Bandopadhyay, B., Sharma, A., 2004. Comparison of mercury systems and the development of genetic tolerance. *Environmental and Experimental Botany, Review* 52, 199-223.

Pereira, L.B., Tabaldi, L.A., Gonçalves, J.F., Jucoski, J.O., Pauletto, M.M., Weis, S.N., Nicoloso, F.T., Borher, D., Rocha, J.B.T., Schetinger, M.R.C., 2006. Effect of aluminum on δ-aminolevulinic acid dehydratase (ALA-D) and the development of cucumber (*Cucumis sativus*). *Environmental and Experimental Botany* 57, 106–115.

Pérez-Sanz, A., Millán, R., Sierra, M.J., Alarcón, R., García, P., Gil-Díaz, M., Vazquez, S., Lobo, M.C., 2012. Mercury uptake by *Silene vulgaris* grown on contaminated spiked soils. *Journal of Environmental Management*, 95, S233-S237.

Polyakov, N.E., Leshina, T.V., Konovalova, T.A., Kispert, L.V., 2001. Carotenoids as scavengers of free radicals in a Fenton reaction: antioxidants or pro-oxidants? *Free Radical in Biology and Medicine* 31, 398–404.

Rai, L.C., 1979. Mercury toxicity to *Chlorella vulgaris*. I. Reduction of toxicity by ascorbic acid and reduced glutathione (GSH). *Phykos* 18, 105–109.

Ranieri, A., Lencioni, L., Schenone, G., Soldatini, G.F., 1993. Gluthatione-ascorbic acid cycle in pumpkin plants grown under polluted air in open-top chambers. *Journal of Plant Physiology* 142, 286–290.

Rellán-álvarez, R., Ortega-Villasante, C., Álvarez-Fernández, A., Campo, F.F., Hernández, L.E., 2006. Stress Responses of *Zea mays* to Cadmium and mercury. *Plant and Soil* 279, 41–50.

Romero-Puertas, M.C., Palma, J.M., Gómez, M., Río, L.A. del., Sandalio, L.M., 2002. Cadmium causes the oxidative modification of proteins in pea plants. *Plant Cell and Environmental* 25, 677–686.

Sassa, S., 1982. Delta-aminolevulinic acid dehydratase assay. *Enzyme* 28, 133–145.

Sandalio, L.M., Dalurzo, H.C., Gomez, M., Romero-Puertas, M.C., Río, L.A. del., 2001. Cadmium-induced changes in the growth and oxidative metabolism of pea plants. *Journal of Experimental Botany* 52, 2115–2126.

Schützendübel, A., Schwanz, P., Teichmann, T., Gross, K., Langenfeld-Heyser, R., Goldbold, D.L., Polle, A., 2001. Cadmium-induced changes in antioxidative systems, hydrogen peroxide content, and differentiation in Scots pine roots. *Plant Physiology* 127, 887–898.

Segura-Muñoz, S.I., Oliveira, A.S., Nikaido, M., Trevilato, T.M.B., Bocio, A., Takayanagui, A.M.M., Dimingo, J.L., 2006. Metal levels in sugar cane (Saccharum spp.) samples from an area under the influence of a municipal landfill and a medical waste treatment system in Brazil. *Environmental International* 32, 52–57.

Shiyab, S., Chen, J., Han, F.X., Monts, D.L., Matta, F.B., Gu, M., Su, Y., 2009. Phytotoxicity of mercury in Indian mustard (*Brassica juncea* L.). *Ecotoxicology and Environmental Safety*, 72, 619-625.

Sierra, M.J., Millán, R., Esteban, E., 2009. Mercury uptake and distribution in *Lavandula stoechas* plants grown in soil from Almadén mining district (Spain*). Food and Chemical Toxicology*, 47, 2761-2767.

Sinha, S., Gupta, M., Chandra, P., 1996. Bioccumulation and biochemical effects on mercury in the plant *Bacopa monnieri* (L.). *Environmental Toxicology and Water Quality: National International Journal* 11, 105–112.

Sinha, S., Saxena, R., Singh, S., 2005. Chromium induced lipid peroxidation the plants of *Pistia stratiotes* L.: role of antioxidants and antioxidant enzymes. *Chemosphere* 58, 595–604.

Sobrino-Plata, J., Ortega-Villasante, C., Flores-Cáceres, M.L., Escobar, C., Campo, F.F.D., Hernández, L.E., 2009. Differential alterations of antioxidant defenses as bioindicators of mercury and cadmium toxicity in alfalfa. *Chemosphere*, Volume 77, Issue 7, November 2009, Pages 946-954.

Tewari, A.K., Tripathy, B.C., 1998. Temperature-stress-induced impairment of chlorophyll biosynthetic reactions in Cucumber and Wheat. *Plant Physiology* 117, 851-858.

Válega, M., Lima, A.I.G., Figueira, E.M.A.P., Pereira, E., Pardal, M.A., Duarte. A.C., 2009. Mercury intracellular partitioning and chelation in a salt marsh plant, *Halimione portulacoides* (L.) Aellen: Strategies underlying tolerance in environmental exposure. *Chemosphere*, 74, 530-536.

Vitória, A.P., Lea, P.J., Azevedo, R.A., 2001. Antioxidant enzymes responses to cadmium in radish tissues. *Phytochemistry* 57, 701–710.

Webber, J., 1981. Trace metals in agriculture. in: Lepp, N.W., (Ed.), Effect of heavy metal pollution on plants: *Metals in environment, Vol. II*. London and New Jersey: Applied Science and Publications, pp. 159–184.

Wu, Y., Wang, W-X., 2011. Accumulation, subcellular distribution and toxicity of inorganic mercury and methylmercury in marine phytoplankton. *Environmental Pollution* 159, 3097-3105.

Zang, W.H., Tyerman, S.D., 1999. Inhibition of water channels by $HgCl_2$ in intact wheat root cells. *Plant Physiology* 120, 849-857.

Zhou, Z.S., Huang, S.Q., Guo, K., Mehta, S.K., Zhang, P.C., Yang, Z.M., 2007. Metabolic adaptations to mercury-induced oxidative stress in roots of *Medicago sativa* L. *Journal of Inorganic Biochemistry* 101, 1–9.

Zhou, Z.S., Guo, K., Elbaz, A.A., Yang, Z.M., 2009. Salicylic acid alleviates mercury toxicity by preventing oxidative stress in roots of *Medicago sativa. Environmental and Experimental Botany* 65, 27–34.

Zhou, Z.S., Wang, S.J., Yang, Z.M., 2008. Biological detection and analysis of mercury toxicity to alfalfa (*Medicago sativa*) plants. *Chemosphere*, 70, 1500–1509.

Zhou, Z.S., Guo, K., Elbaz, A.A., Yang. Z.M., 2009. Salicylic acid alleviates mercury toxicity by preventing oxidative stress in roots of *Medicago sativa. Environmental and Experimental Botany* 65, 27–34.

Zhu, Z., Wei, G., Li, J., Qian, Q., Yu, J., 2004. Silicon alleviates salt stress and increases antioxidant enzymes activity in leaves of salt-stressed cucumber (*Cucumis sativus* L.). *Plant Science* 167, 527–533.

In: Mercury: Sources, Applications and Health Impacts
Editors: Ki-Hyun Kim and Richard J. C. Brown
ISBN: 978-1-62257-721-7

Chapter 8

MERCURY AND ITS CONNECTION WITH PHYSICOCHEMICAL WATER PARAMETERS (CASE STUDY OF THE RIVERS OF THE NORTHERN EUROPEAN TERRITORY OF RUSSIA)

Yu A. Fedorov *** and A. E. Ovsepyan***
Southern Federal University, Rostov-on-Don, Russia

ABSTRACT

The polar regions of the Earth are most vulnerable to different anthropogenic impacts. Rather severe physico-geographical conditions in these regions predetermine a retarded exchange of the substance and energy in the ecosystems, and, as a result, the pollutants that enter them undergo, to an insignificant extent, transformations and decompose, their accumulation causes long-term complications of the ecological situation. The Arctic regions are characterized by a relatively limited industrial and agricultural development and low urbanization level. At the same time, they present a final point of migration of such pollutants as, in particular, heavy metals (HM) brought here with the transboundary atmospheric transport as well as with the river waters from the neighboring regions that are more developed.

The mercury concentration levels in the water of aquatic ecosystem of the Northern Dvina River delta are defined. The interrelations of mercury concentrations with physicochemical parameters of the water body, namely with pH, Eh, temperature, salinity, oxygen and organic matter content are studied. The dominant forms of mercury migration and its presence in the water are revealed.

Regularities of changes and spatial variability of the mercury concentrations in the estuarine area of the river, peculiarities of mercury compound migration and transformation in the barrier zone of the Northern Dvina River–Dvina Bay of the White Sea are studied.

The expeditionary investigations were conducted in a warm season, characterized by certain peculiarities in distribution of temperature, salinity, pH, Eh values, oxygen

* Corresponding author Address: Russia, Rostov-on-Don, Serjantova str., 4, ap. 14, Head of Department of Physical Geography, Ecology and Environmental Protection, Southern Federal University, Rostov-on-Don, Russia>Honors: Doctor of Science.

content. Variations of total dissolved mercury in surface layer were within the limits from 0.005 to 0.18 μg/l. Dissolved form of mercury occurred in water samples of all stations under study. Content of total dissolved mercury greatly changed in waters of the Northern Dvina along the river length and in vertical section. In the most number of samples collected a content of dissolved form of mercury exceeded mercury content in suspended matter.

Keywords*:* Mercury behavior, Northern Dvina River, concentrations, surface waters

Metallic mercury and its combinations belong to prioritive pollutants of natural waters. High migratory ability of mercury and its tendency to the formation and transformation of various forms of existence, among which the most toxic are the combinations with organic matter-methyl-and dimethyl mercury, make a problem of mercury pollution, on the one hand, urgent and, on the other hand, utterly complex task to study. Physico-chemical conditions of a medicum: salinity (mineralization), content of chlorine and oxygen, temperature, values of oxidation-reduction potential (Eh) and acidity (pH) play a leading part in the fate of this metal together with hydrological and, especially, biogeochemical factors and processes. As a rule, under natural conditions the above-mentioned factors have a dependent in time and space effect on mercury behavior. Therefore, to consider this problem by means of looking into closeness of connection between various properties on the one hand and content of gross and dissolved forms of mercury – on the other hand is of great theoretical and utilization interest.

The behavior of total dissolved form of mercury is considered in the present article. Dissolved form of mercury is the only form of mercury existence in natural waters in the Russian Federation for which the sanitary-hygienic standards of maximum permissible concentrations (MPC) are elaborated. In waters in dissolved form an atomic Hg^0 and ionic (univalent and bivalent) mercury, inorganic and organic complexes (fulvic et al.), monomethyl mercury, dimethyl mercury and other mercury organic combinations migrate. The investigations of such purposefulness and comprehension are rather rare-especially for the regions of the Far North on the whole, and the Northern Dvina River basin specifically. A choice for the investigation of the mouth area of the Northern Dvina River is not chance. It is caused, first of all, by the presence here at one time of technogenic sources of mercury, which was detected in waters in the concentrations exceeding MPC essentially [1-8].

The Northern Dvina and its tributaries relate to the territory of the White Sea basin. The river length is 744 km, annual runoff is 110 km^3. Peculiarities of climatic conditions are little solar radiation in winter, predominance of precipitations over evaporation. Winter lasts 5-6 months. In summer precipitation amount is 400-500 mm. Feeding of rivers proceeds by precipitations and thawing snow waters. Considerable bogging of the Northern Dvina basin causes an input of large amount of humus substances into the river system. Up to 50-100 mg/l of organic substances of humic origin can be dissolved in bog water, in addition permanganate oxidability can vary from 30 to 70 O/l [1, 9, 10].

The problem of the HM-pollution of the north of European Russia arouses particular concern. For example, the scientists warn about a possible increase in the mercury concentration in the environment of the northern regions due to the intensified melting of ice and permafrost [10]. It ought to be mentioned that extremely high mercury concentrations in the water and bottom sediments were observed earlier in the region under study [1-8]. The content of the water dissolved form was 6.4–7.4 μg/l; the total mercury content in the bottom sediments was 1.54 μg/l [6, 7]. The inspection of the supposed pollution source, the Open Joint Stock Company (OJSC) Arkhangel'sk Pulp-and-Paper Plant (Arkhangel'sk PPP), organized by the Arkhangel'sk Committee for Nature Protection, found that a chloride manufacturing plant of the PPP discharged mercury-bearing waste waters [6, 7]. Such situation led to negative consequences in the ecosystem, in particular, the researchers revealed that the mercury content in the muscles of the bream caught from the Northern Dvina River, in the zone under the effect of the Arkhangel'sk PPP, exceeded the permissible standards. Furthermore, the phytoplankton biomass in the polluted river sections was three times less, that of zooplankton, more than 2.5 times less, and of zoobentos, 10 times less compared to the data of the middle of the 20th century [2].

Methods of Taking, Preparing, and Analyzing Samples

Several expeditions were organized for revealing peculiarities of mercury migration and transformation under natural conditions in the north of European Russia during the summertime of 2004–2011. The Northern Dvina River, the strongest source of many pollutants brought to the White Sea water area including mercury compounds, was the object of investigations [5].

In order to achieve the target goal, the samples were taken from the near-surface and near-bottom layers of the water column and bottom sediments at 27 stations in the estuarine area of the Northern Dvina River (Fig. 1). The total length of the site studied between the head, the Pinega River mouth, and Mud'yugskii Island within the estuarine coastal water amounted to 135 km. The surveys were performed in the Pinega River mouth, within the village of Ust'-Pinega, upstream and downstream of the towns of Novodvinsk and Arkhangel'sk, in Maimaksa and Kuznechikha channels, and in the watercourses crossing Arkhangel'sk: Yuras and Solombalka, and in the upper reaches of the arms (Nikol'skii, Murmanskii, and Korabel'nyi).

The associates of the Northern Dvina River estuarine station under the Arkhangel'sk Center for Hydrometeorology and Monitoring of the Environment with Regional Functions participated in the work.

With sampling the authors used clean collection techniques.

Water samples for mercury quantification were taken in accordance with the modified methodology tested earlier in different water bodies of European Russia [12-17]. A total of 460 water samples were collected. They were analyzed for the mercury content in the Laboratory of the Southern Federal University with the help of the atomic-absorption spectroscopy in the cold vapor. The detection limit for the cold vapor atomic-absorption method ranged 0.004-0.005 μg/l.

Accuracy of the analysis also confirmed that analysis was performed of the analyst Anikanov A.M. He was took part in an international intercalibration experiment for determination of mercury in standard water samples, the laboratory held AMOS-Proficiency Test, Water Analysis. On the accuracy of determinations performed 30 laboratories from around the world, his findings did not come from the interval of permissible error values.

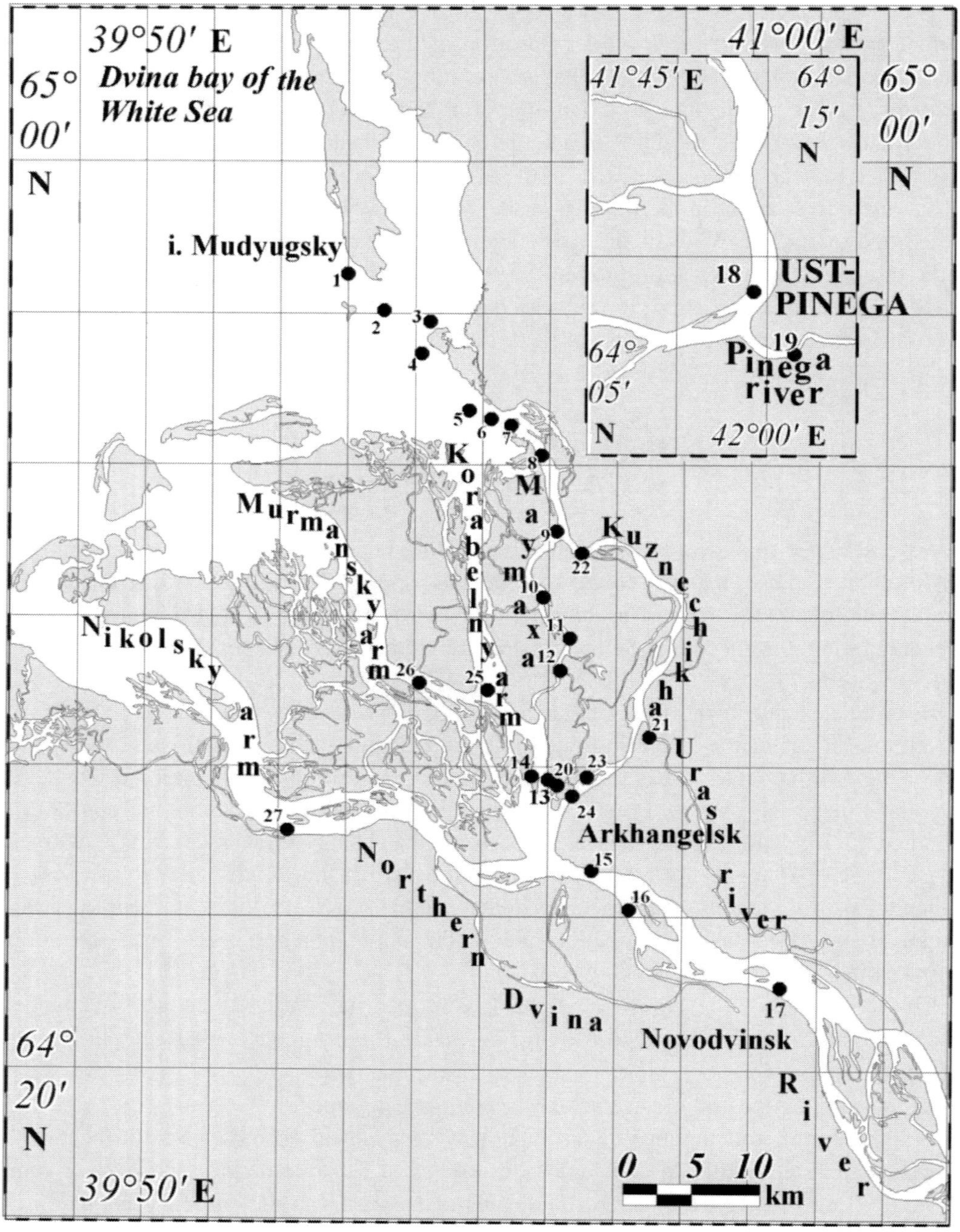

Figure 1. Scheme of location of sampling points. 1 - i. Mudyugsky, 2 - The Dry Sea, 3 - i. Lebedin, 4 - Ustyansk turning sites, 5 - Murov turning sites, 6 - v. Lapominka, 7 - v. Chizhovka, 8 - Staraya Izhma, 9 - Port Ekonomya, 10 - Timber mill No. 24, 11 – Nizhnye Povrakulskye sites, 12 - Hydrolytic plant, 13 - i. Solombala, 14 - above i. Molodezhny, 15 - Top of delta, 16 - Bakaritsa Port, 17 – Novodvinsk, 18 – v. Ust-Pinega, 19 - Pinega River mouth, 20 - Solombalka channel, 21 – Uras River mouth, 22 - Kuznechikha ch., Timber mill No. 29, 23 - Kuznechikha ch., downstream of SPPM, HGS, 24 - Kuznechikha ch., upstream of SPPM, HGS, 25 - Korabelny arm, 26 - Murmansky arm, 27 - Nikolsky arm.

The reductant that was used to convert the dissolved mercury into gas phase mercury was $SnCl_2$ with preliminary destruction related forms of mercury [14-17]. The blank levels for the reagents and filtration process associated with the analysis were below the detection limit. According to method the filters were precleaned by boiling in 1% nitric acid solution [14-17]. It was taken precautions according to method for not contaminated the samples during collection and analysis. The samples of water for mercury determination were collected into the specially prepared sample bottles. To prepare the bottles - it was filled with pre-diluted 1:1 nitric acid, and after the 24 hours was carefully washed from the acid and rinsed by bidistilled water. When sampling the bottles at first were rinsed with small amount of water, and then 100-150 ml of sample were added into them. To measure total dissolved for of mercury the water (400 ml) was filtered through 0.45 μm membrane filters. According to method the water samples were preserved by adding of 20 % solution of $K_2Cr_2O_7$ to nitric acid of particularly pure mark diluted 1:1 by bidistilled water. The preserving solution was added counting 5 ml for 1 l of water. The bottles with samples were corked by glass or plastic caps. A total mercury content, total dissolved mercury (Hg^f_d or Hg^{tot}_d) migration form, and suspended mercury were measured in water. At the same time, 5% of the total samples were analyzed in the Regional Laboratory Center of the Yuzhgeologiya Federal State Unitary Geological Enterprise. The data obtained were within the analysis error. Simultaneously with the sample analysis for the mercury content, the concentration of oxygen, as well as salinity *S*, water temperature, pH, and Eh were measured.

Particular attention was paid to the investigation of the river section in the urban area of Arkhangelsk city. Maymaxa, Kuznechikha channels, the Uras River were studied, and samples were collected in the arms in their upper waters. Stations 19-13 are located at the Northern Dvina River in the main bed before division into arms. Stations 12-1 represent a river section along Maymaxa channel from Hydrolytic plant to i. Mudyugsky situated in Dvina bay. Stations 22-24 are situated in Kuznechikha channel receiving waste waters from Solombalsky Pulp and Paper Mill (SPPM) and Heat Generating Station (HGS). Stations 25-27 are located in the upper waters of three arms of the Northern Dvina. Station 20 is situated in Solombalka channel crossing i. Solombala, 21 – in the mouth of the Uras River, flowing into Kuznechikha channel. More than 150 measurings of mercury content were carried out. A modified methodology for sampling, sample preparation and determination of mercury content in them [5, 12, 17, 18] that had been used before at various water bodies of European territory of Russia (ETR) was used.

The samples of bottom sediments were taken with a bottom scoop and/or a tube of a State Oceanographic Institute design aboard ship Aisberg-2 in the surface (0–5-cm depth) and subsurface (5–10 cm) sediment layers. More than 150 samples were taken at 18 monitoring stations. Most mercury measurements were performed with the help of an atomic absorption method in the cold vapor of the certified Laboratory of the Southern Federal University. The frequency control of the results obtained was performed in the Federal State Unitary Geological Enterprise (FSUGE) Yuzhgeologiya and Hydrochemical Institute of Roshydromet. The measurement error was 10–15%. pH and Eh, methane and organic carbon content were also measured in bottom sediments. Mercury concentration in bottom sediments of the estuarine area of the Northern Dvina River and the Dvina Gulf of the White Sea varied within 0.02–0.80 μg/g of dry mass amounting on average to 0.135 μg/g of dry mass [5, 19].

Analysis and Description of Physico-Chemical Parameters' Behavior

The expeditionary investigations were conducted in a warm season, characterized by certain peculiarities in distribution of temperature, salinity, pH, Eh values, oxygen content (Table 1).

Table 1. Variations of some hydrochemical properties

property		surface layer	near bottom layer
t, °C	l	17.4-19.4 18.5	14.2-19.1 16.5
	h	14-17.6 16.2	12.4-17.2 14.6
O_2, mg/l	l	6.49-7.97 7.29	6.49-7.97 7.11
	h	6.22-8.26 6.9	5.62-8.11 6.84
Eh, mV	l	78...140 107.5	60...136 106.4
	h	-34...92 60.07	-17.9...90 62.49
pH	l	7.58-7.93 78	7.42-8.12 7.75
	h	7.8-8.09 7.9	7.76-7.98 7.8
S ‰	l	0.05-7.72 2.8	0.09-19.08 9.9
	h	0.4-11.98 4.4	1.43-17.89 10
Hg_d^{tot} µg/l	l	0.005-0.18 0.065	0.005-0.12 0.055

In a numerator the minimum-maximum value is presented, in a denominator – the average content.
pH values were on the interval of 7.42-8.12.

Temperature of water in longitudinal section of the river changed within the range from 19.4°C at Ust-Pinega st. to 12.4°C at i. Mudyugsky st. during the period of investigation. Total pattern of spatial temperature distribution is a decrease of values when approaching the sea edge of delta [5, 12].

In vertical section the temperature of surface water layer has exceeded the temperature of near bottom horizon on the average by 3 °C in a «low water» phase and by 2 °C in a "high water" phase. Vertical gradient of temperature distribution was 0.18 °C/m during falling tide and 0.14 °C/m during rising tide.

In surface and near bottom layers pH greatly varied along the river length [5, 12]. Minimum values were detected at the station of i. Solombala and i. Mudyugsky. Maximum values both in near bottom and in surface layers of water were measured at the seaside section between the stations of Nizhnye Povrakulskye sites and Port Ekonomya. At the section i. Lebedin – i. Mudyugsky an abrupt difference of pH values (from 7.6 to 7.9) was observed. Just at this section a penetration of sea water occurs and a hydrofront is rather clear traced. In «low water» phase pH values were smaller on the average (7.78) then in "high water" phase (7.9). Vertical gradient of values' distribution was 0.003 during falling tide and 0.008 during rising tide.

Oxygen content. Amplitude of variations of O_2 content was 5.62-8.26 mg/l. Along the river length if is distributed comparatively uniformly. A tendency to O_2 concentration increase as the approach to the sea edge of delta was noticed.

During rising tide an oxygen content in the surface layer varied from 6.22 to 8.26 mg/l, and in near bottom layer from 5.62 to 8.11 mg/l. During falling tide – from 6.49 to 7.97 mg/l. An increase of oxygen content values was noticed in the mixing zone of sea and river waters (from 7.3 to 7.7 mg/l at 2 kilometer section). Minimum concentrations were determined in near bottom water layer during "high water" phase at station 9 (Port Economya). The greatest uniformity of oxygen distribution was noticed for low water (lesser spread in values, minimum of differences between near bottom and surface layers).

Salinity varies along the river length increasing as the approach to the mouth [5, 12]. Minimum values are characteristic of the river section from station Ust-Pinega to Port Economya and vary from 0.05-0.09‰ to 1.24‰. In the mixing zone with sea water at st. i. Mudyugsky a salinity reaches maximum value of 19.08‰. Salinity distribution in depth and hydrological phases is manifested fairly precise.

The greatest values are determined for near bottom water layer in the mixing zone during rising tide (table 1). Correspondingly, the least ones characterizing the sea edge of delta were revealed for surface layer during "low water". On vertical a salinity increases from surface to near bottom layer approximately by 0.6‰ during "low water" phase, and 0.4 per meter during "high water".

Eh values along the river length varied in a wide range. Minimum values were determined for st. Top of delta, they were – 34 mV, maximum once were 140 mV at st. Staraya Izhma in the mixing zone of river and sea waters. General tendency permits to reveal that in surface layer Eh values, on the average, are higher than in near bottom one. During "low water" Eh values are higher than in "high water" phase (106 and 60 respectively).

Analysis and Description of Total Dissolved Mercury Behavior

Dissolved form of mercury occurred in water samples of all stations under study. Content of total dissolved mercury greatly changed in waters of the Northern Dvina along the river length and in vertical section (Fig. 2).

Variations of total dissolved mercury in surface layer were within the limits from 0.005 to 0.18 μg/l (Table 2). It will be noted that concentration of 0.005 μg/l is a background one for the given region [5, 7, 10].

Mercury content at station 19 (the Pinega River) was 0.03 μg/l. At the section from st. 18 (Ust-Pinega) to st. 16 (Top of delta) it increases to 0.11 μg/l. Then mercury content abruptly drops to 0.005 μg/l at st. 14 (above i. Molodezhny) having achieved minimum value for surface layer at the river section of mouth area. Further, following downstream an increase of concentrations of mercury dissolved in water at st. 13 (i. Solombala), were it increases to 0.12 μg/l was revealed.

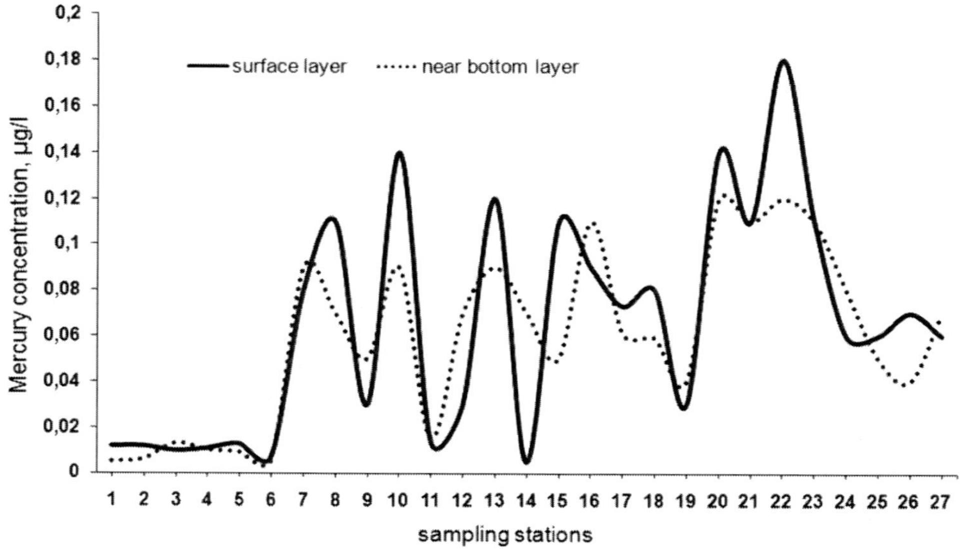

Figure 2. Dynamics of the content of total dissolved forms of mercury in the mouth are waters of the north Dvina River.

On ship channel Maymaxa in section from st. 12 (Hydrolytic plant) to st. 1 (i. Mudyugsky) the values of total dissolved mercury vary from 0.007 µg/l to 0.14 µg/l. Maximum value was determined in the water samples taken at st. 10 (Timber mill No. 24), minimum – in samples from st. 6 (v. Lapominka). During "low water" in the area of st. 6 (v. Lapominka) hydrofront appears. A sharp decrease of mercury concentrations is observed here, but in water downstream from st. 5 (Murov turning sites) it increases 2 times up to 0.013 µg/l and keeps approximately at such level till outlet to Dvina bay. In the samples of st. 4 (Ustyansk turning sites) it is 0.011µg/l, st. 3 (i. Lebedin) – 0.01 µg/l, st. 2 (the Dry sea), st. 1 (i. Mudyugsky) – 0.012 µg/l.

Table 2. Variations in average content of different mercury forms in the water of the estuarine area of the Northern Dvina River

Horizon	Mercury form		
	total, µg/l	dissolved, µg/l	suspended in water, µg/l
	Low water		
Surface	0.008–0.27 0.09	0.005–0.18 0.065	0.004–0.13 0.025
Near-bottom	0.006–0.15 0.074	0.005–0.12 0.055	0.004–0.06 0.019
	High water		
Surface	0.012–0.16 0.075	0.007–0.1 0.04	0.004–0.15 0.036
Near-bottom	0.01–0.19 0.08	0.007–0.12 0.046	0.004–0.08 0.035

Note: The numerator indicates the interval of changes in concentrations, the denominator, an average value.

In Kuznechikha channel 4 sites at 4 stations were investigated. Maximum for surface layer of watr mercury content of 0.18 μg/l was revealed for st. 22 (Timber mill No. 29). This point is situated downstream from SPPM and HGS below the Uras River mouth. The least mercury content in Kuznechikha channel was measured at inlet to the channel – st. 24 (upstream of SPPM, HGS) – 0.006 μg/l.

In the arms of the river delta a mercury content in surface water layer is at a level of 0.06 (Korabelny, Nikolsky) – 0.07 μg/l (Murmansky). High concentrations of mercury were measured in the Solombalka River water – 0.14 μg/l.

Thus, from a graph of mercury concentration variation in surface layer of the Northern Dvina River the least values were measured in mixing zone – from v. Lapominka to i. Mudyugsky, in the Pinega River mouth (background station for mouth region), Maymaxa channel – st. 12 (Hydrolytic plant). The greatest values were measured at st. 22 (Kuznechikha ch., Timber mill No. 29) – 0.18 μg/l, st. 10 (Maymaxa ch., Timber mill No. 24) – 0.14 μg/l, st. 20 (Solombalka ch.) – 0.14 μg/l. An average content was 0.065 μg/l, that is 6.5 times higher than fishery MPC (value is 0.01 μg/l) [3].

25 stations was sampled in near bottom horizon water (at the Uras River and Solombalka ch. are insufficiently deep). Mercury content varied from 0.005 μg/l to 0.12 μg/l.

Dynamics of mercury replicates its variation in surface layer on the whole. It will be noted some shift of extremums (peak and minimum concentrations) in space.

At st. 19 (the Pinega River) a content of dissolved mercury was 0.04 μg/l, downstream it increases and for waters of st. 16 (Bakaritsa Port) it is already 0.11 μg/l. At the next st. 14 (i. Molodezhny) situated downstream it decreases to 0.07 (that is considerably higher than mercury content at this section near surface). In water of st. 13 it somewhat increases (to 0.09). It is interesting to note that in surface layer a range of mercury content change in water of these stations is much wider (0.005-0.12 μg/l).

In the water samples of st. 11 a metal content again decreases to 0.015 μg/l, further an increase is observed – and near v. Chizhovka (st. 7) it is 0.09 μg/l and to st. 6 (v. Lapominka) it decreases synchronously with surface layer up to minimum value of 0.005 μg/l. Moving forward into Dvina bay it increases to 0.013 μg/l at st. 3 (i. Lebedin), then decreases to 0.005 μg/l at st. 1 (i. Mudyugsky).

In Kuznechikha channel a pattern of distribution of various concentrations of mercury replicates its graph for surface layer. However, an average and maximum content is lower here. Minimum concentration is measured at st. 24 (upstream of SPPM, HGS) – 0.08 μg/l; maximum – at st. 22 (Timber mill No. 29) – 0.12 μg/l. And in water of st. 23 (downstream of SPPM, HGS) a content is 0.11 μg/l (as in surface layer). As in surface layer a maximum mercury content for st. 22 is determined in Kuznechikha channel.

In arms a mercury concentration varies from 0.04 μg/l in Murmansky to 0.07 μg/l in Nikolsky (in Korabelny – 0.05 μg/l).

Thus, in near bottom layer the minimum values were measured at st. 1 (i. Mudyugsky) and 6 (v. Lapominka) – 0.005 μg/l, the highest values – in the water samples at st. 22 (0.12 μg/l), 16, 23 (0.11 μg/l). Characterizing variation of mercury concentration on the whole, it should be noted that an average content (0.055 μg/l) is lower here than in surface layer (0.065 μg/l). Maximum value is also lower in near bottom water horizon – 0.12 μg/l. Both in near bottom layer and in surface layer a maximum value is revealed at st. 22 (Timber mill No. 29, Kuznechikha channel). Sharp change in mercury content in water is characteristic also of st. 6 (v. Lapominka), where it decreases from 0.09 μg/l to 0.005 μg/l.

The extremums and minimums of metal concentrations in various horizons coincide in spatial distribution on the whole. Sometimes they are shifted against each other downstream (for example, extremum of mercury content equal to 0.11 μg/l in surface water layer is characteristic of st. 15, and in near bottom horizon – of st. 16). The lowest concentrations as in surface layer were revealed at outlet into Dvina bay.

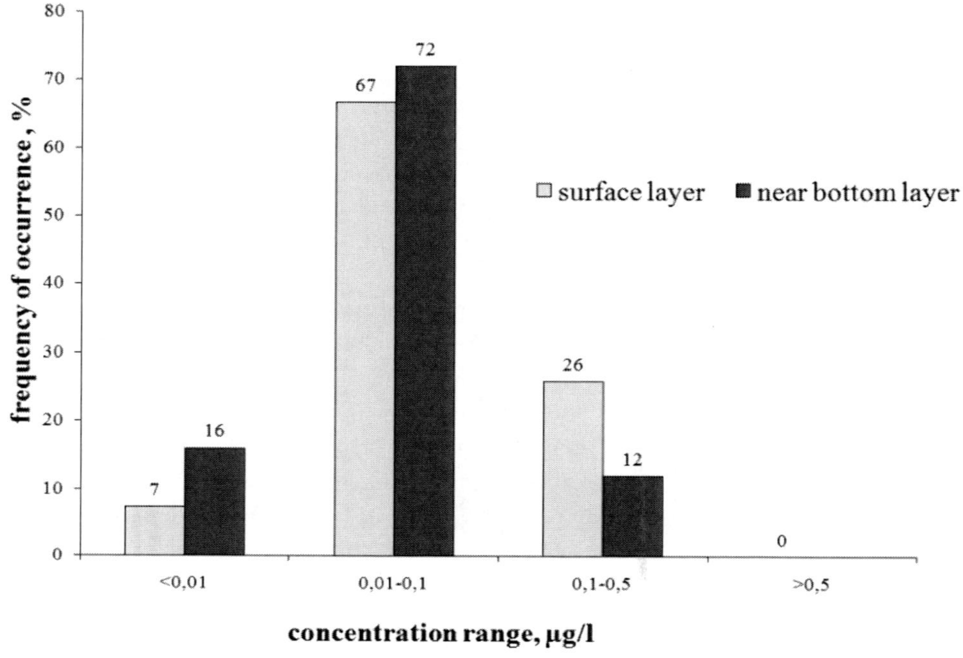

Figure 3. Frequency of occurrence of various concentrations of total dissolved mercury.

Analyzing a frequency of occurrence of various concentrations of total dissolved mercury (Fig. 3) it should be noted that in total mass of data the values exceeding MPS for fishery water bodies (0.01 μg/l) prevailed. In surface water horizon such concentrations were revealed in 93 % of measuring, in near bottom – in 84 %. In surface layer a percent of occurrence of the highest metal concentrations in a range of 0.1-0.5 μg/l, it is 26 %, is also higher; in near bottom horizon a mercury content has exceeded 0.1 μg/l only in 12 % of cases. Large prevalence of high mercury concentrations in surface layer can indicate an occurrence of continuously acting sources of input of this metal into water mass. It may be both industrial enterprises (PPM, HGS, timber mills) and plane washing away of mercury-containing substances from drainage area.

The results obtained are comparable to the dissolved mercury content in other regions of the globe. However, mean concentrations of the metal and its compounds in the water body under study are somewhat above those characteristic of the water of the north of European Russia and are at the level comparable, for example, to its concentration in waters of the Don and Kuban rivers [11, 13, 21, 27-30].

We have calculated a relative content of total dissolved form of mercury (Fig. 4). In the most number of samples collected a content of dissolved form of mercury exceeded mercury content in suspended matter.

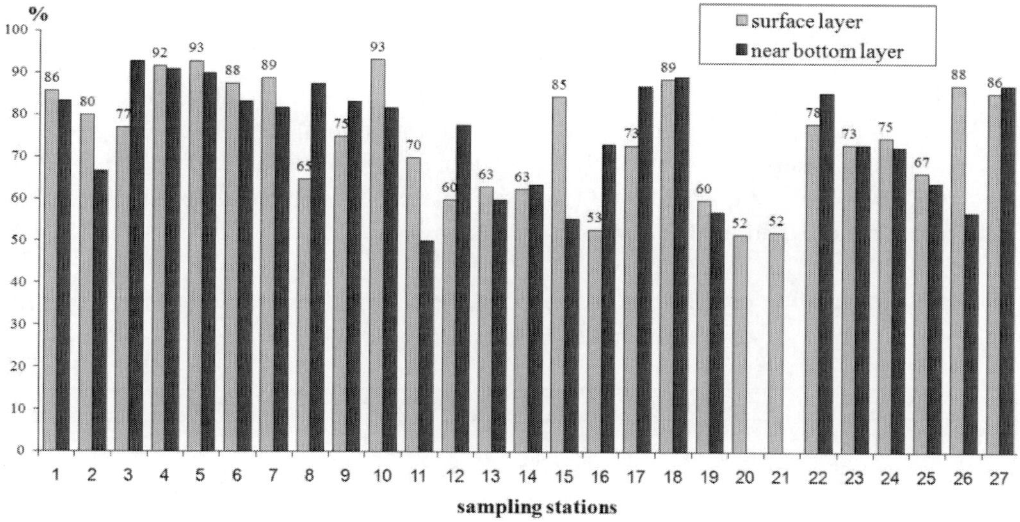

Figure 4. Relative content of total dissolved form of mercury.

In surface layer in 17 cases out of 27 a relative dissolved mercury content exceeded 70 %. In 27 cases out of 27 it exceeded 50 %. At some stations a relative content of dissolved mercury was 92-93 % (10, 4, 5). The least values of a relative content of total dissolved mercury – 52-53 % is characteristic of st. 16, 20, 21.

The Solombalka and Uras Rivers are shallow, they are actively used by local population as the places for parking of small launches and motor boats. One can draw a conclusion that during a process of their exploitation an intensive turbation occurs here (also as a result of turbulent mixing of well heated water mass) and a content of suspended substances is increased, correspondingly a probability of mercury combination with them increases too.

A relative content of total dissolved mercury in the Pinega River, Murmansky and Nikolsky arms and also at a section from v. Chizhovka to i. Mudyugsky is high. That is at outlet into Dvina bay a prevalence of dissolved form of mercury increases. At stations 12-14 a relative content of dissolved mercury was 60-63 %. In Kuznechikha channel a distribution is quite uniform – 73-78 %. The largest is at st. Timber mill No. 29 – 78 %, the least – at st. Kuznechikha ch., downstream of SPPM, HGS – 73%.

Relative Content of Mercury Dissolved in Near Bottom Water Layer

Comparing near bottom and surface water layers it will be noted that on the whole a relative content of total dissolved form of mercury in near bottom layer is smaller than in surface one (in 15 cases out of 25-60 %). That is a percent of mercury content in suspended substances here is though slight but higher. Only in 40 % of cases a relative content of mercury in near bottom layer exceeded or coincided with the same in surface water layer. In some cases a relative content of mercury in near bottom horizon is considerably higher than in surface one, namely at stations 3, 8, 9, 12, 16, 17, 22, 27. In Kuznechikha channel there is also such station – st. Timber mill No. 29. Here a relative content of mercury in surface layer is 78 %, in near bottom – 86 %. Low values of a relative mercury content are recorded at stations 11, 15, 19, 26. At stations 16 (Top of delta) and 11 (Lower Povrakulsk sites) a considerable difference with surface layer where a relative content of metal is higher (85 and 56 %; 70 and 50 % respectively) is observed. For mouth of the Pinega River a low content in both water horizons (60 % - in surface and 57 % - in near bottom) is typical.

The highest relative contents of dissolved mercury are measured at stations 3 (i. Lebedin), 4 (Ustyansk turning sites), 18 (Ust-Pinega). It should be noted that such behavior is typical also for surface horizon of waters at these stations.

Table 3. Correlation coefficients between content of dissolved form of mercury and some hydrochemical properties

Correlation C	S	O_2	t	pH	Eh	CH_4
Hg_d^{tot} surface layer	-0,59	-0,41	0,65	0,2	-0,38	-0,54
Hg_d^{tot} near bottom layer	-0,72	-0,77	0,65	0,03	0,03	0,67

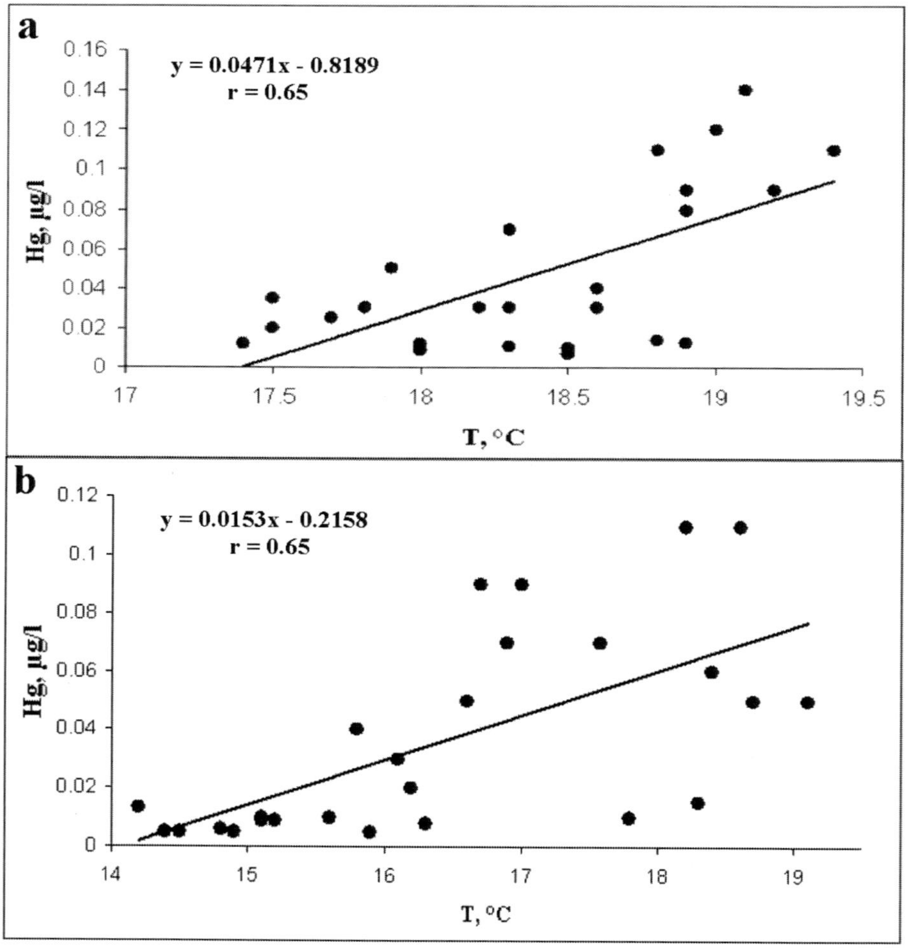

Figure 5. Dependence of the content of total dissolved mercury from the temperature of water: a - surface layer, b – bottom layer.

Thus, one can draw a conclusion that mercury migration in mouth area of the Northern Dvina River occurs mainly in dissolved form, and at outlet into Dvina bay this form prevails more greater over suspended one. It is also necessary to note that a relative mercury content in suspended substance increases with depth.

In the course of the treatment of results obtained during expedition we got correlation coefficients between total dissolved form of mercury and some hydrochemical properties (Table 3).

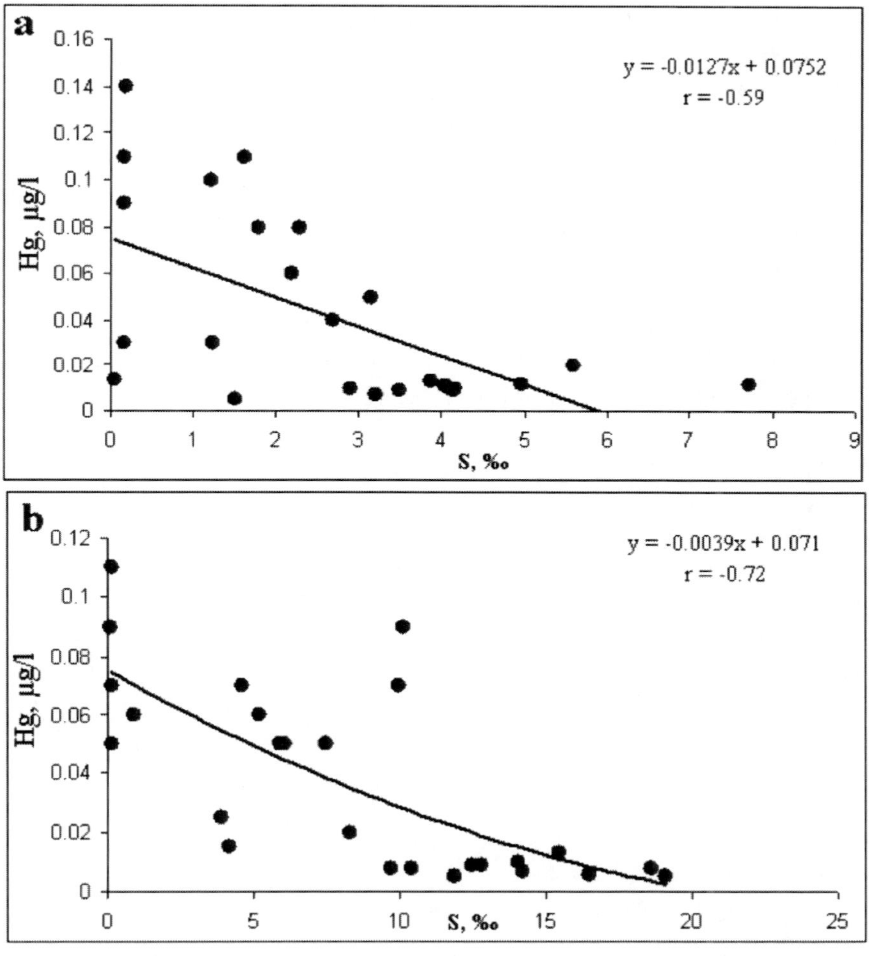

Figure 6. Dependence of the content of total dissolved mercury from the water salinity: a - surface layer, b - bottom layer.

Significant coefficient of positive correlation is calculated for temperature. It is equal to 0.65 (R^2=0,42; p≤0,05) both in surface and in near bottom layer (Fig. 5). Expeditional works were carried out in a warm season when water temperature for this area was quite high and varied in a wide range of values (in surface layer from 14.2 to 19.4 and in near bottom – from 12.4 to 17.6 °C) [5, 12].

In the preceding works we showed that water temperature can promote biogenic and abiogenic transformation of organic substances in bottom sediments, and as a result an activization of the processes promoting transition of mercury from them into water mass [5, 20, 21]. It is corroborated by more early and present investigations having demonstrated a close directly proportional relation between concentrations of mercury and temperature of water [5, 12, 19, 21, 22].

When temperature increasing it is possible an intensification of mercury methylation processes. Thus, E. Bacci (1989) showed that methylation rate can increase 2-5 times when temperature rising from 3 to 13 [23]. It is extremely important in theoretical and practical respect. Especially in the last case since when assessing mercury run off and a degree of pollution of water bodies it is necessary to take into account a temperature factor too.

Significant coefficients of negative correlation were revealed between content of total dissolved form of mercury and salinity (Fig. 6). In near bottom water layer a relation is more close ($r=-0.72$; $R^2=0,51$; $p\leq 0,05$) while in surface layer a correlation coefficient is also rather high ($r=-0.59$; $R^2=0,34$; $p\leq 0,05$). It will be noted that salinity values themselves in near bottom horizon much exceed a value of this property in surface one. In previous papers we considered possible ways of effect of various values of salinity on mercury behavior in different hydrological phases [5, 24]. Changes in this parameter will lead to shift of interfaces of dissolved mercury from one form of occurrence into another changing also a level of toxicity of metal combinations.

When analytically treating a distribution of oxygen and dissolved form of mercury in total mass of data a weak negative correlation was revealed between these elements. In near bottom water layer this relation becomes more close while in surface layer a relationship between concentration of oxygen and mercury is exhibited either rather weakly or absent at all.

An analytical treatment carried out on the basis of the investigation results in the section “i. Solombala – i. Mudyugsky” along the Maymaxa channel revealed significant correlation coefficients for surface ($r=-0.41$) and near bottom ($r=-0.77$) horizon (Table 3). The stronger inversely proportional relation between content of oxygen and mercury is probably concerned with the existence in near bottom layer more reducing condition which during falling tide thanks, first of all, to increased content of bacteria here – sulfate redactors and methane genes stimulates a formation and emission of methyl mercury from bottom sediments into water [5, 19, 24, 25].

A relation between content of total dissolved form of mercury and pH values is not practically exhibited. Correlation coefficients were 0.2 in surface and 0.03 in near bottom layer. Partly it may be connected with relatively slight changeability of pH values. Higher correlation coefficient in surface water layer, where a range of pH changes is wider, evidences it indirectly.

It is known that a prevalence of one or the other form of mercury occurrence depends on pH and chlorides content (pCl) [5, 26]. According to the facts available at the present one may state about $Hg(OH)_2$ prevalence in the upper part of the mouth area and $HgCl_2$ at the outlet into Dvina bay – in Maymaxa, Kuznechikha channels. Moreover, in near bottom horizon a part of $HgCl_2$ increases. However, the results obtained need more in-depth comprehension and study.

Precise relation between Eh values and content of total dissolved form of mercury was not revealed by us. Analytically treating the data obtained in Maymaxa section only in surface layer a comparatively high coefficient of negative correlation equal to -0.38 was determined; for near bottom horizon a coefficient 0.03 not being significant as determined. It will be noted that a factor of effect of oxidation – reduction potential plays a great part in mercury behavior for bottom sediments [5, 19, 26].

CONCLUSION

1. A content of total dissolved mercury along the Northern Dvina River channel varies unevenly. Maximum values of mercury in the river water are characteristic of that areas where it flows in the limits of c. Arkhangelsk. On the whole we have a trend of concentration decrease of total dissolved mercury in the direction of st. Ust-Pinega – st. Lapominka ("river - sea" mixing zone). St. Lapominka is also characteristic of the fact that here during "low water" a hydrofront runs along. Then at the section of st. Lapominka – st. island Mudyugsky some light mercury concentration increase takes place. Apparently, the above mentioned behavior of a metal is caused, on the one hand, by removal of dissolved mercury by its combining with suspended matter precipitated at mechanical barrier (st. Lapominka) and, on the other hand, by the prevalence of a process of desorption over sorption as a result of more long contact of suspended matter with waters of high salinity (st. i. Mudyugsky).

2. Concentration of total dissolved mercury is characteristic of wider intervals and high average concentration in surface layer in comparison with near bottom water layer. It is an indirect corroboration of presence of wider gamut of sources (natural and anthropogenic) of mercury deliver is surface water layer relative to near bottom layer during the period when water is not ice-covered. Perhaps, in winter period this pattern of distribution will change.

3. Total dissolved mercury is contained in concentrations from trace to 0.18μg/l. Such wide variations corroborate its continuous occurrence in waters of the Northern Dvina River and correctness of applied methods for sampling, preparation of samples and determination of this metal content in them. A main reason of relative high concentrations of mercury is anthropogenic factor which effects on the processes of migration, accumulation and dispersion of mercury doth directly and indirectly. Direct effect is a mercury input in a composition of waste waters, with atmospheric precipitations, aerosols and waters of surface runoff polluted by mercury; indirect effect is a consequence of intensification of mercury removal from soils where it is contained in comparatively low quantities under influence of "acid" rains with water of surface runoff or as a result of erosion through an atmospheric channel. Our investigations show a probability of mercury input from bottom sediments, a corroboration of it is an increase of relative mercury content in suspended matter in near bottom horizon. Mercury migrates mainly in dissolved form that considerable differs from its behavior in surface waters of ETR south [5, 13, 22]. Dominating form of mercury occurrence is fulvic complexes, that is corroborated by a close rectilinear relationship between content of mercury and fulvic acids [5, 26].

4. The analysis of the mercury distribution in the frequency of occurrence demonstrated that throughout the entire period under investigation the concentration of the total dissolved mercury in most samples exceeded MPC for fresh-water fishery reservoirs. In the surface water horizon, concentrations of total dissolved mercury exceeded MPC in 93% of measurements, in the near-bottom layer, in 84% of cases. In 26% of cases, the values of 10 MPC of the dissolved mercury form were exceeded in the surface water horizon, in 12% of cases, in the near-bottom layer.

5. Between a content of dissolved mercury and temperature of water a close directly proportional relation is detected that is corroborated by theoretical calculations and experimental data confirming an increase of emission of this metal from bottom sediments into water mass with temperature rise.

6. A weak inversely proportional relation between Eh values and a mercury content is observed. This, on the one hand, shows more favourable conditions for mercury transition from bottom sediments into water mass at lower values of ORP than at higher ones, and, on the other hand, evidences that an effect of this parameter is not dominating or is masked by other factors.

ACKNOWLEDGMENTS

The work is carried out with financial support of the projects MK-1453.2011.5 and NSH-5658.2012.5, P1102

REFERENCES

[1] Brekhovskikh V. F., Volkova Z. V., and Kolesnichenko N. N. Problemy kachestva poverchnostnyh vod v basseine Severnoi Dvini *[Surface Water Quality Problems of the Northern Dvina Basin]*. Nauka, Moscow, 2003. 233 p. [in Russian].

[2] Ecological State of Fish Ponds in Russia, State Report, http://www.ecocom.ru/ *Gosdoklad/Section5.htm, 2004.*

[3] Generalized list of maximum permissible concentrations (MPC) and the estimated safe levels of exposure (ASLE) of harmful substances for the water of fish ponds. M.: *Glavrybvod*. 1990. 400 p.

[4] Izrael Yu. A., *Ecology and Control of the Natural Environment* (Gidrometeoizdat, Moscow, 1984) [in Russian].

[5] *Mercury in the Northern Dvina River Estuarine Area* / A.E. Ovsepyan, Yu. A. Fedorov. Rostov-on-Don - Moscow: ZAO Rostizdat, 2011. 198 p. [in Russian].

[6] Olenicheva A. V. Ob opyte raboty Severnogo UGMS po obsledovaniju rajonov jekstremal'no vysokogo zagrjaznenija (JeVZ) prirodnoj sredy [*About the Experience of the Northern Administration of Hydrometeorological Service (NAHS) in the Field of Studying Extremely Polluted Regions of the Natural Environment*]. Informacionnoe pis'mo SUGMS №1 (164). Arhangel'sk, 1997. pp. 17-22. [in The NAHS Information Letter No. 1(164) (Severnoe UGMS, Arkhangel'sk, 1997. pp. 17-22].

[7] Ovanesyanc A. M., Krasil'nikova T. A., Ivanov A. B. O zagrjaznenii prirodnoj sredy i radiacionnoj obstanovke na territorii Rossijskoj Federacii v marte 2008 g. [About the pollution of the environment and radiation situation in the Russian Federation in March 2008]// *Meteorologija i gidrologija,* 2008. № 6. pp. 98-104.

[8] Vasiliev L.Yu. A problem of the North Dvina River pollution and ways of its solution. *Heads of Reports of VI All-Rus*. Hydrological Congress. St.-Petersburg, 2004. pp.130-131

[9] Gidrologija ust'evoj oblasti Severnoj Dviny [Hydrology of the Northern Dvina mouth area]. Moscow: *Gidrometeoizdat*, 1965. 376 p. [in Russian].

[10] Skibinskii L. E., “The Role of Geochemical Barriers and Geochemical Barrier Zones in Forming of the Hydrochemical Regime of Coastal Waters of the Arctic Seas,” in *Proceedings of the XII Congress of the Russian Geographical Society* (St. Petersburg, 2005), Vol. 5., pp.113-117

[11] AMAP Assessment 2002: *Heavy Metals in the Arctic. Arctic Monitoring and Assessment Program (AMAP)* report, Oslo. 2005. 265 p.

[12] Fedorov Yu. A., D. N. Gar’kusha, A. E. Ovsepyan, and A. N. Kuznetsov, *“Principal Results of Expedition Investigations of the Northern Dvina and the Dvina Bay of the White Sea,”* Izv. vuzov, Severo-Kavkazskii Region, Estestvennye Nauki, No. 3 (2005) [Izv. Higher Educational Institutions, Northern Caucasian Region, Natural Sciences, No. 3 (2005)].

[13] Fedorov Yu. A. et al. Distribution and levels of mercury concentration in atmosphere and water bodies of the Sea of Azov. In book: *Ecosystem investigations of the Sea of Azov and a coast*. V. IV, Apatity: publ. house KHTS RAN, 2002, pp. 150-166.

[14] RD 52.24.479-95. *The method of measurement of mass concentration of mercury in waters by atomic absorption cold vapor.* - Moscow: Publishing House of Hydromet. 1995. P. 25. [in Russian].

[15] Varshal Y.M., Koscheeva I.Ya., Khushvakhtova S.D. et al. // *Geochemistry*, 1999, № 3, pp. 269-275.

[16] Zelyukova Yu. V., T. O. Didorenko, and S. I. Kozlova, “Using Non-flame Atomic-Absorption Spectrophotometry of Mercury for Detecting Trace Amounts of Mercury Compounds,” in *Abstracts to Presentations to the V All-Union Conference on Analytical Chemistry of Organic Compounds* (Moscow, 1984).

[17] Laperdina T.G. Determination of mercury in natural waters. *Novosibirsk,* 2000, P. 222 [in Russian].

[18] *Methodological Instructions for Measuring Pollutants in Marine Bottom Sediments*, Ed. by S. G. Oradovskii (Gidrometeoizdat, Moscow, No. 43, 1979) [in Russian].

[19] Fedorov Yu. A., Ovsepyan A. E., Academician Lisitzin A. P., Dotsenko I. V., Novigatskii A. N., and Shevchenko V. P. *Patterns of Mercury Distribution in Bottom Sediments along the Severnaya Dvina–White Sea Section Doklady Earth Sciences,* 2011, Vol. 436, Part 1, pp. 51–54.

[20] Folsom B.R., Wood J.M. Elsevier *Oceanography Series*, 43. 1986. P. 43.

[21] Moor J. and S. Ramamurti, *Heavy Metals in Natural Waters* (Mir, Moscow, 1987) [in Russian].

[22] Fedorov Yu. A. Mercury in suspended material of water ecosystem. 4th *European Meeting on Environmental Chemistry. Plymouth, England*, 10-13 December, 2003. P.60.

[23] Bacci E., “Mercury in the Mediterranean,*” Marine Pollution Bulletin, No. 2,* P. 20 (1989)

[24] Fedorov Yu.A., A.E. Ovsepyan, V.B. Korobov, Peculiarities of Mercury Distribution, Migration,and Transformation in the Estuarine Areaof the Northern Dvina RiverRussian *Meteorology and Hydrology*, 2010, Vol. 35, No. 4, pp. 289–294.

[25] Fedorov Yu. A., Ovsepyan A.E. About interrelation of concentration of mercury with physical and chemical parameters of water (on an example of Nothern Dvina bay of

White Sea) // *Heads of Reports of XVI Internat. School of Sea Geology.* M., 2005, P. 253

[26] Fedorov Yu.A., Ovsepyan A.E. On an effect of natural and antropogenic organic matter on mercury distribution according to forms of migration in the river waters of ETR North // Proceedings of II scientific-practical conference *"Ecological problems. Outlook in the future"*. Rostov-on-Don, 2005. pp. 130-133.

[27] Bloom N. S., L. M. Moretto, P. Scopece, and P. Ugo, "Seasonal Cycling of Mercury and Monomethylmercury in the Venice Lagoon (Italy)," *Mar. Chem.*, No. 1–4, P. 91 (2004).

[28] Coquery M., D. Cossa, and J. Sanjuan, "Speciation and Sorption of Mercury in Two Macrotidal Estuaries," *Mar. Chem.*, P. 58 (1997).

[29] Kongchum M., I. Devai, R. D. DeLaune, and A. Jugsujinda, "Total Mercury and Methylmercury in Freshwater and Salt Marsh Soils of the Mississippi River Deltaic Plain," *Chemosphere*, P. 63 (2006).

[30] Kudo A., H. Nagase, and Y. Oce, "Proportion of Methylmercury to the Total Amount of Mercury in River Waters in Canada and Japan," *Water Res.,* No. 6, P. 16 (1982).

In: Mercury: Sources, Applications and Health Impacts ISBN: 978-1-62257-721-7
Editors: Ki-Hyun Kim and Richard J. C. Brown

Chapter 9

MERCURY CONTAMINATION IN REPTILES: AN EMERGING PROBLEM WITH CONSEQUENCES FOR WILD LIFE AND HUMAN HEALTH

Larissa Schneider[1], William Maher[1], Aaliyah Green[2] and Richard C. Vogt[3]

[1]Ecochemistry Laboratory. Institute for Applied Ecology,
University of Canberra, Bruce, Australia
[2]Education Specialist/Research Scientist,
Southeastern Natural Sciences Academy, Augusta GA, US
[3]Instituto Nacional de Pesquisas da Amazônia.
Coordenação de Pesquisa em Biodiversidade, Manaus, AM. Brazil

ABSTRACT

Methyl mercury is a persistent environmental contaminant that threatens the health of organisms in ecosystems throughout the world. Since methylmercury bioaccumulates over time and biomagnifies at each trophic level, long-lived, carnivorous species such as reptiles are at greatest risk. Reptiles have often been used as bioindicators of local mercury contamination, and many species have been shown to accumulate large concentrations of mercury in affected habitats. Annually worldwide, millions of reptiles, both wild and farm-raised, are sold for human consumption. Consuming wildlife contaminated with mercury poses a serious threat to human health, particularly for pregnant women and children. In this chapter, we review data of mercury concentrations in reptile species harvested for food. Mercury concentrations in reptile species are compared considering their trophic status and origin (wild-caught or farm-raised) to assess how different groups of organisms respond to mercury contamination. Also, the different origins of Hg are considered in order to understand the effects of different bioaccumulation routes. The issue of legal and illegal trade of reptiles for food in both developed and developing countries is described here to demonstrate the potential health risks to humans. In order to assess the contamination risk per species, we compared Hg concentrations in reptiles with consumption limits developed by the U.S. Environmental Protection Agency and Food and Drug Administration to evaluate mercury consumption through ingestion of fish. Few studies have examined health risks associated with reptile consumption. Although reptiles are not consumed in high numbers in most developed countries, they are a major source of protein in many developing countries where consumers may be at high risk from mercury-related health problems. To conserve endangered species, non-invasive techniques to measure mercury concentrations, by use of skin, shell and blood of reptiles, are described. Shell samples from turtles are especially important because they are composed of layers of keratin deposited throughout an individual's life span, thus providing a history of mercury contamination in turtles.

1. INTRODUCTION

In this chapter, we review data from studies of mercury (Hg) concentrations in reptile species worldwide, with an emphasis on species harvested for food. We review the environmental and biological factors that are potential determinants of mercury bioaccumulation and how the Hg concentration in organisms from undisturbed environments usually reflect the local environmental variability. Moreover, we discuss the issue of legal and illegal trade of reptiles for food and the need to evaluate the health risk of consuming reptiles using consumption limits proposed for Hg concentrations in fish tissues by the U.S. Environmental Protection Agency and Food and Drug Administration. As reptile populations are declining, non-invasive techniques that use skin, shell and blood of reptiles for assessing Hg contamination in reptiles are also discussed.

The class Reptilia includes the orders: *Testudines* (tortoises and turtles), *Crocodylia* (alligators and crocodiles), *Rhynchocephalia* (tuatara), and *Squamata* (lizards and snakes). Reptiles are vertebrates that live in aquatic and terrestrial habitats and have received relatively little attention with respect to Hg bioaccumulation and consequences compared to other animal groups. Reptiles are characterized by having dry skin with keratinized epidermal scales and for producing cleidoic eggs. The eggs are amniotic and have a shell that allows them to develop in a manner less reliant on water than amphibians [1]. All turtles and crocodilians produce shelled eggs, as do most snakes and lizards, but some species of lizards and snakes are ovoviparous, retaining the unshelled embryos within their body cavity and bearing live young. Very few species are actually viviparous, where the embryo receives nutrients from the mother.

Reptiles are important in natural ecosystems, as prey, predators, grazers, seed dispersers and commensal species and serve as bioindicators of environmental health [2, 3]. Reptile population declines are similar in taxonomic breadth, geographic scope and severity as those currently observed for amphibians [4]. Reptilian declines are due to habitat loss and degradation, pollution, disease and climate change [4-6]. Freshwater turtles are highly threatened (46–57% of turtle global populations are threatened) [7]. Studies have also recently reported local declines in snake and lizard populations [8, 9].

Mercury has become a major contaminant in aquatic and terrestrial ecosystems worldwide as its release into these systems has increased over the last 50 years [10, 11]. Mercury is persistent in the environment and bioaccumulates in the tissues and organs of biota, especially higher trophic level animals [12, 13]. Thus the risk to reptiles and humans consuming reptiles needs to be evaluated.

2. MERCURY: SPECIES AND SOURCES

2.1. Mercury Species and Toxicity

Mercury can occur in the environment in different chemical species (Figure 1). Elemental Hg (Hg^0), in the form of a gas, can be efficiently transported in the atmosphere around the world [14]. Yet inorganic Hg [Hg(II)] and methylmercury (CH_3Hg^+), are the main forms of Hg present in water. Inorganic mercury can be in its ionic form (Hg^{2+}) or can be combined forming mercury chloride ($HgCl_2$), mercury sulfide (HgS) and organic acids. Organic forms of Hg contain Hg covalently bound to carbon, for example methylmercury CH_3Hg^+ and ethylmercury ($CH_3CH_2\,Hg^+$).

The most toxic Hg species found in aquatic environments is CH_3Hg^+ and can accumulate readily in biota [15, 16]. As Hg has an affinity to adipose tissue, CH_3Hg^+ tends to bioacumulate and biomagnify in aquatic ecosystems [17, 18]. In order to better understand these processes, research has focused primarily on aquatic environments where inorganic Hg is more efficiently transformed to CH_3Hg^+ [19]. The key factor that determines Hg concentration in the biota is the concentration of CH_3Hg^+ present in water, which is ultimately controlled by the relative efficiency of the methylation and demethylation processes [14].

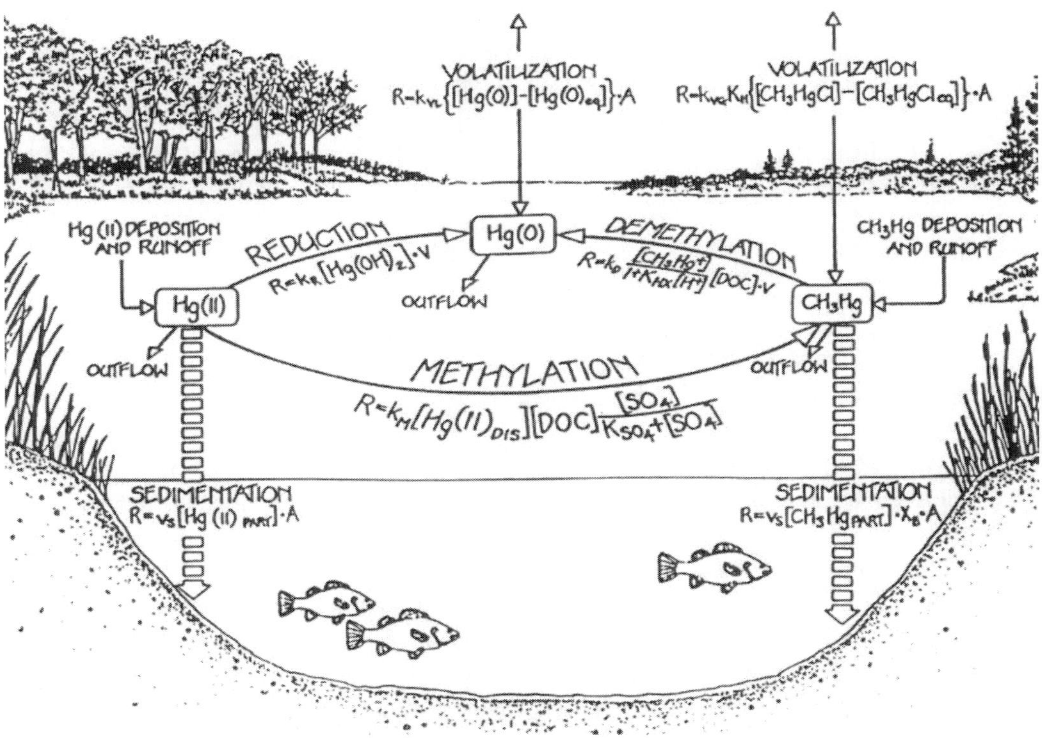

Figure reproduced from Bowles (1998) [192].

Figure 1. Mercury cycling pathways in a lacustrine environment.

2.2. Mercury Origins

Mercury can originate from natural or anthropogenic sources. Natural sources include volcano emissions, natural weathering of geologic deposits and volatilization from ocean surfaces [20]. Anthropogenic sources include agricultural activities, fossil-fuel fired power plants, ferrous and non-ferrous metal manufacturing facilities, caustic soda production plants, ore processing facilities, incinerators for urban, medical and industrial wastes, cement plants and chemical production facilities [21, 22]. Human activities have substantially altered the global Hg cycle so that anthropogenic sources now dominate natural inputs [23, 24]. Anthropogenic Hg emissions to the atmosphere can occur as Hg^0, which undergoes global transport, or as oxidised Hg (e.g. Hg^{2+}) or particulate mercury, which are short-lived and deposited close to source [20]. Thus mercury deposition to a given region results from a combination of global and local sources, with elevated Hg concentrations now found in many remote locations such as the Amazon, Arctic and Antarctic [25-28]. Point sources, such as industrial discharges, can also be important sources at some sites but many of these can be eliminated or at least curtailed by environmental controls [29].

Regardless of the source, the form of Hg that accumulates in biota is mostly CH_3Hg^+, and elevated Hg concentrations in biota are typically associated with environments that enhance methylation [30]. In animals, Hg causes adverse effects such as behavioral and endocrine disruption and at high concentrations can be lethal [31-33]. Toxicity depends on whether the Hg is present as CH_3Hg^+ or in an inorganic form. Methyl mercury is a severe problem because of its propensity to affect the nervous system, its long retention in the body, and its effect on developing organs through transplacental passage [22, 34, 35]. Hg toxicity to reptiles is of particular concern because many reptilian species are already experiencing population declines [36]. Hg contamination is a potential threat to all reptiles but of greater concern for species that forage in aquatic environments because of the potential for rapid movement of Hg into watercourses with Hg stored in sediments, providing ongoing exposure of wildlife to Hg [37].

2.3. Considerations of Mercury Origins and Effects on the Use of Wildlife as Biomonitors

The origin of Hg should be taken into account when using reptiles as biomonitors of Hg contamination in a given area. When studying reptiles in an uncontaminated environment, one has to consider the potential determinants of Hg bioaccumulation such as environmental factors (Hg leaching rate, pH, dissolved organic carbon and temperature), ecological factors (productivity and trophic position) and physiological factors (respiration and assimilation). In a contaminated environment, Hg bioaccumulation in reptiles depends more heavily on the form of Hg being discharged than environmental factors, particularly in lotic ecosystems. In the case of anthropogenic Hg contamination, health effects may initially be more obvious than in animals subjected to long term exposure to natural sources of Hg, due to high levels of exposure over a relatively short period.

A good example of an area with naturally occurring Hg is the Rio Negro Basin of the Amazon. Studies conducted in the Rio Negro basin have shown that soils from the Amazon Central basin are exceptionally rich in Hg and this Hg is predominantly of natural origin [25].

On examination of the Hg concentrations found in this area, it is clear that soil is the main Hg reservoir in the Amazon region and the most probable source of Hg for the fluvial system. It is also evident that the quantity of Hg naturally present in soil is much larger than that introduced into the region by gold mining activities. It was estimated that soils in the Rio Negro Basin have 40 times more Hg than released by gold mining activities during the last 30 years and more than 10 times the quantity of Hg released by the gold and silver mining in the whole of South America during the last 500 years [25, 38]. When considering the influence of environmental factors such as pH, dissolved organic carbon and the relative abundance of potential methylation sites (floodable areas and hydromorphic soils) on the levels of total Hg in two non-migratory predatory fish species, *Cichla spp.* and *Hoplias malabaricus* [30], DOC, pH, and percent floodable area upstream from the sampling site explained most of the variation (proximally 85%) in Hg concentrations in *H. malabaricus*.

In areas where anthropogenic input of Hg is significant, repeated evaluation and temporal measurements of Hg burdens are important to provide information about changes in contaminant load to a system and also to understand the bioaccumulation of Hg by reptiles. Combining laboratory-based *in vitro* assessment of the toxicity of environmental media from a contaminated site with assessment of *in vivo* markers from resident wildlife populations will provide a foundation leading to more timely and efficient efforts to reduce the affects of environmental exposure in areas with direct human disturbances. An example of anthropogenic Hg contamination is in the Republic of Azerbaijan [39], which contains some of the most heavily contaminated sites in the world. The city of Sumgayit is of particular significance because of the number of chemical factories established there by the Soviet Union. Widespread contamination has occurred in and around the city as a result of a lack of regulation of manufacturing processes, waste disposal and resources to maintain facilities. For example, a chlor-alkali plant which is still operational and produces chlorine gas by use of metallic Hg as a catalyst, has spilled an estimated 1,566 tons of Hg into the environment. Turtles were successfully used as a sentinel species to monitor Hg contamination here because they are relatively common in the area, are easy to collect and are carnivorous, eating a variety of invertebrates, amphibians and carrion. Hg concentrations were positively correlated with DNA damage as measured by micronucleus frequency.

2.4. Why Some Reptiles are Good Mercury Bioindicators?

One of the reasons to use reptiles as biomonitors of Hg in the environment (and perhaps the reason for the increase in the number of publications on this topic in recent years) is that reptiles are long-lived animals, which results in accumulation of Hg in their tissues over time from exposure to Hg in sediments, water and food. Moreover, they have a varied diet, depending on their position in the trophic web and where they occur. They reflect well the state of Hg contamination in their environments. However, to be useful biomonitors they also must exhibit a correlation between their tissue Hg concentrations and the Hg concentration in the surrounding water and/or sediment, and this correlation must be comparable among individuals of the species, among all sites and under all conditions [36]. In this sense, reptiles have been successfully used as biomonitor species as they accumulate Hg, are relatively sessile, are ubiquitous, are easy to collect and have sufficient tissues for analysis [29, 40, 41]. Lizards are particularly suitable biomonitors because they are common and widely-dispersed.

Turtles have carapaces composed of hard keratinized plates which store Hg, giving a reliable history of Hg contamination over time. Alligators are sensitive indicators of Hg contamination because they are top predators.

Besides exhibiting the requirements of a good biomonitor, reptiles have other traits that make them better biomonitors than other wildlife species. First, reptiles have very high energy conversion efficiencies (i.e. the amount of ingested energy converted to biomass). High energy conversion efficiency should be associated with higher rates of Hg accumulation but this prediction has not been tested [42]. Second, some reptiles ingest large meals (>30% of their body mass) at infrequent intervals (weeks or months between meals), a dietary pattern that would result in pulse exposure to contaminants not experienced by most other wildlife groups. Third, the enzymatic detoxification system of reptiles is thought to be less developed than in their endothermic counterparts [42]. Available information suggests that reptiles have the major components of the vertebrate mixed function oxygenase system, but the concentrations and activity of these components are sometimes lower in reptiles than in other vertebrates [42]. In addition to surviving in regions with climatic extremes, reptiles brumate or aestivate and these two strategies can remobilize stored contaminants, resulting in latent toxicity. During this time reptiles decrease their activities and feeding and reduce their metabolism and water loss. For example, freshwater turtles at the northern extremes of their range have a period of dormancy that may account for half of their life [42]. This is also true of some tropical species of turtles, *Platemys platycephala*, a small species which lives in rainwater pools in the Amazon forests, is active only when these pools have water, they aestivate in the leaf litter from June until the pools are filled again in December [43]. When reptiles rely on stored energy reserves, such as metabolizing lipids of livers which contain Hg, this will be remobilized during these periods.

3. Mercury Accumulation in Reptiles: A Health Issue

3.1. Historical Use of Reptiles for Human Consumption

Perhaps one of the strongest reasons to describe Hg concentrations in reptiles as an emerging problem is the use of these animals as a food source. Consumption of reptiles has increased over the last decade. An important source of proteins in the past, consumption of these animals is still part of traditional cultures around the world. In many modern cultures, consumption of exotic reptiles is a symbol of wealth.

Reptiles have been in the diet of humans perhaps since the beginning of time; indigenous people all over the world capture and eat turtles, crocodiles, lizards and to a lesser extent snakes. Although very few of these groups rely on reptiles as their sole source of animal protein intake, people living in developing areas, such as along rivers in the Amazon Basin, rely on these animals as a source of protein [44], with alligators and turtles a mainstay in their diets [45]. These people continue to consume reptiles today, possibly to a lesser extent than they did a century ago, mainly because reptiles populations are lower and the widespread availability of other commercial sources of animal protein (e.g. chicken, beef and pork).

For hundreds of years, turtles were a mainstay in the diets of sea voyagers, and perhaps were responsible for keeping crews from dying of starvation or thirst. Galapagos tortoises, sea turtles, and large Amazon river turtles could be kept alive in hulls of boats for months, and slaughtered when other food was scarce. The water in their bladders provided a source of uncontaminated water [46]. Today with refrigeration, canned foods, and freeze dried foods, turtles are no longer necessary for mariners. However, indigenous people still collect tortoises and freshwater turtles in the Amazonian Basin and hold them for weeks or months in corrals or pits until they eat them, without the need of refrigeration or drying the meat.

Turtles have become a luxury food item in different parts of the world [47]. The descendents of river people that have moved to the large metropolitan areas of the Amazon- Manaus, Belem, Porto Velho, and Santarem, enjoy a journey into the past of their forefathers, and pay a premium price to poachers; $300-400 US for a large *Podocnemis expansa*. Consumption of turtles is used as a status symbol at birthday or other parties and on special holidays. Large and small species of turtles are consumed, as well as eggs and hatchlings [48].

In Mexico, the Central American River Turtle, *Dermatemys mawii,* has been driven to the status of Critically Endangered according to the IUCN/SSC Tortoise and Freshwater Turtle Group. Although there is sufficient habitat for this species to recover, local poachers continue to hunt and sell these turtles in cities such as Villahermosa in Tabasco. There are no longer large hunting expeditions for this turtle, as there are no large populations left to hunt. Turtles found by fisherman are sold in the city, as one turtle represents a month of wages [49].

This is also a problem in developed countries. In the USA, in the 1930s, the Diamondback terrapin, *Malaclemmys terrapin,* almost went extinct because of the demand for turtle soup in east coast restaurants. Only when populations of this species dipped so low that it was no longer economically feasible to collect them did this practice stop. In the 1960s-80s, the Alligator snapping turtle, *Macrochelys temmincki*, also became endangered throughout its range in the southern USA [50] because trappers collecting this species sold them for canned turtle soup. Only after this species reached the brink of extinction did the production of canned soup cease.

Throughout the world, sea turtles have been overharvested for their meat and eggs. Since sea turtles are large, long-lived animals, their potential for accumulating Hg or depositing Hg in their eggs is high.

In Latin America, sea turtles and their eggs are often thought to be aphrodisiacs (Figure 2) [51], when in reality because of the accumulation of metal contaminants, polychlorinated biphenyls (PCBs), dichlorodiphenyltrichloroethanes (DDT), and other hydrocarbons, they actually may be harmful to one's health and lower the libido of people who consume them. Only through strong conservation efforts and legislation were sea turtles kept from going extinct [52]. Many populations have remained stable and others have increased 500 to 800% after 40 years of protection [53].

Guidelines for consumption of reptiles based on Hg concentration do not exist and often fish guidelines are used to advise on the allowable amount of reptile meat that can be consumed.

There is a urgent need to develop consumption guidelines specific for reptiles as the risk from eating wildlife depends not only on the quantity of animals eaten but also on the species eaten as Hg concentrations vary by species (Table 1).

Figure 2. Advertising campaign to challenge the use of sea turtle eggs as an aphrodisiac in Latin America, by members of the group Los Tigres del Norte joined by a multitude of environmental organizations. *"Mi hombre no necesita comer huevos de tortuga"* meaning "My man doesn't need turtle eggs." "*Por que sabe que no lo hacen más potente",* "Because he knows they do not make him more potent."

Table 1. Mean Hg concentration in reptiles reported in muscle, liver, skin/carapace, blood, kidney and egg tissues. Maximum meals numbers were calculated for muscle tissues based on the limits determined by the U.S. Environmental Protection Agency (2000)

Order	Common name	Species	Year	Authors [Ref]	Location	Hg muscle (µg/g)	Hg liver (µg/g)	Hg skin/ carapace (µg/g)	Hg Blood (µg/g)	Hg Kidney (µg/g)	Hg eggs (µg/g)	Maximum meals/month
Crocodilia	Alligator	*Alligator mississippiensis*	2011	Chumchal et al [95]	Texas and Louisiana - USA	0.211*	0.711*					4
Crocodilia	Alligator	*Alligator mississippiensis*	2010	Campbell et al [13]	South Carolina - USA	5.680						0
Crocodilia	Alligator	*Alligator mississippiensis*	2002	Rumbold et al [171]	Everglades, Florida - USA	0.640	4.890					1
Crocodilia	Alligator	*Alligator mississippiensis*	2000	Burger et al [189]	Florida - USA	0.080	0.403	0.060				>4
Crocodilia	Alligator	*Alligator mississippiensis*	1999	Elsey et al [198]	Louisiana - USA	0.034*						>4
Crocodilia	Alligator	*Alligator mississippiensis*	1998	Jagoe et al [102]	Everglades, Florida - USA	1.465*	41.030**	5.834**		36.420**		0.5
Crocodilia	Alligator	*Alligator mississippiensis*	1998	Jagoe et al [102]	South Carolina - USA	1.271*	14.9**	5.14**				0.5
Crocodilia	Alligator	*Alligator mississippiensis*	1998	Jagoe et al [102]	Central Florida - USA	0.486*	14.61**	0.52**		12.590**		1
Crocodilia	Alligator	*Alligator mississippiensis*	1998	Jagoe et al [102]	Georgia - USA	0.210*	4.3**	0.290*		4.820*		4
Crocodilia	Alligator	*Alligator mississippiensis*	1997	Heaton-Jones [54]	Florida - USA	2.610	39.990	1.030				0
Crocodilia	Alligator	*Alligator mississippiensis*	1997	Heaton-Jones [54]	Farm raised - USA	0.110	0.100	0.080				>4
Crocodilia	Alligator	*Alligator mississippiensis*	1997	Yanochko et al [169]	Everglades, Florida - USA	1.460*	40.950**	5.725**		36.730**		0.5
Crocodilia	Alligator	*Alligator mississippiensis*	1997	Yanochko et al [169]	South Carolina - USA	1.074*	17.730**	4.580**				0.5
Crocodilia	Alligator	*Alligator mississippiensis*	1993	Ruckel [168]	Georgia - USA	0.480						1

Table 1. (Continued)

Order	Common name	*Species*	Year	Authors [Ref]	Location	Hg muscle (μg/g)	Hg liver (μg/g)	Hg skin/ carapace (μg/g)	Hg Blood (μg/g)	Hg Kidney (μg/g)	Hg eggs (μg/g)	Maximum meals/month
Crocodilia	Alligator	*Alligator mississippiensis*	1988	Delany et al [199]	Florida - USA	0.305						3
Crocodilia	Alligator	*Alligator sinensis*	2006	Xu et al [144]	Zhejiang province-China	0.05*	0.559**			0.902*		>4
Crocodilia	Alligator	*Caiman crocodilus sp*	2012	Schneider et al [251]	Amazon - Brazil	0.291						>4
Crocodilia	Alligator	*Caiman crocodilus yacare*	2011	Vieira et al [200]	Pantanal - Brazil	0.038*						>4
Crocodilia	Alligator	*Crocodilus porosus*	2001	Jefree et al [201]	Northern Territory - Australia	BDL						>4
Crocodilia	Alligator	*Crocodylus acutus*	2007	Rainwater et al [202]	Puntarenas - Costa Rica			0.090				N/A
Crocodilia	Alligator	*Crocodylus moreletii*	2007	Rainwater et al [202]	Northern Belize - Belize			0.090				N/A
Crocodilia	Alligator	*Crocodylus moreletii*	2002	Rainwater et al [208]	Northern Belize - Belize						0.110	N/A
Crocodilia	Alligator	*Crocodylus niloticus*	2005	Almli et al [203]	Kafue and Luangwa rivers - Zambia		3.600			1.730		N/A
Crocodilia	Alligator	*Melanosuchus niger*	2012	Schneider et al [251]	Amazon - Brazil	0.194						4
Squamata	Lizard	*Anolis sagrei*	2004	Burger et al [204]	Florida - USA	0.079						>4
Squamata	Lizard	*Not identified*	2006	Hsu et al [205]	Kenting National Park - Taiwan	5.41**						N/A
Squamata	Lizard	*Podarcis sicula*	2009	Marsili et al [206]	Val d'Agri - Italy	0.070**						N/A
Squamata	Lizard	*Psammodromus algirus*	2009	Marquez-Ferrando et al [207]	Guadiamar Basin - Spain	0.166**						N/A
Squamata	Snake	*Agkistrodon piscivorus*	2011	Chumchal et al [95]	Texas and Louisiana – USA	3.292**	7.456**					N/A
Squamata	Snake	*Agkistrodon piscivorus*	2005	Rainwater et al [209]	Texas - USA	0.163	0.739			0.211		N/A

Order	Common name	Species	Year	Authors [Ref]	Location	Hg muscle (µg/g)	Hg liver (µg/g)	Hg skin/carapace (µg/g)	Hg Blood (µg/g)	Hg Kidney (µg/g)	Hg eggs (µg/g)	Maximum meals/month
Squamata	Snake	*Akistrodon piscivorus*	2006	Burger et al [211]	South Carolina - USA	0.900**			0.1**			N/A
Squamata	Snake	*Elaphe guttata*	2006	Jones and Holladay [210]	Laboratory experiment - USA	0.403**						N/A
Squamata	Snake	*Nerodia fasciata*	2006	Burger et al [211]	South Carolina - USA	0.600**			0.4**			N/A
Squamata	Snake	*Nerodia sipedon*	2005	Burger et al [154]	Tennessee - USA	0.673	0.103	0.423	0.427	0.696	0.061	1
Squamata	Snake	*Nerodia sipedon*	2005	Campbell et al [212]	Tennessee - USA	0.662	1.077	0.433	0.404	0.752		1
Squamata	Snake	*Nerodia sipedon*	1980	Heinz [213]	Lake Michigan - USA	0.450						2
Squamata	Snake	*Nerodia spp.*	2007	Burger et al [214]	New Jersey - USA	0.357	0.303	0.159	0.128	0.136		2
Squamata	Snake	*Nerodia taxispilota*	2006	Burger et al [211]	South Carolina - USA	0.700**			0.7**			N/A
Squamata	Snake	*Not identified*	2006	Hsu et al [205]	Kenting National Park - Taiwan	5.410**						N/A
Squamata	Snake	*Pituophis melanoleucus*	1992	Burger [215]	New Jersey - USA	0.130**		0.28**				N/A
Squamata	Snake	*Thamnophis gigas*	2009	Wylie et al [216]	California - USA	0.082	0.393					>4
Squamata	Snake	*Thamnophis sauritus*	2007	Albrecht et al [217]	Alabama - USA	0.580**						N/A
Squamata	Snake	*Thamnophis sirtalis*	1980	Heinz [213]	Lake Michigan - USA	0.324						2
Testudines	Turtle	*Caretta caretta*	1998	Storelli et al [218]	Adriatic Sea - Italy	0.177*	0.168**			0.65**		4
Testudines	Turtle	*Caretta caretta*	1998	Storelli et al [219]	Adriatic Sea - Italy	0.210	0.700					4
Testudines	Turtle	*Caretta caretta*	2011	Ley-Quinonez et al [220]	Baja California Sur - Mexico				BDL			N/A
Testudines	Turtle	*Caretta caretta*	2010	Day et al [221]	Southeastern coast - USA				0.129			N/A

Table 1. (Continued)

Order	Common name	*Species*	Year	Authors [Ref]	Location	Hg muscle (µg/g)	Hg liver (µg/g)	Hg skin/ carapace (µg/g)	Hg Blood (µg/g)	Hg Kidney (µg/g)	Hg eggs (µg/g)	Maximum meals/month
Testudines	Turtle	*Caretta caretta*	2010	Jerez et al [222]	Murcia - Spain	0.036*	0.390**		0.020**	0.078**		>4
Testudines	Turtle	*Caretta caretta*	2009	Deem et al [223]	Georgia - USA				0.0003			N/A
Testudines	Turtle	*Caretta caretta*	2007	Day et al [108]	Southeastern coast - USA				0.029			N/A
Testudines	Turtle	*Caretta caretta*	2006	Kampalath et al [224]	Baja California Sur - Mexico	0.026****	0.15225 ****			0.0995* ***		>4
Testudines	Turtle	*Caretta caretta*	2005	Day et al [100]	Southeastern coast - USA	0.155	0.594	0.941	0.099	0.214		4
Testudines	Turtle	*Caretta caretta*	2005	Maffucci et al [225]	Western Mediterranean - Italy	0.102*	1.100**			0.900**		>4
Testudines	Turtle	*Caretta caretta*	2005	Storelli et al [226]	Eastern Mediterranean Sea - Italy	0.180	0.430			0.160		4
Testudines	Turtle	*Caretta caretta*	2000	Sakai et al [101]	Cape Ashizuri - Japan	94.4	400	43.2		237	5.54	0
Testudines	Turtle	*Caretta caretta*	1999	Godley et al [227]	Mediterranean Sea - Cyprus	0.123*	2.41**			047**		>4
Testudines	Turtle	*Caretta caretta*	1998	Gordon et al [228]	Queensland - Australia		0.015			0.045		N/A
Testudines	Turtle	*Caretta caretta*	1995	Sakai et al [229]	Cape Ashizuri - Japan	0.108	1.510			0.247		>4
Testudines	Turtle	*Chelodina parkeri*	2010	Green et al [172]	New Guinea	0.329	0.593	0.667		0.361		N/A
Testudines	Turtle	*Chelonia mydas*	2011	Komoroske et al [230]	California - USA			0.0475**	0.001			N/A
Testudines	Turtle	*Chelonia mydas*	2006	Kampalath et a [224]	Baja California Sur - Mexico	0.021****	0.091** **			0.0892* ***		>4
Testudines	Turtle	*Chelonia mydas*	2001	Anan et al [231]	Yaeyama Island - Japan	0.010*	0.420**			0.030**		>4
Testudines	Turtle	*Chelonia mydas*	2000	Sakai et al [101]	Cape Ashizuri - Japan	0.004535	0.18885	0.00241		0.045	0.0013 5	>4

Order	Common name	*Species*	Year	Authors [Ref]	Location	Hg muscle (µg/g)	Hg liver (µg/g)	Hg skin/carapace (µg/g)	Hg Blood (µg/g)	Hg Kidney (µg/g)	Hg eggs (µg/g)	Maximum meals/month
Testudines	Turtle	*Chelonia mydas*	1999	Godley et al [227]	Mediterranean Sea - Cyprus	0.023*	0.55**			BDL		>4
Testudines	Turtle	*Chelonia mydas*	1994	Aguirre et al [232]	Hawaii - USA		BDL			BDL		N/A
Testudines	Turtle	*Chelus fimbriatus*	2011	Schneider et al [109]	Amazon - Brazil	0.432	3.582	0.324	0.158			2
Testudines	Turtle	*Chelydra serpentina*	2011	Turnquist et al [233]	New York - USA	0.481		2.851				1
Testudines	Turtle	*Chelydra serpentina*	1993	Meyers-Schone et al [234]	Tennessee - USA	0.135				0.820		4
Testudines	Turtle	*Chelydra serpentina*	1983	Helwig and Hora [235]	Minnesota - USA	0.150						4
Testudines	Turtle	*Chelydra serpentina*	2001	Golet and Haines [236]	Connecticut - USA	0.165	1.531	1.809	0.127			4
Testudines	Turtle	*Chelydra serpentina serpentina*	2004	Ashpole et al [237]	Great Lakes - Canada						0.218	N/A
Testudines	Turtle	*Chelydra serpentina serpentina*	1998	Bishop et al [153]	Great Lakes - Canada						0.065**	N/A
Testudines	Turtle	*Chelydra serpentina serpentina*	1995	Bonin et al [238]	St. Lawrence River - Canada						0.120	N/A
Testudines	Turtle	*Chelydra serpentine*	2007	Bergeron et al [239]	Virginia - USA				0.956			N/A
Testudines	Turtle	*Chinemys reevesii*	2010	Green et al [172]	Obtained from Chinese food markets	0.037	6.323	1.141		0.240		>4
Testudines	Turtle	*Chinemys reevesii*	2010	Fu et al [240]		0.116*						>4
Testudines	Turtle	*Chrysemys picta*	2007	Bergeron et al [239]	Virginia - USA				0.457			N/A
Testudines	Turtle	*Clemmys marmorata*	2003	Henny et al [241]	Oregon - USA	***0.072*						>4
Testudines	Turtle	*Cuora amboinensis*	2010	Green et al [172]	Malaysia	0.022	0.227	0.241		0.069		>4

Table 1. (Continued)

Order	Common name	*Species*	Year	Authors [Ref]	Location	Hg muscle (µg/g)	Hg liver (µg/g)	Hg skin/carapace (µg/g)	Hg Blood (µg/g)	Hg Kidney (µg/g)	Hg eggs (µg/g)	Maximum meals/month
Testudines	Turtle	*Dermochelys coriacea*	2008	Guirlet et al [242]	Northwest coast - French Guiana				0.011		0.012	N/A
Testudines	Turtle	*Dermochelys coriacea*	1990	Davenport and Wrench [243]	Cardigan Bay - UK	0.031*	0.390**					>4
Testudines	Turtle	*Dermochelys coriacea*	2011	Perrault et al [244]	Florida - USA		0.017		0.0007		0.048	N/A
Testudines	Turtle	*Dermochelys coriacea*	2011	Harris et al [245]	California - USA				0.021			N/A
Testudines	Turtle	*Emys orbicularis*	2003	Swartz et al [39]	Ali Bairamly - Republic of Azerbaijan	1.342*						2
Testudines	Turtle	*Emys orbicularis*	2003	Swartz et al [39]	Sumgayit - Republic of Azerbaijan	5.102*						N/A
Testudines	Turtle	*Eretmochelys imbricata*	2001	Anan et al [231]	Yaeyama Island - Japan	0.010*	1.300**			0.487**		>4
Testudines	Turtle	*Eretmochelys imbricata*	1998	Gordon et al [228]	Queensland - Australia		0.042			0.036		N/A
Testudines	Turtle	*Geochelone elegans*	2010	Green et al [172]	Obtained from pet trade	0.002	0.116	0.010				>4
Testudines	Turtle	*Geoemyda spengleri*	2010	Green et al [172]	Obtained from pet trade	0.017	0.130	2.289		0.060		>4
Testudines	Turtle	*Heosemys spinosa*	2010	Green et al [172]	Malaysia	0.010	1.379	0.743		0.243		>4
Testudines	Turtle	*Kinosternon flavencens*	2010	Green et al [172]	Obtained from pet trade	0.030	0.605	1.893				>4
Testudines	Turtle	*Lepidochelys kempii*	2008	Innis et al [246]	Massachusetts - USA		0.067		0.024			N/A
Testudines	Turtle	*Lepidochelys olivacea*	2011	Paez-Osuna [247]	Oaxaca - Mexico				0.0006**		0.0145**	N/A
Testudines	Turtle	*Lepidochelys olivacea*	2006	Kampalath et al [224]	Baja California Sur - Mexico	0.05****	0.214****			0.143****		>4

Order	Common name	*Species*	Year	Authors[Ref]	Location	Hg muscle (µg/g)	Hg liver (µg/g)	Hg skin/carapace (µg/g)	Hg Blood (µg/g)	Hg Kidney (µg/g)	Hg eggs (µg/g)	Maximum meals/month
Testudines	Turtle	*Leucocephalon yuwonoi*	2010	Green et al [172]	Sulawesi, Indonesia	0.004	0.078	0.056		0.013		>4
Testudines	Turtle	*Macrochelys temminchii*	2008	Chaffin et al [248]	Georgia and Florida - USA				0.363****			N/A
Testudines	Turtle	*Malaclemys terrapin*	2007	Blanvillain et al [249]	South Carolina and Georgia- USA			0.252	0.045			N/A
Testudines	Turtle	*Malaclemys terrapin*	2002	Burger [40]	New Jersey - USA	0.172	1.139				0.035	4
Testudines	Turtle	*Malaclemys terrapin terrapin*	2010	Green et al [172]	New York - USA	0.054	1.493	0.647		0.090		>4
Testudines	Turtle	*Pelochelys cantorii*	2010	Green et al [172]	Obtained from Chinese food markets	0.118	0.691	0.081**		0.399		>4
Testudines	Turtle	*Pelodiscus sinensis*	2010	Green et al [172]	Obtained from Chinese food markets	0.020	0.102	0.027**				>4
Testudines	Turtle	*Peltocephalus dumerilianus*	2011	Schneider et al [109]	Amazon - Brazil	0.106	0.579	0.327	0.028			>4
Testudines	Turtle	*Podocnemis erythrocephala*	2011	Schneider et al [109]	Amazon - Brazil	0.033	0.469	0.067	0.002			>4
Testudines	Turtle	*Podocnemis expansa*	2011	Schneider et al [109]	Amazon - Brazil	0.062	0.287	0.188	N/A			>4
Testudines	Turtle	*Podocnemis sextuberculata*	2011	Schneider et al [109]	Amazon - Brazil	0.061	1.712	0.088	0.007			>4
Testudines	Turtle	*Podocnemis unifilis*	2011	Schneider et al [109]	Amazon - Brazil	0.034	0.335	0.329	N/A			>4
Testudines	Turtle	*Pseudemys rubriventris*	2007	Bergeron et al [239]	Virginia - USA				0.082			N/A
Testudines	Turtle	*Pyxidea (Cuora) mouhotii*	2010	Green et al [172]	Obtained from Chinese food markets	0.050	0.638	0.857		0.362		>4
Testudines	Turtle	*Rhinoclemmys pulcherrima manni*	2010	Green et al [172]	Obtained from pet trade	0.005	1.109	0.148		0.107		>4
Testudines	Turtle	*Siebenrockiella crassicollis*	2010	Green et al [172]	Malaysia	0.011	1.337	0.210		0.183		>4

Table 1. (Continued)

Order	Common name	*Species*	Year	Authors[Ref]	Location	Hg muscle (µg/g)	Hg liver (µg/g)	Hg skin/carapace (µg/g)	Hg Blood (µg/g)	Hg Kidney (µg/g)	Hg eggs (µg/g)	Maximum meals/month
Testudines	Turtle	*Sternotherus odoratus*	2007	Bergeron et al [239]	Virginia - USA				0.922			N/A
Testudines	Turtle	*Trachemys scripta*	2011	Chumchal et al [95]	Texas and Louisiana - USA	0.058*	0.778**					>4
Testudines	Turtle	*Trachemys scripta*	1998	Burger and Gibbons [250]	South Carolina - USA						0.04**	N/A
Testudines	Turtle	*Trachemys scripta*	1993	Meyers-Schone et al [234]	Tennessee - USA	0.065				0.380		>4

* Original data were reported as dry mass. For this table, alligators dry concentrations were divided by a factor of 3.8 (the wet/dry mass ratio calculated by Jefree et al, 2001), for turtles dry concentrations were divided by a factor of 3.9 (the wet/dry mass ration calculated by Eggins et al *In press*) for comparison.

** concentrations reported on a dry mass basis.

*** Mean calculated including values that were below detection limit concentrations which were set as 0.025 µg/g.

**** Wet or dry mass not specified.

N/A: For turtles and alligators, Hg concentrations were not reported in muscle. For snakes and lizards, no calculation was possible as the dry mass could not be transformed into wet mass.

BDL: Hg concentration below detection limit.

3.2. Legal and Illegal Trade of Reptiles for Food and Their Potential as Sources of Mercury to Humans

As an alternative to wildlife consumption, many species of reptiles, including turtles, are being successfully raised in captivity for human consumption to reduce the stress on wild populations and to also produce a healthier product for the consumer.

Farm-raised animals can provide protection of wild populations against overexploitation as well as protecting consumers from dietary exposure to the high Hg concentrations often found in wild animals. Heaton-Jones et al [54] compared the Hg concentrations of American alligators (*Alligator mississippiensis*) captured in the Everglades, Florida USA and in farm-raised animals. Mercury concentrations from farm-raised animals (0.11 μg/g wet mass) were much lower than wild animals (2.70 μg/g wet mass). In the Amazon, farmed turtles (*Podocnemis expansa*) also have lower Hg concentrations (0.034 μg/g wet mass) than wild animals (0.062 μg/g wet mass) (Schneider, unpublished data).

The illegal trade of reptiles, however, is a lucrative activity for the seller and a cheaper option for the consumer, therefore, farm-raised animals are only a solution in developed countries.

In the USA, alligators are farmed and there are regulated harvests of wild animals, therefore large breeding adults are taken from the wild much less frequently because of the lower demand and greater risk of obtaining them. In developing countries, there are few barriers to hunting alligators.

In the Piagaçu-Purus Sustainable Development Reserve, located in the lower Purus River in the Brazilian Amazon, at least 50 tons of dried-salted caiman meat is illegally harvested annually, which represents the largest amount of illegal trade of caiman in Latin America [55, 56].

Reptiles in developing countries are under threat of over exploitation because wildlife is the primary source of meat and income for millions of people. This also represents a large risk of Hg ingestion for human consumers [48].

Globalization is another threat faced by reptiles from countries without enforced illegal wildlife trade laws or with a culture of lack of respect for the current laws. With globalization, wildlife is being harvested worldwide, independent of a countries' economic status. China, in particular, is purchasing turtles from all over the world for use as food, pets and medicinal purposes. An upward shift in the Asian economy has facilitated this demand. As a result, most of the turtles in Southeast Asia are threatened with extinction because of China's quest for reptiles [57-59].

Recently, commercial harvesting laws in the United States have been enacted to protect species of turtles that were being exported to China. The World Chelonian Trust estimates that about 31.8 million turtles were shipped from the United States between 2002 and 2005. A single supplier from Texas shipped 250,000 wild-caught turtles from 2002–2005 [60]. Therefore, reptile consumption has increased significantly in both developed and developing countries. Globalization is leading to an increase of Hg release by industrial activities. This widespread contamination and increase in reptile consumption has led to an increased risk of dietary Hg exposure.

4. Environmental and Biological Factors Affecting Mercury Bioaccumulation

4.1. Environmental Factors

Bacteria within fine-grained organic sediments can transform elemental Hg and Hg^{2+} into CH_3Hg^+ [61]. This process generally occurs under specific environmental conditions. Research has shown that elevated water temperatures, low pH, anaerobic conditions, and high dissolved organic carbon concentrations (DOC) increase the rates of methylation of mercury [62, 63]. Methylmercury is readily absorbed or ingested by organisms, and it is transported to all organs, particularly affecting the nervous system. Because CH_3Hg^+ bioaccumulates in organisms, Hg concentrations in fish tissue can be orders of magnitude higher than Hg concentrations found in water and sediments.

4.1.1. Dissolved Organic Carbon

Mercury concentrations in water is often strongly positively correlated with DOC concentrations [64-66] because Hg binds to organic particles being leached by water [67]. The organic material reduces the ultraviolet radiation into the water, decreasing the transformation of Hg^{2+} into Hg^0, consequently increasing the quantity of Hg^{2+} available to be transformed into $MeHg^+$ [68, 69]. In addition, organics are used as substratum for the methylating bacteria [61, 67, 70]. DOC, however, can also have a negative influence on reptile Hg concentrations because it can also bind Hg, preventing biomethylation and reducing the availability of CH_3Hg^+ for accumulation [61, 71, 72].

In the Rio Negro Basin of the Amazon, Hg concentrations in *Hoplias sp.* significantly increased with increased DOC concentrations [30]. Strong correlations between DOC and Hg concentrations in fish have been found in lakes and streams with extensive areas of wetland and peatland in their upstream drainage basins, from which these two are co-exported [64, 65, 73]. The extensive areas of wetlands and hydromorphic soils in the Negro basin provide a similar source of Hg and DOC.

4.1.2. pH

In temperate areas, more efficient transformation of inorganic Hg to CH_3Hg^+ occurs at low pH [61, 74, 75]. Acidic conditions increase the adsorption of Hg to organic material in flooded soils [74], favoring the transformation of Hg into CH_3Hg^+ [67, 69, 75]. Therefore, pH affects the supply of CH_3Hg^+ [72] and promotes the bioaccumulation of Hg in fish in natural environments [76]. For example, in New York, the acid neutralizing capacity (ANC) of water was measured to determine if it was correlated with total Hg concentrations in snapping turtles. ANC is defined as the difference between cations of strong bases and anions of strong acids, or dynamically as the amount of acid needed to change the pH value from the sample's value to a chosen different value. The results showed that high Hg concentrations found in snapping turtles were correlated with ANC.

In tropical waters, the opposite trend has been reported with a positive correlation between pH and Hg methylation [77]. For example, in the Pantanal area of Brazil, methylation increased with increased pH, reaching the greatest methylation rate between pH 6- 7 [78]. The same pattern was observed in the Rio Tapajos in the Amazon [79].

4.1.3. Dissolved Oxygen

Several laboratory studies have shown that anoxic conditions favor methyl mercury production in sediment and water [80-82]. In field studies, methylation has been shown to occur predominantly in anoxic environments [83-85] such as wetlands [86], floodplain forests, aquatic macrophyte beds [79] and hydromorphic soils [87].

The effects of aerobic/anaerobic conditions on the rate of Hg methylation have been successfully tested in the laboratory [88]. When freshwater systems were made anaerobic and microbial activity was stimulated by the addition of tryptic soybroth (TSB), the level of CH_3Hg^+ in the water increased by two orders of magnitude. There was also an increase in the concentration of total Hg in the water, which was greater than the increase in CH_3Hg^+, indicating release of both CH_3Hg^+ and other Hg species from sediment into water. In the systems that were kept aerobic, there was no increase in either total Hg or CH_3Hg^+ after stimulating microbial activity with TSB.

4.1.4. Availability of Methylation Sites

Wetlands are important environments in promoting Hg biomethylation. When wetlands are inundated, they are characterized by frequent periods of anoxia and high concentrations of organic matter [64, 89]; factors that favor methylating bacteria and methylation of Hg [70, 72].

In addition to being methylation sites, alluvial wetlands can also be significant sources of Hg, due to the leaching of Hg stored in recently flooded soils [70]. In the Tapajós River Basin of the Amazon, high Hg concentrations were found in undisturbed soils with levels in the mineral horizon up to 120 times higher than those commonly found in temperate zone soils [90]. Exceptionally high Hg concentrations were also found in soils of the Negro River basin, including soils collected from alluvial wetland areas [91]. Leaching and release of Hg from these soils during seasonal flooding represents a major source of Hg to the river system [92]. As for pH, anoxia also promotes Hg methylation by bacteria. In the Amazon, this condition is found in flooded areas, such as *igapós* and *várzeas* and in hydromorphic soils. In the same area it was found that the aquatic macrophytes and the flooded forest are important areas of methylation in the region.

4.1.5. Selecting a Suitable Focal Species

It is important to consider the following factors when investigating the influence of spatial variation in water chemistry and other environmental characteristics on wildlife Hg concentration:

- a non-migratory species can better reflect the characteristics of the environment.
- predatory species are especially useful in this context, since they biomagnify more mercury through the food chain and are, thus, more sensitive indicators of bioaccumulation.
- the concentration of mercury in most species increases with age and size of the animal This factor must be taken into account when investigating the effects of environmental variability on animals.

In the Amazon, pH, dissolved organic carbon and availability of potential methylation sites (floodplain forests and hydromorphic soils) are important environmental factors influencing Hg concentrations in the wildlife. In order to check the influence of environmental factors on wildlife Hg uptake, Belger and Forsberg [30] selected two non-migratory fish species as focal species: *Cichla spp.* and *Hoplias malabaricus*. They investigated the influence of pH, dissolved organic carbon and the relative abundance of potential methylation sites (floodable areas and hydromorphic soils) on the levels of total Hg in these two species in the Rio Negro Basin. Their results indicated that the Hg concentrations in *H. malabaricus* were more sensitive to environmental variables than *Cichla spp.*, probably because this species is more territorial and sedentary; DOC, pH, and percent floodable area upstream from the sampling site explained most of the variation in Hg concentrations in *H. malabaricus* (approximated 85%).

On the other hand, when using a turtle species (*Podocnemis erythrocephala*) from the same basin to investigate the influence of the same environmental factors on Hg concentrations, none of the environmental factors studied had a significant relation with blood, muscle, liver and carapace [197]. This is perhaps a result of the extensive movement patterns of the turtle within the basin and because Hg concentrations in *P. erythrocephala* were relatively low due to its exclusively herbivorous diet. In open and interconnected aquatic systems that are characterized by high limnological variability, such as the Amazon Basin, a more locally distributed sedentary species might be a better indicator of the influence of environmental characteristics on mercury biomethylation and accumulation. In addition, the low variation in Hg concentration was insufficient to track differences in the physical characteristics of the environment. For a species to be a good bioindicator of Hg concentration it needs to be relatively sedentary to represent a specific habitat and to be a predatory species so the bioacumulated Hg concentrations are sufficient to track differences in the physical characteristics of the local environmental conditions.

4.2. Biological Factors

Although production of CH_3Hg^+ is an important factor controlling bioaccumulation in reptiles, the biology of reptile is the prime determinant of Hg bioaccumulation.

4.2.1. Diet

In vertebrates, food is the dominant pathway for Hg uptake but within a trophic category, species possess a variety of feeding strategies that will influence $MeHg^+$ acquisition [93]. A wide variation in Hg bioaccumulation among reptiles is expected because of the diversity of feeding strategies and hence diets involving mobility, foraging location, migratory traits, as well as differences in Hg assimilation of each food type [94]. In spite of a wide range of Hg concentrations between and within reptile species, studies clearly show that the dominant feature in Hg assimilation is an increase in muscle-Hg with increasing trophic levels [95]. Some reptiles may change food acquisition strategies during their life cycle [96]. For example, the Amazonian alligators, *Melanosuchus niger,* reach a greater size than *Caiman crocodilus,* thus it would be expected that *M. niger* can prey on larger animals than *C. crocodilus* and therefore accumulate higher Hg concentrations. However, when *M. niger* reaches a larger size they start preying on terrestrial mammals such as capybaras

(*Hydrochaeris hydrochaeris),* which have lower Hg concentrations and alligator size does not reflect an increase in Hg concentration with age.

Turtles also change their diet during their life cycle. Young turtles of many species are carnivores; it is only when they reach adulthood that they begin to vary their diet. For example, mature green sea turtles (*Chelonia mydas*) prefer algae, occasionally taking crustaceans, mollusks, and jellyfish. The loggerhead sea turtle (*Caretta caretta*) consumes mostly fish, bivalves, and cephalopods, while sponges are the main dietary component of the hawksbill sea turtle (*Eretmochelys imbricata*) [97]

Mercury concentrations are lower in the smaller, non-predatory vertebrates and can increase many-fold in higher trophic level predatory invertebrates [98]. Species like the alligator snapping turtle are sit and wait predators, while matamata turtles stalk their prey and swallow small fish whole by a strike gape and suck technique. Snapping turtles also stalk and strike at crayfish swallowing them as the neck is recoiling. Many species of *Podocnemis* feed using neustophagia, sucking up small insects and floating seeds and later expelling the water, while the food particles remain in the digestive tract by erecting the throat papillae [99].

4.2.2. Gender

Differences may exist in Hg accumulation among male and female individuals of the same species because of reproduction effects. Differing reproductive behaviour in mature adults during the mating and nesting season affects short-term Hg accumulation. For example, during reproductively active periods, both male and female Loggerhead turtles decrease or stop feeding [43]. Loggerhead males in North America resume normal foraging after the mating season (March to April). Females have a more prolonged period of reproductive activity lasting until their last clutch is laid, typically in mid or late July. Day et al [100] calculated the Index of Recent Exposure (IRE), representing recent Hg intake relative to the historical exposure for these turtles in July and August. As expected, the mean IRE in females was over three times lower than the IRE in males (-0.0032 and -0.001 respectively).

Reproduction can also result in the elimination of Hg in females by maternal transfer of Hg to their eggs, thus Hg concentrations can be lower in females because they are able to transfer Hg from soft tissues to their eggs [101]. Jagoe et al [102] estimated the Hg body burden of adult female American alligators (*Alligator mississippiensis*) lost by laying eggs. This estimation was based on clutch sizes of 25-40 eggs. A typical female alligator loses about 2.5 % of her body burden during reproduction. Given a body burden of 7-25 mg Hg and a clutch burden of 0.4 mg, a female alligator could lose 1.6 to 5.7 % of its total Hg body burden each year.

4.2.3. Age and Size

Given constant exposure to Hg, individuals tend to accumulate Hg with age as a result of the slow elimination of $MeHg^+$ and increased Hg intake due to changes in diet that often occur when animals grow to larger sizes (i.e., consumption of predator animals and larger prey items). Larger animals feed on larger or more prey which are expected to have higher Hg concentrations. For example, larger alligators prey on predatory fish species. In addition, older reptiles generally have had a longer period over which to accumulate Hg. Therefore, older vertebrates typically have higher tissue Hg concentrations than younger vertebrates of the same species [30].

Although growth (e.g. length as snout-to-vent length [SVL], curved carapace length [CCL] and body weight) and other growth related measures may be used to predict longevity, interpreting growth measurements in reptiles is not always straightforward [103]. Growth is associated with, and responsive to, environmental factors such as temperature, abiotic conditions of microhabitats and biological interactions such as predation and competition [104]. In addition, food will be consumed and assimilated; energy will then be used to support basic physiological processes. Energy and materials remaining after maintenance costs are satisfied, are then available for growth, reproduction or storage. All reptiles are principally ectothermics but the effect on Hg bioaccumulation is unknown.

Since reptile age is indeterminate, size can be used to some extent as a surrogate measure for age [43]. Turtle carapaces are a good indicator of age and can be used to study Hg bioaccumulation in turtles.

The carapace of chelonians is composed of a deep, bony dermal-endoskeleton and an external corneous epidermis subdivided into rows of scutes [105]. These scutes are the hard keratinized plates that armor the carapace and they originate from growing regions at the basal infolding of the epidermis. The scutes form a series of non-living keratin layers deposited one on the top of the other [106]. Mercury is deposited in a metabolically inactive keratin matrix or is bound in relatively immobile inorganic Hg complexes. Consequently, since Hg is very stable in non-living proteinaceous tissue [107, 108], a correlation between Hg concentration and turtle size can be obtained by analyzing the scutes. In studies of Hg concentrations in *Podocnemis erythrocephala,* a herbivorous turtle from the Amazon, no correlation of Hg concentration and size was found for blood, liver or muscle tissues but a strong correlation was found when carapace Hg concentrations were used [109]. Likewise, female and male reptiles can substantially vary in size, where size may be influenced by sex as well as age.

5. MERCURY BIOACUMULATION: DIET AND TROPHIC LEVEL COMPARISONS

Mercury bioaccumulation occurs within a trophic level because of assimilation from food and the surrounding environment. In aquatic ecosystems, bacteria convert inorganic Hg to CH_3Hg^+ which is readily assimilated by organisms. Organisms at the base of the food web such as phytoplankton and periphyton concentrate CH_3Hg^+ and, to a lesser extent, inorganic Hg directly from water, whereas food represents the major route of exposure for invertebrate and vertebrate consumers [110, 111].

Aquatic food webs tend to have more trophic levels than terrestrial webs, where wildlife predators rarely feed on each other, and biomagnification of Hg in aquatic food webs typically reaches higher values [112].

As CH_3Hg^+ biomagnifies, the number of trophic levels in a food web is one of the most important factors in determining the ultimate concentration of Hg in the tissues of consumers. Therefore, greater Hg accumulation is expected for aquatic reptiles which occupy the highest trophic levels [95, 113].

5.1. Mercury Bioaccumulation in Reptiles

Information on Hg accumulation in reptiles dates from the 1980s (Figure 3) with studies reporting Hg concentration in alligators, turtles and snakes. Although some studies date from the 1980s, it was in the 1990s that the number of studies increased significantly especially in the USA. Turtles are the most studied reptiles, while lizards are the least studied with the first articles dating from the early 2000s (Figure 3).

Mercury concentrations reported in reptile tissues found in the literature are in the order liver > kidney > muscle > eggs. Mercury concentrations in carapaces (for turtles) and skin (for alligators) for some individuals are lower than muscle tissues; however, for the majority of species, carapace and skin had higher Hg concentrations than muscle tissues (Table 1, Figure 4).

Mean Hg concentrations in reptile muscle tissues are shown in Figure 5. From our bibliographic survey, the highest mean concentrations of Hg in reptiles were reported in the American alligator (*Alligator mississippiensis*) from the Florida Everglades, USA. The mean Hg concentrations were 2.70 μg/g in muscle, 25.8 μg/g in kidney, 39.99 μg/g in liver and 1.03 μg/g wet mass in tail scale [54].

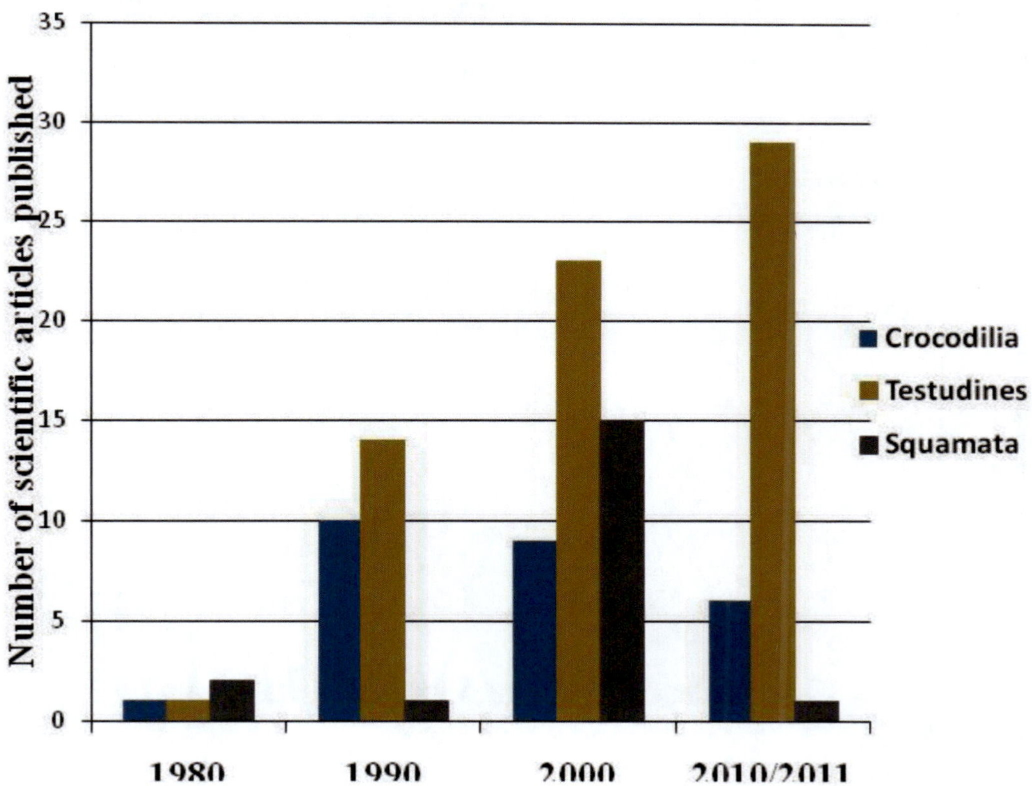

Figure 3. Publications on Hg concentration in reptiles from 1980 to 2012.

For turtles, the highest muscle Hg concentrations reported are for the carnivorous species *Chelus fimbriatus* [41], with a mean Hg concentration of 0.432 μg/g wet mass. The snake with the highest muscle Hg concentrations reported is *Agkistrodon piscivorus* with a mean Hg concentration of 0.329 μg/g dry mass [114]. For lizards, *Psammodromus algirus* is the species with highest Hg concentrations reported in muscle tissues, with 0.166 μg/g, dry mass.

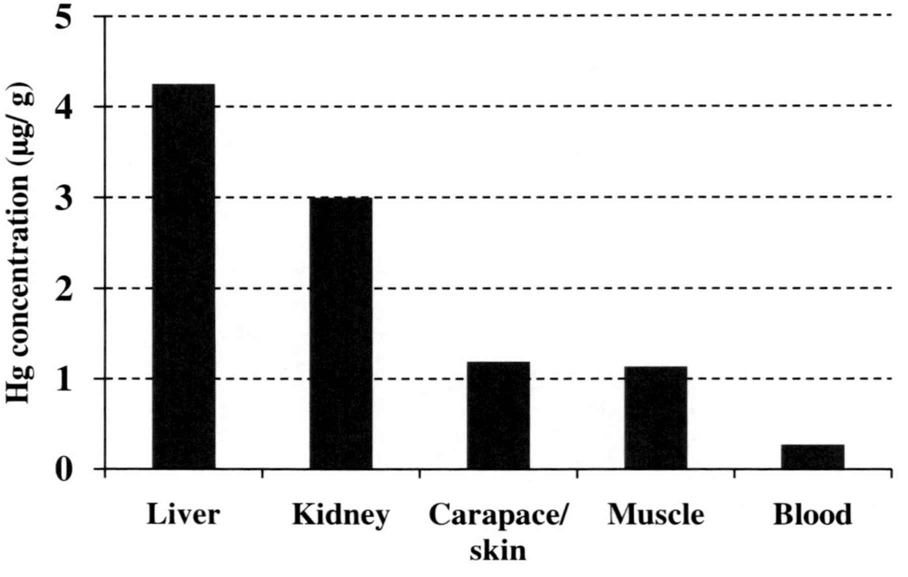

Figure 4. Mean Hg concentrations in tissues of reptiles (wet mass) based on Table 1 data.

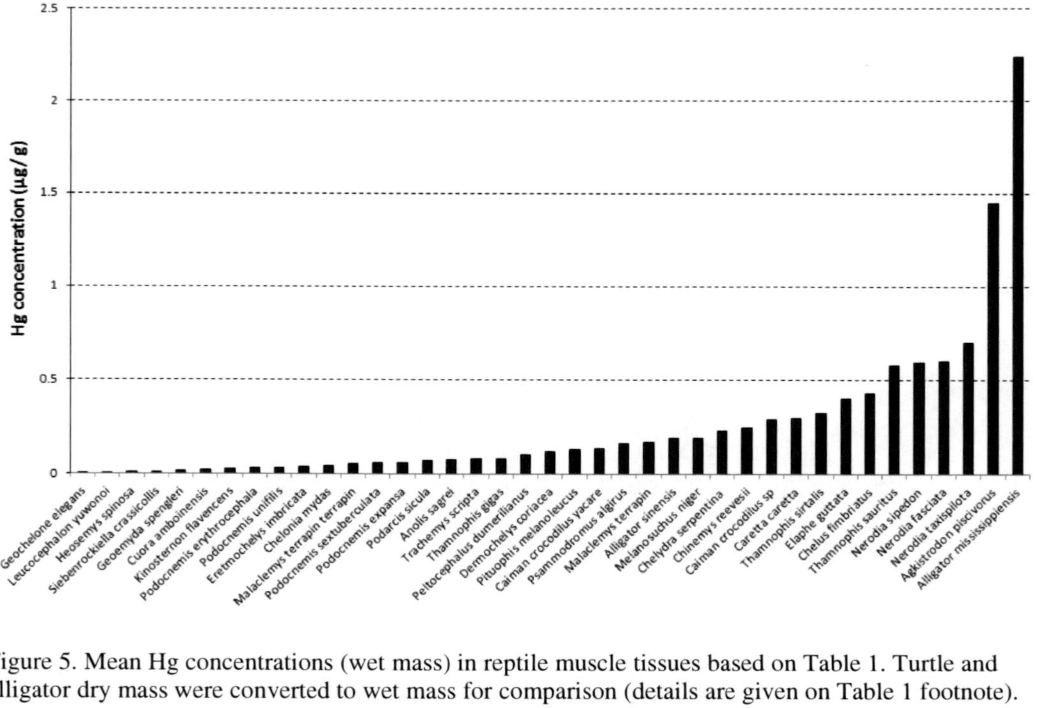

Figure 5. Mean Hg concentrations (wet mass) in reptile muscle tissues based on Table 1. Turtle and alligator dry mass were converted to wet mass for comparison (details are given on Table 1 footnote).

Mercury contamination in the environment may pose an actual or potential threat to animals because of the bioaccumulation and biomagnification of Hg in a reptile's food web and subsequent effects. From our knowledge, there are only two representative studies that measured Hg concentrations in reptiles and in their food webs in which organisms from different trophic levels (predators, herbivores, omnivores and carnivores) are included. Thome-Souza et al [113] examined a food web in the Rio Negro Basin of the Amazon (Brazil) including turtle species and Chumchal et al [95] examined a food web in Lake Caddo on the border of Texas and Louisiana (USA) including snake, turtles and alligator species. Reptiles have evolved and adapted to survive a number of environmental challenges over many years, but Hg bioaccumulation in reptiles could be a *tipping point* for these animals in a world where human exploitation, diseases, invasive predators, drought, and other threats have been affecting reptile populations. Increased exposure to Hg within the past century may seriously impair their ability to continue to survive because Hg has a long term, cumulative and toxic effects. Many reptile species around the world suffer from a variety of threats and the cumulative effects eventually may exceed some species ability to sustain viable populations.

5.2. Reptile Diets

Diet provides a foundation for understanding the relative importance of both foraging habitat and trophic position in the transfer of Hg through successive trophic levels. Table 2 shows the trophic position of selected turtles, snakes, lizards and alligators as a reference for reptiles in general

Most reptiles are carnivorous, meaning they eat other animals. There are a small number of reptile species, however, that eat plants for survival Most reptiles have highly developed predatory instincts as well as reflexes to aid them in catching their prey.

Most land tortoises are nearly 100% herbivorous, although some, such as red foot and yellow foot tortoises, will take carrion when found in the jungle [43, 115, 116]. Many tortoises graze on a variety of grasses, but also consume fruits and seeds [115, 117].

Sea turtles vary from being nearly entirely herbivorous as in the case for the green turtle *Chelonia mydas*, feeding on sea grasses, to carnivorous, as in the leatherback *Dermochelys coriacea* which feeds almost exclusively on jellyfish [115, 117]. Other species specialize in consuming crustaceans such as the loggerhead turtle (*Caretta caretta*), while the hawksbill turtle (*Eretmochelys imbricata*) consumes sponges [115, 117].

Most freshwater turtles are omnivorous, opportunistic feeders, feeding on what is locally abundant, and subsisting on fruits and seeds when protein rich sources are not readily available [118]. There are exceptions, however. The Central American river turtle, *Dermatemys mawii* is nearly 100% herbivorous from hatchling to adult, and has specialized gut flora that allow them to digest the cell walls of plants [49, 119]. The cooters of the southern USA (*Pseudemys nelsoni*, *Pseudemys floridana*, *Pseudemys concinna*, *Pseudemys alabamensis*) are also herbivorous when adults, but may consume small animals when young [120]. Some species of map turtles (*Graptemys geographica*, *Graptemys barbouri*, *Graptemys pulchra*) are specialists consuming molluscs and depending on the amount of molluscs consumed, develop larger heads and larger alveolar crushing surfaces. Most of the Australian and South American chelids are omnivores to some extent; they eat a substantial amount of animal material, but readily feed on fruits and seeds [121]. Some species of Kinosternidae, *Kinosternon carinatus*, *Staurotypus triporcatus*, and *Staurotypus salvini,* are also specialists consuming molluscs, but when molluscs and crustaceans are scarce they will also feed on hard seeds and fruits [122]. This is similar to the large headed *Peltocephalus dumerilianus* of the Amazon Basin. Unlike other Pelomedusids, it specializes in consuming molluscs and crustaceans, but like *Staurotypus* spp. feeds on hard seeds and fruits when these animals are unavailable [43].

Table 2. Trophic position of selected turtles, snakes, lizards and alligators

Taxa	Trophic position	Juvenile	δN[e]	δN baseline[e]
Adult Turtles				
Yellow-bellied slider (*Trachemis scripta*)[a]	3.3	3.5		
Florida cooter (*Pseudemys floridana*)[a]	2.3	2.3		
Stinkpot (*Sternotherus adoratus*)[a]	3.6			
Eastern mud turtle (*Kinosternon subrubrum*)[a]	4			
Florida softshell (*Apalone ferox*)[a]	3.8	3.2		
Snapping turtle (*Chelydra serpentina*)[a]	3.5			
Snakes				
Florida green watersnake (*Nerodia floridana*)[a]	3.4			
Banded watersnake (*Nerodia fasciata*)[a]	3.8			
Cottonmouth (*Agkistrodon piscivorous*)[a]	3.2			
Mud snake (*Farancia abacura*)[a]	4.6			
Chequered Keelback (*Xenochrophis piscator*)[b]	3.4		9.9	1.7
Chequered Keelback (*Xenochrophis piscator*)[b]		3	8.7	1.7
Rice paddy snake (*Enhydris plumbea*)[b]		3	8.5	1.7
Brahminy blind snake (*Ramphotyphlops braminus*)[b]	2.9		8.1	1.7
Red-tailed pipe snake (*Cylindrophis ruffus*)[b]		3	8.6	1.7
Small-spotted coral snake (*Calliophis maculiceps*)[b]	3.6		10.5	1.7
Indo-Chinese ra snake (*Ptyas korros*)[b]	3.5		10.2	1.7
Crocodilian				
American alligator (*Alligator mississippiensis*)[a]	3.3	2.7		
Lizards				
Side-blotched lizard (*Uta stansburiana*)[c] (coastal)	3.3			
Side-blotched lizard (*Uta stansburiana*)[c] (inland)	2.7			
Sail fin lizard (*Hydrosaurus pustulatus*)[d]	1.3		1.7	0.6
Panay monitor (*Varanus mabitang*)[d]	1.5		2.2	0.6
Reeves' smooth skink (*Scincella reevesii*)[b]	2.6		7	1.7
Speckled forest skink (*Mabuya macularia*)[b]	2.5		6.8	1.7

Note: Trophic position: Primary producers = 1, primary consumers = 2, secondary consumers = 3, tertiary consumers = 4 or greater. If δN is presented, no trophic position was listed, but was calculated *post hoc*.

[a] Aresco and James (2005)

[b] Kupfer et al (2006)

[c] Barrett et al (2002)

[d] Ulrich et al (2002)

Table reproduced from de Solla et al [196].

There are few species that are truly carnivorous turtles. The matamata (*Chelus fimbriata*) from the Amazon Basin is, perhaps, the only 100% carnivorous turtle, feeding primarily on small fish which they capture by striking out and simultaneously sucking in the water surrounding the fish [123]. They consume, to a lesser extent, tadpoles and aquatic insects as well [43]. The chopontil (*Claudius angustatus*) is a small voracious predator, feeding on crustaceans, molluscs, and aquatic insects [124]. The snapping turtles of the Americas (*Chelydra* spp.) and the alligator snapping turtle (*Macrochelys temmincki*) have been mistakenly labeled as large carnivorous predators of fish and birds. However it has been shown that the northern snapping turtle is an omnivore, often feeding on duckweed (*Lemna* spp.), tubers, and other aquatic plants. Carrion is a dietary staple while crayfish seem to be its preferred prey [115]. Even though the *Macrochelys temminckii* reaches nearly 100 kg, much of its diet is made up of acorns and molluscs, although they do eat whatever animal comes to nibble on the worm-like lure they possess on the floor of their mouth, including fish and other turtles [125]. *Kinosternon leucostomum*, a small mud turtle from southern Mexico, is an opportunistic omnivore; if protein rich items such as tadpoles, snails, and crayfish are available they will gorge on these items, but in other populations may feed only on fallen fruits and seeds. Thus it is important to not only know what a species of turtle normally eats, but also what the population being sampled eats [126].

Most lizards are carnivorous, and specifically insectivorous [117]. A few such as the green iguana, (*Iguana iguana*), which ranges from Mexico to the Amazon, are 100% herbivorous [127]. The young follow subadults and consume their faeces to inoculate their guts with the flora necessary to digest cellulose [128]. Transfer of fermentation microbes occurs between generations in this herbivorous lizard. The marine iguana of the Galapagos specializes in feeding on algae which grows on submersed rocks in the sea [117]. The Central American genus, *Ctenosaura* spp., spiny-tailed Iguanas, also feed to a great extent on plants, but will take insects when young [117]. Similarly the desert iguana (*Dipsosaurus dorsalis*) and the chuckawallas (*Sauromalus* spp.) of the North American deserts are almost exclusively herbivorous [117]. In the same deserts reside the leopard lizards (*Gambelia* spp.) which specialize in feeding on other lizards. The tegus (*Tupinambis* spp.) of South America and the monitors (*Varanus* spp.) of Asia and Australia are carnivorous, feeding on eggs and nestling animals. The largest of the monitors, the Komodo Dragon, consumes deer and wild pigs, and is large enough to track down and eat humans as well [117]. The two venomous lizards of the North American Sonoran Desert, the gila monster (*Heloderma suspectum*) and the beaded lizard (*Heloderma horridum*) principally consume eggs and nestling animals [117]. No plants have ever been found in their diet. The horned lizards (*Phrynosoma* spp.) from the same deserts are specialized in consuming ants [117].

Snakes as a group are carnivorous. Small snakes feed on small invertebrates, such as scorpions, crickets, earthworms, and beetle larvae (*Geophis* spp., *Tantilla* spp., *Storeria* spp., *Diadophis* spp.) [117]. Coral snakes of the New World (*Micrurus* spp. and *Micruroides* spp.) specialize in feeding on other snakes, as do cobras (*Naja* spp.), king snakes (*Lampropeltis* spp.), and cribos (*Drymarchon* spp.) [117]. Many medium sized snakes consume amphibians (*Heterodon* spp., *Thamnophis* spp., *Nerodia* spp., *Natrix* spp.) [117], which they overpower and consume alive. Larger constricting snakes consume birds and mammals (*Elaphe* spp., *Pseustes* spp., *Spilotes* spp., *Epicrates* spp.). Some large snakes such as the racers (*Coluber* spp.) and whipsnakes (*Masticophis* spp.) are generalists and consume amphibians, lizards, small mammals and birds, simply by overpowering and swallowing them. The Old World vipers and New World pit vipers feed to a large extent on warm blooded prey. However, the water moccasin (*Agkistrodon piscivorus*) feeds on fish [117]. Sea snakes also feed to a large extent on fish [117]. The largest constricting snakes, pythons, boas and anacondas principally feed on large mammals and birds, although some pythons and anacondas devour crocodilians. The anaconda (*Eunectes* spp.) will also broadly feed on fish [117].

5.3. Mercury Concentrations in Reptiles and Their Position in Food Webs

Few studies have analysed CH_3Hg^+ concentrations in complete food webs. The study of Chumchal et al [95] has reported total Hg and CH_3Hg^+ concentrations in muscle tissues and whole bodies of species in the Caddo Lake food web in the USA, including the Red-eared slider turtle, the cottonmouth snake and the American alligator (Figure 6). This study confirms the hypothesis that vertebrates bind CH_3Hg^+strongly as nearly 100 % of Hg that accumulates in the muscle of predator vertebrates is CH_3Hg^+ [98, 129]. Most of the CH_3Hg^+ in vertebrate tissues is covalently bound to protein sulfhydryl groups which results in a long half-life of elimination (about two years; [130]). As a consequence, there is a selective enrichment of CH_3Hg^+ relative to inorganic mercury as one moves from one trophic level to the next higher trophic level. In Caddo Lake, the Hg concentrations are lower in the smaller, non-predatory vertebrates and increase many-fold up the food chain (Figure 6).

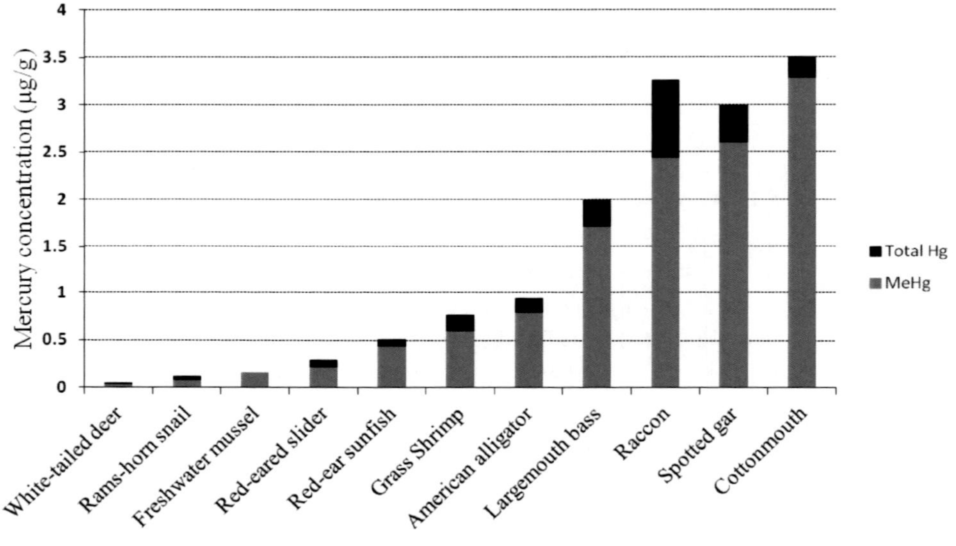

Figure 6. Mean total and methyl Hg concentrations (dry mass) in muscle tissues and whole bodies of Caddo Lake food web species, USA [95]. In this food web three reptile species are shown: red-eared slider turtle, American alligator and cottonmouth snake.

Another complete food web study, but only reporting total Hg concentration, is the one by Thome-Souza et al [113] in which Hg concentrations were analysed in organisms from different trophic levels in the Rio Negro Basin of the Amazon. Figure 7 shows four species of turtles from different trophic levels in comparison with other food web elements. Similar to Chumchal et al [95] study, the Hg concentrations were lower in the smaller, non-predatory vertebrates and increased many-fold up food chains.

Although Hg levels largely correspond to reptiles trophic level, when comparing them to other taxonomic groups, one has to consider that reptiles are principally ectothermics and it isn't just their trophic level that dictates their relative Hg burdens or exposure. Given that reptiles have lower metabolic rates and thus energetic demands compared to mammals and particularly birds, they in turn have a lower daily food requirement, and thus dietary exposure to contaminants [252]. The Field Metabolic Rate for birds is generally higher than mammals, and much higher than for reptiles. Nagy [252] found that endothermic mammals and birds have FMRs about 12 and 20 times higher, respectively, than FMRs of equivalent-sized, but ectothermic, reptiles. So, assuming that diet is the main source of Hg, assuming a similar trophic position and body size, reptiles should have lower burdens than birds or mammals.

6. Toxicokinetics of Mercury in Reptiles

Although all species of Hg can accumulate to some degree, $MeHg^+$ is assimilated and accumulates to a greater extent than other species (refer to section 2; Figure 1).

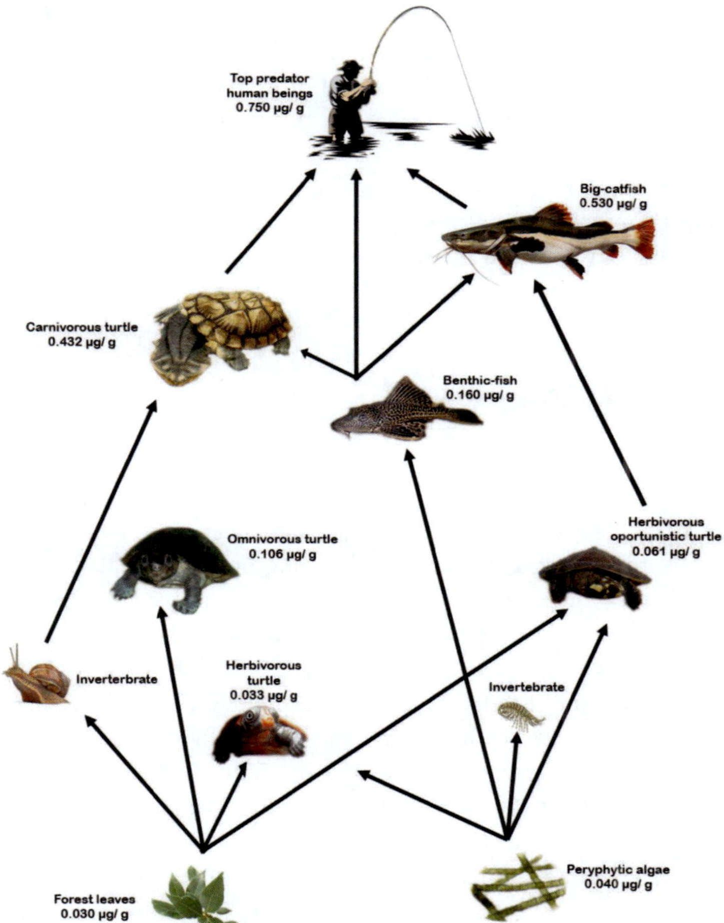

Figure 7. Mean values of total Hg concentrations (wet mass) for a Rio Negro Basin food web indicating trophic levels based on $\delta^{13}C$ according to Thome-Souza et al [113] and gut content based on Schneider et al [41].

In vertebrates, CH_3Hg^+ and inorganic Hg are absorbed through the gut with assimilation efficiencies varying among species [131, 132]. After gut absorption, CH_3Hg^+ is transferred to blood and distributed to other body tissues and most CH_3Hg^+ is eventually redistributed to skeletal muscle tissue, where it accumulates bound to sulfhydryl groups in proteins [133]. Relative to CH_3Hg^+, a larger proportion of inorganic Hg is retained in the gastrointestinal tract [132]. Inorganic Hg can be sequestered by binding to proteins in the liver [134, 135], and demethylation may occur in the liver, resulting in the accumulation of inorganic Hg [136, 137]. As a result of these processes, most Hg in the muscle of vertebrates is CH_3Hg^+ ([138, 139], but the liver can contain elevated concentrations of inorganic Hg [137, 140].

The major routes of Hg exchange in adult reptiles are ingestion and elimination through the digestive (particularly gastrointestinal) system barriers. In addition, Hg uptake and elimination may occur via the integument system, particularly for reptiles with comparatively high transcutaneous water and gas exchange rates [141]. However, studies on the transfer of Hg through reptilian skin and lungs have not been undertaken.

For terrestrial reptiles, besides the dietary ingestion of Hg [142], there are few studies documenting its ingestion through incidental ingestion of soil, sediment or rocks containing Hg [143]. Even though reptiles ingest Hg, not all of it will be absorbed depending on factors mediating Hg absorption and elimination. Mercury that is ingested but not absorbed in the vertebrate gut is eliminated through the faeces, which may sometimes contain high Hg concentrations. For example, a study of metal concentrations in the faeces and eggs of the Chinese alligator *(Alligator sinensis)* suggested that Chinese alligators could reduce their body burden of metals by excreting them to faeces and/or sequestrating them into eggs to a lesser extent [144].

6.1. Fate of Hg within Organs

No data is available on the fate of assimilated Hg in reptiles and we have to extrapolate from data available for other vertebrates. Hg tends to have organ-specific affinities and organs tend to serve as Hg-specific locations of accumulation [141].

Mercury is absorbed in different ways, depending on the animal and the form of Hg. As mentioned in Section 2, there are three common species of Hg: Hg^0, inorganic Hg salts and CH_3Hg^+. In human beings these three species have different toxicokinetic profiles. Elemental Hg is well absorbed through the lungs via inhalation or through the skin, yet is poorly absorbed through the gastrointestinal tract [141]. Elemental Hg is distributed primarily to the kidney and the brain and is excreted in urine and faeces. Inorganic Hg species are absorbed through the gastrointestinal tract and through the skin. The kidney is the main site of deposition of inorganic Hg, with elimination occurring through both the urine and faeces. Methyl mercury is absorbed through the gastrointestinal tract, the lungs and to a small extent through the skin [141]. In mammals, the main route of exposure to Hg^0 and inorganic mercury is inhalation and ingestion, respectively. In fish, Hg is absorbed through their gills as they swim and through their digestive tracts as they feed.

Mercury absorbed through the alimentary tract and transferred primarily to the liver via the hepatic portal vein system and metabolized, stored or excreted before entering the systemic circulation (Figure 8). Once Hg enters the bloodstream, it is distributed throughout the body bound to plasma proteins and blood cells. At this point Hg can be eliminated, transferred to different tissues where it might be sequestered or it is stored in a form unavailable to the individual for their whole lifetime. Redistribution of Hg from storage depots into circulation may occur through an animal's lifetime, particularly during reproduction and during the fasts common to hibernation [42, 145].

Mercury may also be directly transferred to the reptile kidneys and undergo renal presystemic excretion, a mode of elimination, has not yet investigated. In addition, Hg elimination occurs through presystemic or systemic enteral excretion routes (via bile fluid and gastrointestinal mucosa). The biliary route has been observed to be the important elimination route of CH_3Hg^+ in vertebrates [146], but in the American alligator (*Alligator mississippiensis*) Hg concentrations were higher in liver and kidney tissues but low in the bile [54], suggesting this route of elimination is of secondary importance. Elimination of Hg may also occur through sequestration of these constituents into gonodal or integument tissues. Additional studies are required to understand the relative significance of these elimination routes in reducing whole body burden of Hg in reptiles.

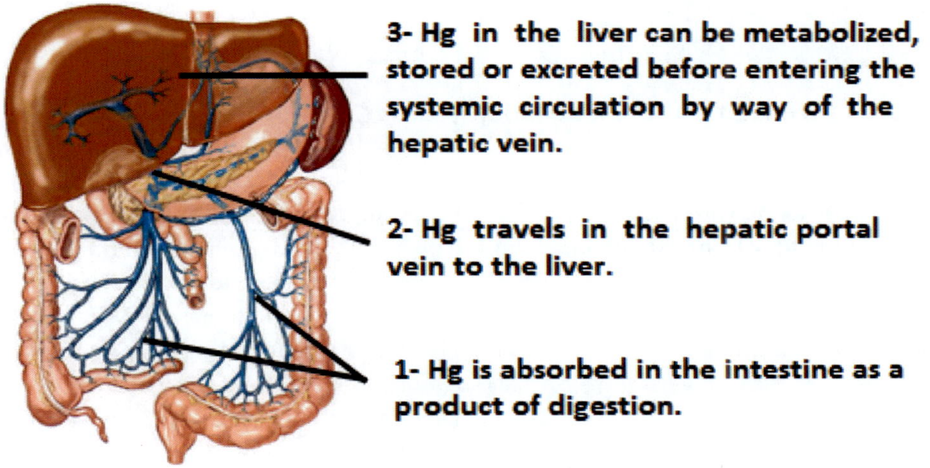

Figure 8. Fate of Hg in the vertebrate's hepatic portal vein system illustrating the circulation of mercury.

Mercury may also be directly transferred to the reptile kidneys and undergo renal excretion. Mercury concentrations are normally higher in liver tissues because this organ is responsible for detoxification. Muscle Hg concentrations may also be particularly high but still less than liver concentrations. The liver plays a central role in the storage and redistribution of recently ingested Hg, while muscle represents a sink for accumulated Hg with a long half-life [147]. Most studies, therefore, focus on Hg concentrations in muscle tissues because these measurements best represent Hg accumulation over time and muscle tissues are also consumed by humans. Thus liver Hg concentrations represent dietary exposure, whereas muscle concentrations represent long-term integration of Hg exposure and assimilation over time [19].

Carapace and skin Hg concentrations in relation to muscle tissue concentrations vary among species. This is because differences in carapace and skin morphology may result in differences in Hg concentrations since they have different patterns of non-living keratin layers and Hg can be stored and metabolically unavailable in these tissues [100, 107, 148].

Blood is metabolically the principal transport medium that links all tissue compartments within an organism. Except for initial phases of high-level exposure, blood Hg concentrations will be well below those in tissues [19]. Mercury shows distinct partitioning among blood compartments being preferentially bound to blood cells or plasma proteins [141]. Although Hg accumulation in brain tissues may result in neurotoxicity, Hg concentrations in brain tissues are not frequently analysed. For all reptiles which have had their brain tissues analysed, Hg was measurable. In turtles (*Caretta caretta*) Hg was present in brain tissues at concentrations of 0.039 μg/g wet mass [101]. For the alligators, the species *Alligator mississipiensis* had a brain Hg concentration of 0.46 μg/g wet mass [149] and 1.37 μg/g wet mass [54]. Mercury concentrations in brain tissues were lower than in liver tissues but similar to muscle tissue concentrations. Intake of Hg into the brain is restricted by the blood-brain barrier and the blood-cerebral spinal fluid barrier, which can be easily passed only by lipid-soluble species such as CH_3Hg^+ and Hg^0. Mercury has rarely been analysed in bone but when analysed levels were comparatively low [141].

6.2. Fate of Mercury at the Subcellular Level

The fate of Hg at the subcellular level is the key to understanding the toxicity of Hg. Detoxification of Hg (or increased toxicity) may occur through changes in Hg oxidation state and through methylation and demethylation. However, there is limited information available on the biotransformation of Hg in reptiles. Based on vertebrate systems, the distribution of Hg into tissues is mediated by the unselective carrier system metallothioneins (MTs) [141]. These are low-molecular mass, cysteine-rich, high affinity metal-binding intracellular peptides comprising a variety of subcategories and isoforms. MTs have the capacity to bind Hg through the thiol groups of cysteine residues, which represents nearly 30% of the amino acids [150]. MTs provide protection against Hg toxicity and are involved in protection against oxidative stress. MTs are present in all tissues, but the capacity for MT induction is greatest in tissues that are involved in Hg uptake, storage and excretion such as gills, intestine, liver and kidney tissues [141].

Besides MTs, melanin also plays an important defensive role in many organisms because it is capable of binding oxygen-derived radicals and cations. Melano-macrophages are also considered to be important in detoxification and have been described in the liver of various reptile species [151]. In particular for reptiles, keratins, the major epidermal structural proteins, are sulfur-rich metal-binding proteins and concentrate thiol-reactive metal ions and metalloid elements such as Hg in the epidermis and its structure, such as scutes and nails. Keratin may immobilize Hg preventing toxicity [141].

7. CONSEQUENCES FOR WILDLIFE

The presence of Hg in ecosystems can have deleterious effects on aquatic organisms as Hg is persistent and can cause toxicity. Compared with studies of other taxa, the analyses of reptilian Hg contamination induced effects are sparse. However, studies of birds and mammals have provided insights into the possible effects of Hg contamination on reptiles.

Here we summarize the information currently available on Hg contamination in reptiles within the context of what is known for non-reptilian species.

7.1. Embryos

Because of the direct effects of Hg contamination on eggs and thus reproductive success, studies examining maternal transfer of Hg and transfer of Hg from nesting substrates are needed. Embryonic reptiles are exposed to Hg via two primary mechanisms: transfer from mother to offspring and absorption from the egg's surroundings. Most of what is currently known about embryonic exposure to Hg comes from surveys of reptilian egg residues (Figure 9), but the actual route of exposure is often not identified. In many cases, maternal transfer is probably the most important route of exposure, but studies need to account for other sources of exposure routes of Hg before reaching this conclusion [42].

Maternal transfer is initiated early in vitellogenesis when females begin synthesizing large quantities of lipoproteins that are crucial for supplying nutrients, essential trace metals and hormones to the developing embryo [153]. Lipoproteins may serve to transport Hg from the mother to the offspring [42]. As the early stages of embryonic development direct the developmental trajectory of the individual, maternal transfer of Hg represents an important vulnerability for developing offspring and ultimately for the mother's fitness [154]. In some cases, the quantity of Hg transferred can be adequate to disrupt early development [42].

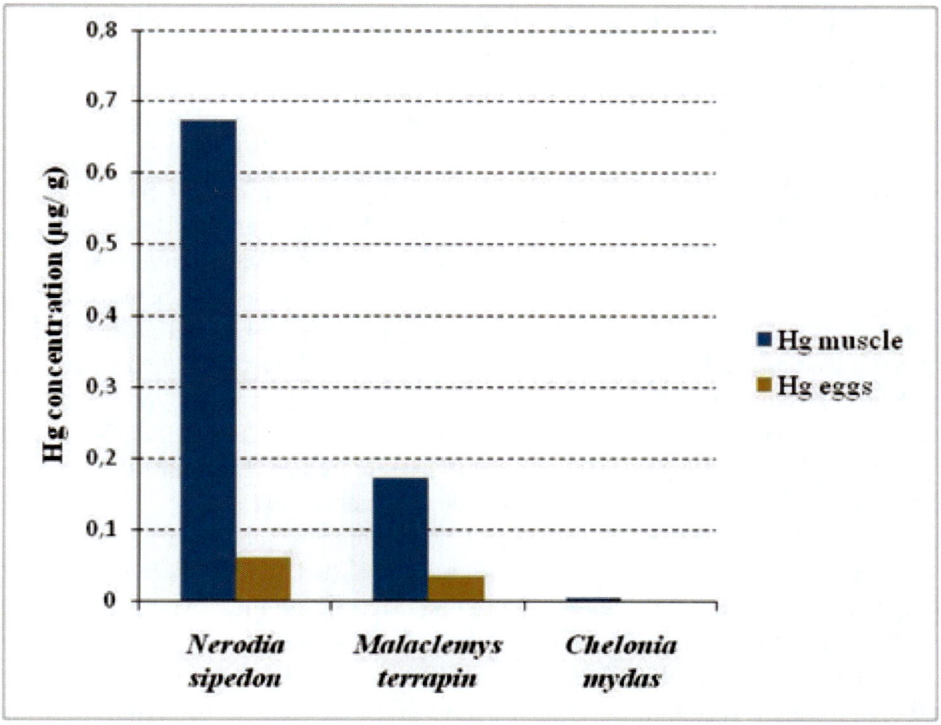

Figure 9. Mercury concentrations (wet mass) in adults and eggs of the snake *Nerodia sipedon* [152] and the turtles *Malaclemys terrapin* [40] and *Chelonia mydas* [101].

The importance of Hg (and contaminants in general) transfer from nesting substrate to the egg has not been adequately investigated. Reptiles generally deposit their eggs in soil, leaf litter or other debris where the potential exists for dissolved Hg species to transverse the eggshell. However, the structure and calcification of the eggshell varies widely among reptiles and is likely to affect Hg transfer [42]. Some turtles and most snakes and lizards have parchment-like eggs that are very permeable to water during development that can result in a two fold to four fold increase in egg mass. In contrast, crocodilians and some geckos and turtles have hard-shelled eggs that are generally less permeable to water [42].

Whereas the distribution of Hg among tissues in reptiles is generally similar to those in other vertebrates, reptiles possess a unique set of traits that may influence Hg accumulation patterns. High energy conversion efficiency, a less developed enzymatic detoxification system, dietary patterns and behavioural traits are all factors that may influence the uptake and accumulation of Hg in reptiles.

7.2. Juveniles and Adults

Detrimental effects of Hg have been reported in numerous studies of birds, fish, and mammals. Neurologic symptoms are well- documented in dose-response studies at high levels of Hg exposure. Lower, environmentally-relevant concentrations of Hg have resulted in sublethal effects on reproduction, behavior, and physiology. Although few studies have focused on the effects of Hg accumulation in reptiles, the breadth of research on other vertebrate groups suggests that reptiles are not immune to the harmful sublethal effects of long-term environmental Hg exposure. Methylmercury's chemical properties allow it to persist in biological tissues as well as in the environment. Its tendency to bioaccumulate over time and biomagnify with trophic position places long-lived predatory species at highest risk from exposure [155, 156].

An extensive literature search revealed a scarcity of experimental feeding studies of reptiles and Hg. Table 3 summarises some of the health effects reported in the literature for reptiles and other vertebrates as potential effects for reptiles. Most studies of dietary Hg effects on reptiles involve analysis of tissue Hg concentrations compared with various reproductive, behavioral, and physiological endpoints observed in exposed populations. Experimental feeding studies conducted on other vertebrate groups provide insights into how dietary Hg exposure may directly influence reptile physiology. Several studies have investigated the effects of sublethal Hg exposure in birds. Three generations of mallards (*Anas platyrhynchos*) fed 0.5 μg/g Hg in feed, offered *ad libitum*, from nine days of age until euthanization at one year laid a significantly greater percent (5.4% - 8.2%) of eggs outside of the nestbox. Shells of viable eggs were significantly thinner than those laid by unexposed hens. Ducklings had muted responses to maternal calls and were hypersensitive to fright stimuli compared to controls. Ducklings also gained significantly less weight during the first week of life than controls [157]. Common loons (*Gavia immer*) consuming fish with whole body concentrations exceeding 0.300 μg/g Hg (wet mass) had significantly reduced productivity [158]. Great egret nestlings (*Ardea alba*) that consumed 0.500 μg/g Hg in fish *ad libitum* for 14 weeks showed a marked decrease in grooming behavior [159].

Sublethal effects of dietary Hg have also been documented in fish. Juvenile male walleye (*Sander vitreus*) fed catfish containing 0.137 – 0.987 μg/g of Hg three times per week for six months had significantly impaired growth and testicular atrophy [160]. Female juvenile fathead minnows (*Pimephales promelas*) fed 0.080 – 0.850 μg/g Hg from hatching to sexual maturity also displayed reduced gonadal development, which led to a decrease in reproductive effort and spawning success [161]. Golden shiners (*Notemigonus crysoleucas*) fed 0.500 μg/g and 1 μg/g Hg daily for 90 days displayed abnormal schooling behavior and decreased predator avoidance behavior [162].

Table 3. Potential adverse health effects caused by Hg exposure in reptiles and other vertebrates

Vertebrate Class	Species	Hg exposure (μg/g)	Length of Exposure	Effects	Reference
Reptile	Loggerhead sea turtle (*Caretta caretta*)	0.5 - 1.0	120 hours	Significant decrease in blood cell viability	Day et al [108]
Reptile	Yellowbelly slider (*Trachemys scripta*)	not given	1 and 2 hours	Inhition of Na^+ transport by the turtle bladder	Schwartz and Flamenbaum [193]
Reptile	Corn snake (*Elaphe guttata*)	1.0 - 6.0	20 weeks	Reduced growth rate	Bazar et al [194]
Reptile	Common garter snake (*Thamnophis sirtalis*)	200	no exposure data	No effects	Wolfe et al (1998) [155]
Bird	Mallard (*Anas platyrhynchos*)	0.50	1 year	*Behavioral*: More eggs laid outside of nestbox, muted response to maternal calls, hypersensitivity to fright stimuli *Physiological*: thin eggshells, reduced weight gain in ducklings	Heinz [157]
Bird	Common loon (*Gavia immer*)	0.30	no exposure data	Reduced productivity	Barr [158]
Bird	Great egret (*Ardea alba*)	0.50	14 weeks	Marked decrease in grooming behavior	Spalding et al [159]
Fish	Walleye (*Sander vitreus*)	0.137 - 0.987	6 months	Impaired growth, testicular atrophy (males)	Friedmann et a*l* [159]
Fish	Fathead minnow (*Pimephales promelas*)	0.08 - 0.85	hatching to sexual maturity	Reduced gonadal development (females)	Hammerschmidt et al [160]
Fish	Golden shiner (*Notemigonus crysoleucas*)	0.50	90 days	Abnormal schooling behavior, decreased predator avoidance behavior	Webber and Haines [161]
Mammal	Mink (*Mustela vision*)	1.1 - 15.0	93 days	Anorexia, fine head tremors, ataxia, clonic convulsions, blindness, lateral recumbency	Wobeser et al [195]

Mercury can also act as an endocrine disruptor when it mimics natural hormones, inhibits the action of hormones, or alters the normal functions of the endocrine system as well as the immune and nervous systems [163]. This is an expanding area of research not well emphasized to date. A classical review for Hg as an endocrine disruptor is the one of Tan et al [31] which summarizes the existing literature on the effects of Hg on the endocrine system and identifies gaps in the knowledge. It is clear that Hg has an affinity for the endocrine system. In several of the reviewed studies, the pituitary gland, thyroid, and testes had accumulated Hg such that concentrations were similar to or not much lower than those in kidney or liver tissues. It is not clear why Hg may target the endocrine system, but one reason may be Hg's affinity for Se, and thus, the selenoproteins. This affinity has specific implications for Hg accumulation in the endocrine system because of the abundance of selenoproteins in the endocrine system, including an importance in hormone metabolism [31].

Endocrine disruptors, in general, may interfere with the synthesis, secretion, transport, binding, action, or elimination of natural hormones in the body that are responsible for homeostasis, reproduction, development, and behaviour.

For each aspect of the endocrine system, both direct mechanisms (e.g. direct inhibition of enzymes in steroidogenesis) and indirect effects (e.g. abnormalities in female cyclicity) of Hg are reported in the literature. Accumulation of Hg in the pituitary gland potentially links each of the hypothalamic–pituitary axes (HPA, HPT, and HPG) to endocrine dysfunction. An example of this can be seen in the decrease in cortisol and TH produced in Hg-exposed catfish [164, 165], or in the trend for increased adrenal and thyroid gland weights in response to Hg-related disruption of feedback mechanisms of the hypothalamic–pituitary axes as described in the sections on the adrenal gland and thyroid gland.

Few studies have investigated the role of Hg as an endocrine disruptor in reptiles and the existing data are based on field observations for systems where other endocrine disrupting chemicals are also present. However, the review of Tan et al [31] clearly shows that Hg has specific endocrine effects within all three hypothalamic–pituitary axes and across humans and wildlife (which may be extrapolated to predict effects in reptiles). Perhaps the review conducted by Tan et al [31] will stimulate interest and research in reptiles.

7.3. Predators of Reptiles

Turtles function in the ecosystems of the world by transforming plant material, seeds, fruits, insects, crustaceans, and fish into protein packages of eggs stored in their nests or bite size morsels as hatchlings. The carnivorous species of turtles clean their systems of Hg through metabolism of lipids in the liver to produce large yolked eggs. Mammal predators, such as foxes, raccoons, opossums, otters, monkeys and jaguars raiding turtle nests can accumulate Hg from the eggs they consume. In turn hatchlings are swallowed whole by catfish and wading birds such as herons and egrets. The tegu lizard (*Tupinambis teguxin*) also specializes on consuming turtle eggs, and would accumulate considerable Hg burdens if enough are eaten [166]. Crocodiles also feed to a moderate extent on turtles of all sizes, even the giant amazon river turtle is not free from predation by the black caiman, (*Melanosuchus niger*) as an adult. Both crocodilians and catfish are regularly consumed by humans and these animals could be potential candidates for high Hg concentrations from feeding on turtles [43].

8. Implications for Human Health

Hg exposure can cause several health problems in humans. Mercury has no known role in human metabolism and may be toxic to the renal, nervous, cardiovascular, and reproductive systems. Chronic effects of long-term, low-level dietary Hg exposure have been examined in several studies of populations with high fish consumption. Examining health effects of humans exposed to Hg through seafood provides valuable insight into how consumption of reptiles with high Hg concentrations can affect consumer health.

No regulatory agency oversees the international trade in reptiles for food and medicine, and reptiles sold for food are not usually tested for contaminants. As a result, people who consume reptiles may be at risk of health consequences associated with dietary exposure to Hg. In order to evaluate these risks, we have used consumption thresholds based on Hg concentrations in fish tissues. In 2004, the Joint Food and Agriculture Organization of the United Nations/ World Health Organization Expert Committee on Food Additives established a tolerable Hg weekly intake of 1.6 μg/kg (0.0016 μg/g) as CH_3Hg^+ for children and pregnant or nursing women. For non-pregnant adults, the committee recommends a tolerable weekly intake of 3.2 μg/kg. In the United States, the Environmental Protection Agency [163] and Food and Drug Administration have partnered to develop risk-based consumption limits for Hg based on a reference dose of 0.1 μg/kg day. To avoid exceeding this dose, it is recommended that no single meal exceed 1 μg/g for children and pregnant or nursing women, and 1.9 μg/g for the general public. To characterize the risk of consuming meals below this threshold, risk-based consumption limits, expressed as the maximum allowable number of meals per month, have been determined using a standard meal size of 0.227 kg (8 oz.) and standard consumer body weight of 70 kg [167].

8.1. Risk-Based Consumption Limits

Daily consumption limits (CR_{lim}) are calculated using Equation 1 from U.S. Environmental Protection Agency (2000):

(Equation 1) $CR_{lim} = (R_fD \times BW)/C_m$

Where CR_{lim} = Maximum allowable consumption rate (kg/day)

R_fD = Reference dose (1×10^{-4} mg/kg/day for Hg)

BW = Consumer body weight (kg)

C_m = Hg tissue concentration (mg/kg)

From this, the monthly consumption limit (CR_{mm}) can be calculated using Equation 2:

(Equation 2) $CR_{mm} = (CR_{lim} \times T_{ap})/MS$

Where CR_{mm} = Maximum allowable consumption rate (meals/month)

T_{ap} = Time averaging period (365.25 days/12 months = 30.44 days/month)

MS = Meal size (kg)

Table 1 lists the Hg concentrations in reptile tissues reported in previous research studies. The high Hg concentrations measured in American alligators are of note. Out of 15 separate study locations, nine had alligators with average muscle Hg concentrations making them unsafe to eat more frequently than once per month [13, 54, 168-171]. Alligators in six locations had average muscle Hg concentrations that made them unsafe for pregnant women to eat [13, 54, 168, 171]. Alligator meat is a popular item in many restaurants in China, Brazil, and the southeastern United States. Children of women who consumed alligator meat while pregnant may be at risk of developmental problems later in life.

The maximum allowable number of meals per month is based on Hg concentrations in muscle tissue, as it is the most frequently sampled tissue type in toxicological studies and, for many species, is also the most likely to be consumed. However, several species of turtles are often consumed whole. For these species, consumption limits based on muscle tissue can obscure the risks of consuming other tissue types, most notably liver. This is worth noting especially in the case of *Chinemys reevesi* [172], a popular turtle in Chinese restaurants, as well as *Podocnemis sextuberculata* from Brazil [41] and northern diamondback terrapins [40, 172].

8.2. Developmental Neurotoxicity

Both longitudinal and cross-sectional studies examining the relationships between maternal fish consumption, maternal Hg concentrations, and neurological development in offspring have been conducted in the Seychelles, Faroe Islands, New Zealand, Poland, Britain, and Massachusetts (USA). Assessment of neurological development involves measurement of subtle endpoints of neurotoxicity, such as visual memory, cognition, manual dexterity, and motor function. Jedrychowski et al[173] reported delayed development associated with high maternal Hg concentrations in children in Krakow, Poland. One year-old infants with normal performance on tests of motor function and mental development were born to mothers with significantly lower maternal and umbilical cord blood Hg concentrations than mothers of infants with delayed performance. The authors determined that the relative risk for delayed performance increased more than threefold for infants born to mothers with cord blood Hg concentrations greater than the median (0.80 μg/L) and nearly threefold with maternal blood Hg concentrations greater than the median (0.50 μg/L).

Much research on subclinical effects of prenatal Hg exposure has examined responses in infants, but a few studies have focused on long-term effects measurable in school-aged children. A New Zealand study [174] found that four- to six-year old children of mothers with maternal hair Hg concentration > 6 μg/g during pregnancy performed significantly poorer on cognitive tests than children exposed to lower Hg concentrations in utero. Grump et al [175] tested scholastic performance and psychological development in six- and seven-year olds from New Zealand. Scores on six of 26 tests were negatively correlated with maternal hair Hg during pregnancy. Grandjean et al [176] followed a cohort of children born 1986-1987 in the Faroe Islands of Denmark. Exposure was characterized by measuring Hg concentrations in cord blood, maternal hair at partuition, and in offspring hair samples at 12 months and at seven years of age. Neurobehavioral examinations were conducted at age seven to assess development. Children displayed dose-related deficiencies in tests of language, attention, and memory.

Conflicting results reported in other studies [177-179] suggest that the relationship between prenatal dietary Hg exposure and developmental neurotoxicity is still not clearly understood. These discrepancies may be explained by the use of different assessments of exposure, such as Hg concentration in umbilical cord blood, maternal blood or hair samples (taken either at different stages of pregnancy, at partuition, or at time of testing), and self-reported seafood intake (through diet logs, written questionnaires, or face-to-face interviews). Additionally, a wide variety of endpoints were used to assess developmental effects. Cognitive, motor, psychological, visual, and mental testing was conducted on children at a range of ages and developmental stages. The use of several different types of tests measuring different areas and stages of neurological development has likely contributed to the uncertainties in this area of research.

8.3. Neurotoxicity in Adults

Extensive research on neurologic effects in adults has been conducted in the Amazon Basin of Brazil, where a history of gold mining and reliance on fish for protein has resulted in high levels of dietary Hg exposure. Lebel et al [180] measured hair Hg concentrations and tested motor skills and visual function in adult residents of a fishing village along the Rio Tapajós. They reported dose-dependent effects of hair Hg on near visual contrast sensitivity, restricted visual fields, muscular fatigue, manual dexterity, and alternating hand coordination. Dolbec et al [181] reported that high Hg concentrations in resident hair samples were significantly correlated with decreased performance on psychomotor tests. A similar study by Yokoo et al [182] examined relationships between hair Hg concentrations and performance on a variety of cognitive and neurobehavioral tests in adult residents of a fishing village near the Rio Cuiaba. Hair Hg concentrations were significantly correlated with lower scores on tests of fine motor speed, dexterity, and response inhibition.

8.4. Cardiovascular Effects

Recently, there has been increased interest in the effects of dietary Hg on cardiovascular health. Fish consumption has traditionally been associated with reduced cardiovascular disease and is recommended as an important part of a heart-healthy diet. Salonen et al [183] identified a population in eastern Finland with high fish intake, but exceptionally high rates of death due to coronary heart disease. Researchers hypothesized that dietary Hg exposure may be a contributing factor. They collected hair, blood, and urine samples from middle-aged men and determined risk for various cardiovascular disorders based on blood chemistry and self-reported risk factors (e.g., smoking, alcohol consumption, etc.). Subjects also recorded food intake for four days prior to giving a blood sample. A dose dependent effect of Hg on risk was reported for a variety of cardiovascular disorders. Most notably, average daily fish intake of ≥ 30 g was associated with a twofold increase in risk of acute myocardial infarction (AMI).

Guallar et al [184] conducted a large-scale study of cardiovascular effects of dietary mercury exposure in men participating in the European Multicenter Case-Control Study on Antioxidants, Myocardial Infarction, and Cancer of the Breast (EURAMIC). Subjects were native male residents of eight European countries and Israel ≤ 70 years of age with a first diagnosis of AMI. Men in the control group were from the same populations and age groups, but had no history of AMI. Mercury concentrations in toenail samples were used to assess long-term chronic exposure, and samples of blood and adipose tissue were used along with self-reported risk factors to calculate risk of AMI. Concentrations of docosahexaenoic acid (DHA, an omega-3 fatty acid present in fatty coldwater fish required for normal brain development in infants and normal brain function in adults) in adipose tissue were strongly correlated with Hg concentrations in toenails. Researchers reported reduced risk of AMI with higher DHA concentratinos (which supports cardiac benefits of eating fatty fish), but only in subjects with relatively low toenail Hg concentrations. At high Hg concentrations, high DHA concentrations were associated with an increase in AMI risk. There was also a positive increase in risk of AMI associated with toenail Hg concentrations above 0.25 μg/g (wet or dry mass not specified). The authors suggest that high concentrations of Hg may counteract the cardiovascular benefits of eating fish.

Virtanen et al [185] compared Hg concentrations in hair and blood samples, levels of eicosapentaenoic (EPA) and docosapentaenoic acids (DPA), DHA in blood, and frequency of fish consumption with risk of acute coronary events, coronary heart disease (CHD) and cardiovascular disease (CVD) in Finnish men participating in the EURAMIC. Subjects were divided into thirds based on hair Hg concentrations. Men in the highest third had significantly increased risk of acute coronary events, CVD death, and CHD death. The authors also observed a significant decrease in the risk of AMI in men with low hair Hg concentrations and high serum DHA and DPA concentrations. No such association was seen in men with high hair Hg concentrations. This finding supports Guallar's conclusion that high Hg concentrations in fish attenuates the beneficial effects of fish consumption on cardiovascular health.

Only one research study has examined the effects of maternal fish consumption on cardiovascular development in offspring. Sorensen et al [186] analysed blood pressure, heart rate, and heart rate variability in seven-year old subjects from the Faroe Islands study. An increase in cord blood Hg concentrations from 1 to 10 μg/L was associated with an increase of 13.9 mm Hg in diastolic blood pressure and an increase of 14.6 mm Hg in systolic blood pressure. No further increase in blood pressure was seen above 10 μg/L Hg. The authors conclude that this result suggests that prenatal Hg exposure affects development of cardiac homeostasis, which can have detrimental consequences later in life.

9. Non Invasive Techniques to Study Mercury Bioaccumulation in Reptiles

Most tissue Hg studies focus on the accumulation of contaminants in tissues (e.g. liver, kidney etc.) and require euthanizing individuals. However, the use of reliable, non-lethal sampling techniques (sampling animals without killing them) is preferable, especially when studying rare species or those with declining populations. Several studies have measured Hg concentrations in reptile blood, skin and carapaces [100, 108, 109, 187, 188], as a nondestructive method of sampling.

In studies with the alligator *Alligator mississippiensis* and snakes *Nerodia sipedon* and *Pituophis melanoleucus*, skin proved to be a useful tissue to predict muscle Hg concentrations [152, 189]. The use of turtle carapace and blood as non-lethal measures of internal exposure to Hg has also been successful [190]. Golet and Haines [19] found that the uniform distribution of Hg in the muscle tissues of *Chelydra serpentina* from different body locations indicated that analysis of small portions of muscle from any tissue will produce representative results. Therefore, a portion of the tail could be removed non-lethally for monitoring Hg concentrations in this species. Mercury concentrations in the carapaces of *Chelydra serpentina* were correlated with muscle concentrations, making non-lethal sampling for this species feasible. Similarly, Jagoe et al [101] demonstrated that measuring Hg concentration in carapaces is a useful non-lethal way of monitoring Hg concentrations in sea turtles. Golet and Haines [19] also suggest the use of carapaces to obtain historical information on Hg concentrations in freshwater aquatic environments by analysing scutes from turtle specimens in museums, as has been done with bird feathers [191].

Even though non-invasive approaches have been shown to be successful, one has to consider the origin and time that the animal has been exposed to Hg when applying these techniques. For example, in the Brazilian Amazon, Schneider et al [109] found that only blood was a reliable indicator of Hg concentrations in liver and muscle tissues for the turtle *Podocnemis sextuberculata,* suggesting that sampling blood from the tail vein is a simple non-invasive (non-lethal) method of obtaining an indication of the body burden of Hg. It was expected that the Hg concentrations in keratinized carapaces would also be correlated to Hg concentrations in other tissues of freshwater turtle species found in the Rio Negro basin. However, Hg concentrations are naturally high in soils [25] and the bioaccumulation of Hg is different in those turtles that derive Hg from anthropogenic sources. Since the turtles of the Rio Negro basin were exposed to Hg for long evolutionary periods, they may have developed defence systems to eliminate Hg which results in different Hg bioaccumulation rates in tissues. Mercury concentrations in muscle tissues only correlated to the Hg concentrations in livers for *P. sextuberculata*. The more Hg ingested, the higher the liver Hg concentrations, and consequently the higher the CH_3Hg^+ bioaccumulation in muscle tissues. Further studies are needed to understand the manner in which Hg bioaccumulation occurs in these species that have always experienced high Hg concentrations.

SUMMARY AND CONCLUDING REMARKS

Mercury has become a major contaminant in aquatic and terrestrial ecosystems worldwide as its release into these systems has increased over the last 50 years. Mercury can originate from natural or anthropogenic sources. Elemental Hg in the form of a gas, can be efficiently transported in the atmosphere around the world thus elevated Hg concentrations are now found in many remote locations such as the Amazon, Arctic and Antarctic. Many studies have used the tissues of field-collected, long lived reptiles as biomonitors of Hg environmental contamination. Reptiles are ideal biomonitors as they accumulate Hg, are relatively sessile, are ubiquitous, are easy to collect and have sufficient tissues for analysis. Some reptiles such as turtles have carapaces composed of hard keratinized plates which store Hg, giving a reliable history of Hg contamination over time. Most studies focus on the accumulation of Hg in tissues (e.g. liver, kidney etc.) and require euthanizing individuals. However, the use of reliable, non-lethal sampling techniques is preferable, especially when studying rare species or those with declining populations. Although tissues from these organisms may be well suited for this purpose, reptiles deserve attention in their own right, not just as tools for monitoring habitat quality.

Mercury is persistent in the environment and bioaccumulates in the tissues and organs of biota, especially higher trophic level animals. Elevated Hg concentrations in biota are typically associated with environments that enhance methylation as CH_3Hg^+ accumulates efficiently in biota. Although production of CH_3Hg^+ (which is controlled by DOC, pH and DO) is an important factor controlling Hg bioaccumulation in reptiles, the biology of reptile (diet, gender, age and size) is the prime determinant of Hg bioaccumulation. In animals, Hg causes adverse effects such as behavioural and endocrine disruption and at high concentrations can be lethal Hg toxicity to reptiles is of particular concern because many reptilian species are already experiencing population declines.

Mercury concentrations in reptiles are an emerging problem, especially in terms of human health, because of the historical use of these animals as food sources. An important source of proteins in the past, consumption of these animals is still part of traditional cultures around the world and has increased over the last decade. Turtles have become a luxury food item in different parts of the world. Guidelines for consumption of reptiles based on Hg concentration do not exist and often fish guidelines are used to advise on the allowable amount of reptile meat that can be consumed. Legal and illegal trade of reptiles for food has increased their potential as sources of mercury to humans. No regulatory agency oversees the international trade in reptiles for food and medicine, and reptiles sold for food are not usually tested for contaminants. As a result, people who consume reptiles may be at risk of health consequences associated with dietary exposure to Hg.

Some general observations about published studies are that:

- Published data on reptiles with relatively low Hg concentrations are lacking. We feel there is a bias for not publishing low or non-detectable concentrations. This makes it difficult to ascertain the true world-wide levels of Hg contamination.
- Natural or anthropogenic sources of Hg are usually not determined thus it is not possible to understand the biogeochemical cycling of Hg at the location being examined.
- The majority of studies on Hg in reptiles only examine Hg concentrations in tissues of individual reptile species and rarely examine different trophic levels thus the biomagnification of Hg in food webs cannot be elucidated.
- Mercury species (Hg^{2+} and CH_3Hg^+) in food webs need to be measured to assist in understanding the bioaccumulation and biomagnification of Hg in food webs.
- There is a lack of experimental data on the effects of Hg on the health of reptiles and postulated effects are extrapolated from other vertebrates. Laboratory and field experiments are required to determine dose-response curves at different developmental stages for a wide range of reptiles. Linkages of species responses to changes in population and community level effects also need to be investigated.
- With new technologies such as radiotelemetry to track reptiles, studies could monitor reptiles before, during and after exposure to Hg to examine dietary patterns, determine Hg uptake and their behaviour and other effects of Hg.
- There is a need to develop human consumption guidelines specific for reptiles as the risk from eating wildlife depends not only on the quantity of animals eaten but also on the species eaten as Hg concentrations vary by species

In conclusion, Hg concentrations in the environment are increasing. As reptiles typically live in environments where Hg biomethylation to CH_3Hg^+ and accumulation is favoured, and thus bioaacumulation and biomagnification in foodwebs is certain to occur, the risks to reptile health needs to be determined. Also as legal and illegal trade of reptiles for food is increasing there is a need to evaluate the human health risk of consuming reptiles.

REFERENCES

[1] Karns, D. Amphibians and reptiles. In: *The Patterned Peatlands of Minnesota*, Wright, H.; Coffin, B.; Aaseng, N. Eds.; University of Minnesota Press: Minneapolis, Minnesota, (1992), pp. 131-150.

[2] Raxworthy, C.J.; Pearson, R.G.; Zimkus, B.M.; Reddy, S.; Deo, A.J.; Nussbaum, R.A.; Ingram, C.M. Continental speciation in the tropics: contrasting biogeographic patterns of divergence in the uroplatus leaf-tailed gecko radiation of Madagascar. *Journal of Zoology*, (2008) 275, 423-440.

[3] Read, J. Are geckos useful bioindicators of air pollution? *Oecologia*, (1998) 114, 180-187.

[4] Gibbons, J.W.; Scott, D.E.; Ryan, T.J.; Buhlmann, T.D.; Metts, B.S.; Greene, J.L.; Mills, T.; Leiden, Y.; Poppy, S.; Winne, C.T. The global decline of reptiles, déjà vu amphibians. *BioScience*, (2000) 50, 653-666.

[5] Cox, N.A.; Temple, H.J. European Red List of Reptiles. Office for Official Publications of the European Communities: Luxembourg (2009).

[6] Todd, B.D.; Willson, J.D.; Gibbons, J.W. The global status of reptiles and causes of their decline. In: *Ecotoxicology of Amphibians and Reptiles*. Sparling, D.; Linder, G.; Bishop, C.; Krest, S. Eds.; CRC Press: Boca Raton, Fl, (2010), pp. 47-67.

[7] Buhlmann, K.A.; Akre, T.S.B.; Iverson, J.B.; Karapatakis, D.; Mittermeier, R.A.; Georges, A.; Rhodin, A.G.J.; Van Dijk, P.P.; Gibbons, J.W. A global analysis of tortoise and freshwater turtle distributions with identification of regional priority conservation areas. *Chelonian Conservation and Biology*, (2009) 8, 116-149.

[8] Reading, C.J.; Luiselli, L.M.; Akani, G.C.; Bonnet, X.; Amori, G.; Ballouard, J.M.; Filippi, E.; Naulleau, G.; D, P.; Rugiero, L. Are snake populations in widespread decline? *Biology Letters*, (2010) 6, 777-780.

[9] Sinervo, B.; Méndez-de-la-Cruz, F.; Miles, D.; Heulin, B.; Bastiaans, E.; Villagrán-Santa Cruz, M.; Lara-Resendiz, R.; Martínez-Méndez, N.; Calderón-Espinosa, M.; Meza-Lázaro, R.; Gadsden, H.; Avila, L.; Morando, M.; De la Riva, I.; Sepulveda, P.; Duarte Rocha, C.; Ibargüengoytía, N.; Puntriano, C. A; Massot, M.; Lepetz, V.; et al Erosion of lizard diversity by climate change and altered thermal niches. *Science*, (2010) 328, 328-333.

[10] Haines, T.A.; Komov, V.T.; Matey, V.E.; Jagoe, C.H. Perch mercury content is related to acidity and color of 26 Russian lakes. *Water, Air, and Soil Pollution*, (1995) 85, 823-828.

[11] Mason, R.P.; Fitzgerald, W.F; Morel, F.M.M. The biogeochemical cycling of elemental mercury: anthropogenic influences. *Geochimica Cosmochimica Acta*, (1994) 58, 3191-3198.

[12] Bowles, K.C.; Apte, S.C.; Maher, W.A.; Kawei, M.; Smith, R. Bioaccumulation and biomagnification of mercury in Lake Murray. *Canadian Journal of Fisheries and Aquatic Sciences*, (2001) 58, 888-897.

[13] Campbell, J.W.; Waters, M.N.; Tarter, A.; Jackson, J. Heavy metal and selenium concentrations in liver tissue from wild American alligator (*Alligator mississippiensis*) livers near Charleston, South Carolina. *Journal of Wildlife Diseases*, (2010) 4, 46-53.

[14] Morel, F.M.M.; Kraepiel, A.M.L.; Amyot, M.. The chemical cycle and bioaccumulation of mercury. *Annual Review of Ecology and Systematics*, (1998) 29, 543-566.

[15] Lindqvist, O.; Rodhe, H. Atmospheric mercury: a review. *Tellus*, (1985) 37B, 136-159.

[16] Mahalingam, R. Interactions between mercury and dissolved organic matter - a review. *Chemosphere*, (2004) 55, 319-331.

[17] Hund, E.G.; Bischoff, A.L. Inimical effects on wildlife of periodic DDT applications to Clear Lake. *California Fish and Game*, (1960) 46, 91-106.

[18] Padovani, C.R.; Forsberg, B.R.; Pimentel, T.P. Contaminação mercurial em peixes do Rio Madeira: resultados e recomendações para consumo humano. *Acta Amazônica*, (1996) 25, 127-135.

[19] Golet, W.J.; Haines, T.A. Snapping turtles (*Chelydra serpentina*) as monitors for mercury of aquatic environments. *Environmental Monitoring and Assessment*, (2001) 71, 211-220.

[20] Jaffe, D.; Strode. Sources, fate and transport of atmospheric mercury from Asia. *Environmental Chemistry*, (2008) 5, 121-126.

[21] Pirrone, N.; Cinnirella, S.; Feng, X.; Finkelman, R.B.; Friedli, H.R.; Leaner, J.; Mason, R.; Mukherjee, A.B.; Stracher, G.; Streets, D.G.; Telmer, K. Global mercury emissions to the atmosphere from anthropogenic and natural sources. *Atmospheric Chemistry and Physics* (2010) 10, 5951-5964.

[22] Nelson, N. Hazards of mercury. *Environmental Research*, (1971) 4, 1-69.

[23] Fitzgerald, W.; Mason, R.; Vandal, G. Atmospheric cycling and air-water exchange of mercury over mid-continental lacustrine regions. *Water, Air, and Soil Pollution*, (1991) 56, 745-767.

[24] Mason, R.P.; Sheu, G.R. Role of the ocean in the global mercury cycle. *Global Biogeochemical Cycles*, (2002) 16, 1093-2007.

[25] Fadini, P.S.; Jardim, W.F. Is the Rio Negro (Amazon) impacted by naturally occuring mercury? *The Science of the Total Environment*, (2001) 275, 71-82.

[26] Forsberg, B.; Silva-Forsberg, M.; Padovani, C.; Sargentini, E.; Malm, O. High level of Hg in fish from the Rio Negro Basin (Brazilian Amazon): natural background or anthropogenic contamination? In: *International Workshop on Environmental Hg Pollution and its Health Effects in the Amazon River Basin*. Kato, H.; Peifer, W. Eds.; National Institute for Minamata Disease, Universidade Federal do Rio de Janeiro Rio de Janeiro, RJ, (1995), pp. 33-40.

[27] Jitaru, P.; Gabrielli, P.; Marteel, A.; Plane, J. M. C.; Planchon, F.A.M.; Gauchard, P.A.; Ferrari, C.P.; Boutron, C.F.; Adams, F.C.; Hong, S.; Cescon, P.; Barbante, C. Atmospheric depletion of mercury over Antarctica during glacial periods. *Nature Geoscience*, (2009) 2, 505-508.

[28] Macdonald, R.W.; Loseto, L.L. Are Artic Ocean ecosystems exceptionally vulnerable to global emissions of mercury? A call for emphasised research on methylation and the consequences of climate change. *Environmental Chemistry*, (2010) **7**, 133-138.

[29] Burger, J.; Cooper, K.; Saliva, J.; Gochfeld, D.; Lipsky, D.; Gochfeld, M. Mercury bioaccumulation in organisms from three Puerto Rican estuaries. *Environmental Monitoring and Assessment*, (1992) 22, 181-197.

[30] Belger, L.; Forsberg, B. R. Factors controlling Hg levels in two predatory fish species in the Rio Negro Basin, Brazilian Amazon. *Science of the Total Environment*, (2006) 367, 451-459.

[31] Tan, S.W.; Meiller, J.C.; Mahaffey, K.R. The endocrine effects of mercury in humans and wildlife. *Critical Reviews in Toxicology*, (2009) 39, 228-269.

[32] Klaper, R.; Rees, C.B.; Drevnick, P.; Weber, D.; Sandheinrich, M.; Carvan, M.J. Gene expression changes related to endocrine function and decline in reproduction in fathead minnow (*Pimephales promelas*) after dietary methylmercury exposure. *Environmental Health Perspectives*, (2006) 114, 1337-1343.

[33] Anway, M.D.; Skinner, M.K. Epigenetic transgenerational actions of endocrine disruptors. *Endocrinology*, (2006) 147, S43-S49.

[34] Goldwater, L. Mercury, a history of quicksilver. York Press: Baltimore, Maryland, (1972).

[35] Clarkson, T.W. Human health risks from methylmercury in fish. *Environmental Toxicology and Chemistry*, (1990) 9, 957-961.

[36] Burger, J.; Mckenzre, D.; Thackston, R.; Demaso, J. The role of farm policy in achieving large-scale conservation: bobwhite and buffers. *Wildlife Society Bulletin*, (2006) 34, 986-994.

[37] Burger, J.; Gochfeld, M. Comparison of arsenic, cadmium, chromium, lead, manganese, mercury and selenium in feathers of bald eagle (*Haliaeetus leucocephalus*) and comparison with common eider (*Somateria mollissima*), glaucous-winged gull (*Larus glaucescens*), pigeon guillemot (*Cepphus columba*), and tufted puffin (*Fratercula cirrhata*) from the Aleutian Chain of Alaska. *Environmental Monitoring and Assessment*, (2009) 152, 357-367.

[38] Nriagu, J.O. Mercury pollution from the past mining of gold and silver in the Americas. *Science of the Total Environment*, (1994) 149, 167-181.

[39] Swartz, C.D.; Donnelly, K.C.; Islamzadeh, A.; Rowe, G.T.; Rogers, W.J.; Palatnikov, G.M.; Mekhtiev, A.A.; Kasimov, R.; McDonald, T.J.; Wickliffe, J.K.; Presley, B.J.; Bickham, J.W. Chemical contaminants and their effects in fish and wildlife from the industrial zone of Sumgayit, Republic of Azerbaijan. *Ecotoxicology*, (2003) 12, 509-521.

[40] Burger, J. Metals in tissues of diamondback terrapin from New Jersey. *Environmental Monitoring and Assessment*, (2002) 77, 255-263.

[41] Schneider, L.; Belger, L.; Burger, J.; Vogt, R.; Ferrara, C. Mercury levels in muscle of six species of turtles eaten by people along the Rio Negro of the Amazon Basin. *Archives of Environmental Contamination and Toxicology*, (2010) 58, 444-450.

[42] Hopkins, W. Use of tissue residues in reptile ecotoxicology: a call for integration and experimentalism. In: *Toxicology of Reptiles*, Gardner, S.; Oberdorster, E. Eds.; CRC Press: Boca Raton, FL, (2006).

[43] Vogt, R. *Amazon turtles*, Grafica Bıblos: Lima, (2008).

[44] Teran, A.F.; Vogt, R.C..; Thorbjarnarson, J. Patterns of use and hunting of turtles in the Mamirauá Sustainable Development Reserve, Amazonas, Brazil. In: *People in*

Nature: Wildlife Conservation in South and Central America. Silvius K., Bodmer R.E., Fragoso J.M.V. Eds.; Columbia University Press: New York, (2004), pp. 363-377.

[45] Bates, H. *The Naturalist on the River Amazon*. Murray: London, (1892).

[46] Pritchard, P.C.H. 1989. The Galapagos tortoises nomenclatural and survival status. *Chelonian Research Monographs*, (1996) 1, 1-85.

[47] VanDijk, P.P.; Stuart, B.L.; Rhodin, A.G.J. Asian turtle trade: proceedings of a workshop on conservation and trade of freshwater turtles and tortoises in Asia. *Chelonian Research Mongraphs*, (2000) 2, 164.

[48] Schneider, L.; Ferrara, C. R.; Vogt, R. C.; Burger, J. History of turtle exploitation and management techniques to conserve turtles in the Rio Negro Basin of the Brazilian Amazon. *Chelonian Conservation and Biology*, (2011) 10, 149-157.

[49] Vogt, R.C.; Polilsar, J.; Moll, D.; Porter, G.P.G. *Dermatemys mawii* Gray 1847-Central American River Turtle, Tortuga Blanca, Hickatee. In: *Conservation Biology of Freshwater Turtles and Tortoises: A Compilation Project of the IUCN/SSC Tortoise and Freshwater Turtle Specialist Group*. Rhodin, A.G.J.; Pritchard, P.C.H.; van Dijk, P.P.; Saumure, R.A.; Buhlmann, K.A.; Iverson, J.B.; Mittermeier, R.A. Eds.; Chelonian Research Monographs, (2011) 5, pp. 058.1-058.2.

[50] Pritchard, P. *The alligator snapping turtle biology and conservation*. Milwaukee Public Museum: Milwauke, Wi, (1989).

[51] Orenstein, R. *Turtles, Tortoises and Terrapins survivals in armor*. Firefly Books: Buffalo, NY, (2001).

[52] Spotila, J. *Sea turtles a complete guide to their biology , behaviour , and conservation*. The John Hopkins University Press: Baltimore, MD, (2004).

[53] Mortimere, J.E.; Von Brandis, R.J.; Liljevik, A.; Chapman, R.; Collie, J. Fall and rise of nesting green turtles (*Chelonia mydas*) at Alsabra Atoll, Seychelles: positive response to four decades of protection (1968-2008). *Chelonian Conservation and Biology*, (2011) 10, 165-176.

[54] Heaton-Jones, T.G.; Homer, B.L.; Heaton-Jones, D.L.; Sundlof, S.F. Mercury distribution in American alligators (*Alligator mississippiensis*) in Florida. *Journal of Zoo and Wildlife Medicine*, (1997) 28, 62-70.

[55] Da Silveira, R., Avaliação preliminar da distribuição, abundância e da caça de jacarés no Baixo Rio Purus. In: *Piagaçu-Purus: Base Cientifica para a Criação de uma Reserva de Desenvolvimento Sustentável*. Deus, C.; Da Silveira, R.D.; Py-Daniel, L. Eds.; Instituto de Desenvolvimento Sustentável Mamirauá, Manaus, AM, (2003).

[56] Marioni, B.; Mhülen, E.V.; Silveira, R.D. Illegal caiman hunting in the Sustainable Development Reserve Piagaçu-Purus, Brazilian Amazonia. 18th *Working Meeting of the IUCN-SSC Crocodile Specialist Group*, (2006).

[57] VanDijk, P.; Stuart, B.; Rhodin, A. *Asian Turtle Trade: Proceedings of a Workshop on Conservation and Trade of Freshwater Turtles and Tortoises in Asia* (2000) 2, 164.

[58] WWF. *The State of Wildlife Trade in China. Information on the Trade in Wild Animalsand Plants in China* 2007. Beijing, China, (2007).

[59] Turtle Conservation Coalition. *Turtles in Trouble: The World's 25+ Most Endangered Tortoises and Freshwater Turtles—2011*. Rhodin, A.; Walde, A.; Horne, B.; VanDijk, P. v.; Blanck, T.; Hudson, R. Eds.; IUCN/SSC Tortoise and Freshwater Turtle Specialist Group, Turtle Conservation Fund, Turtle Survival Alliance, Turtle

Conservancy, Chelonian Research Foundation, Conservation International, Wildlife Conservation Society, San Diego Zoo Global: Lunenburg, MA, (2011).

[60] Lowe, H. The globalization of the turtle trade. In: *Turtle Survival Alliance* (2009).

[61] Miskimmin, B.M.; Rudd, J.W.M.; Kelly, C.A. Influence of dissolved organic carbon, pH and microbial respiration rates on Hg methylation and demethylation in lake water. *Canadian Journal of Fishery Aquatic Science*, (1992) 49, 17-22.

[62] US Environmental Protection Agency. *Mercury Study Report to Congress.* EPA-452/R-97-003 (1997).

[63] Power, M.; Klein, G.M.; Guiguer, K.R.R.A.; Kwan, M.K.H. Mercury accumulation in the fish community of a sub-artic lake in relation to trophic position and carbon sources. *Journal of Applied Ecology*, (2002) 39, 819-830.

[64] Meili, M. Mercury in lakes and rivers. In: *Metal Ions in Biological Systems*. Siegel, A.; Siegel, H. Eds.; Marcel Dekker Inc.: New York, (1997).

[65] Rask, M.; Metsala, R. Hg concentrations in northern pike, *Esox lucius* L., in small lakes of Evo area, Southern Finland. *Water, Air and Soil Pollution*, (1991) 56, 369-378.

[66] Montgomery, S.; Lucotte, M.; Pichet, P.; Mucci, A. Total dissolved mercury in the water column of several natural and artificial aquatic systems of Northern Quebec (Canada). *Canadian Journal of Fisheries and Aquatic Sciences*, (1995) 52, 2483-2492.

[67] Bisinoti, M.; Sargentini-Junior, E.; Jardim, W. Seasonal behavior of mercury species in waters and sediments from the Negro River Basin, Amazon, Brazil. *Journal of Brazilian Chemical Society* (2007) 18, 544-553.

[68] Seller, P.; Kelly, C.A.; Rudd, J.W.M.; MacHutchon, A.R. Photodegradation of methylmercury in lakes. *Nature*, (1996) 380, 694-697.

[69] Serudo, R.L.; de Oliveira, L.C.; Rocha, J.C.; Paterlini, W.C.; Rosa, A.H.; da Silva, H.C.; Botero, W.G. Reduction capability of soil humic substances from the Rio Negro Basin, Brazil, towards Hg (II) studied by a multimethod approach and principal component analysis (PCA). *Geoderma*, (2007) 138, 229-236.

[70] Bodaly, R.A.; Hecky, R.E.; Fudge, R.J. Increases in fish Hg levels in lakes flooded by the Churchill River diversion, Northern Manitoba. *Canadian Journal of Fisheries and Aquatic Sciences*, (1984) 41, 682-681.

[71] Snodgrass, W.J.; Jagoe, C.H.; Bryan, A.L.; Brant, H.A.; Burger, J. Effects of trophic status and wetland morphology, hydroperiod, and water chemistry on Hg concentrations in fish. *Canadian Journal of Fishery Aquatic Science*, (2000) 57, 171-180.

[72] Watras, C. J.; Back, R.C.; Halvorsen, S.; Hudson, R. J. M.; Morrison, K. A.; Wente, S.P. Vioaccumulation of mercury in pelagic freshwater food webs. *Science of the Total Environment*, (1998) 219, 183-208.

[73] Sorensen, J.A.; Glass, G.E.; Schmidt, K.W.; Huber, J.K.; Rapp, J. G. Airborn Hg deposition and watershed chracteristics in relation to Hg concentrations in water, sediments, plankton and fish of eighty Northern Minnesota lakes. *Environmental Science and Technology*, (1990) 24, 1716-1727.

[74] Lodenius, M. The effects of peatland drainage on the Hg contends of fish. *Suo*, (1983) 34, 21-24.

[75] Lodenius, M.; Seppänen, A.; Autio, S. Sorption of Hg in soils with different humus content. *Bulletin of Environmental Contamination and Toxicology*, (1987) 39, 593-600.

[76] Harris, R.; Snodgrass, W. Bioenergetic simulations of mercury uptake and retention in walleye (*Stizostedion vitreon*) and yellow perch (*Perca flavescens*). *Pollution Research Journal of Canada*, (1993) 28, 217-236.
[77] Mauro, J.B.N.; Guimarães, J.R.D.; Melamed, R. Mercury methylation in a tropical macrophyte: influence of abiotic parameters. *Applied Organometallic Geochemistry*, (1999) 13, 631-636.
[78] Guimarães, J.R.D.; Meili, M.; Malm, O.; Brito, E.M.S. Hg methylation and floating meadows of a tropical lake in the pantanal floodplain, Brazil. *Science of the Total Environment*, (1998) 213, 165-175.
[79] Guimarães, J.R.D.; Meili, M.; Hylander, L.D.; de Castro e Silva, E.; Roulet, M.; Mauro, J.B.N.; de Lemos, R.A. Mercury net methylation in five tropical floodplain regions of Brazil: high in the root zone of floating macrophyte mats but low in surface sediments and flooded soils. *Science of the Total Environment*, (2000) 261, 99-107.
[80] Compeau, G.; Bartha, R. Methylation and demethylation of mercury under controlled redox, pH and salinity conditions. *Applied Environmental Microbiology*, (1984) 48, 1203-1207.
[81] Olson, B.H.; Cooper, R.C. Comparison of aerobic and anaerobic methylation of mercury chloride by San Francisco Bay sediments. *Water Research*, (1976) 10, 113-116.
[82] Regnell, O.; Tunlid, A. Laboratory study of chemical speciation of mercury in lake sediment and water under aerobic and anaerobic conditions. *Applied Environmental Microbiology*, (1991) 57, 789-794.
[83] Berman, M.; Bartha, R. Levels of chemical versus biological methylation of Hg in sediments. *Bulletin of Environmental Contamination and Toxicology*, (1986) 36, 401-404.
[84] Regnell, O. The effect of pH and dissolved oxygen levels on methylation and partitioning of mercury in freshwater model systems. *Environmental Pollution*, (1994) 84, 7-13.
[85] Bodaly, R.A.; St.-Louis, V.L.; Paterson, M.J.; Fudge, R.J.P.; Hall, B.D.; Rosenberg, D.M. Metal ions in biological systems. In: *Hg and its effects on environmental and biology*. Sigel; A, S. H. Eds.; Marcekl Dekker (1997). pp 259.
[86] Roulet, M.; Guimarães, J.R.D.; Lucotte, M. Methylmercury production and accumulation in sediments and soils of an Amazonian floodplain: effect of seasonal inundation. *Water, Air, and Soil Pollution*, (2001) 128, 41-61.
[87] Branfireun, B.A.; Heyes, A.; Roulet, N.T. The hydrology and methylmercury dynamics of a Precambrian shield headwater peatland. *Water Resources Research*, (1996) 32, 1785-1994.
[88] Regnell, O. The effect of pH and dissolved oxygen levels on methylation and partitioning of mercury in freshwater model systems. *Environmental Pollution*, (1994) 84, 7-14.
[89] Rudd, J.M.W. Sources of Hg to freshwater ecossystems: a review. *Water, Air and Soil Pollution* (1995) 80, 697-720.
[90] Roulet, M.; Lucotte, M.; Saint-Aubin, A.; Tran, S.; Rhéault, I.; Farella, N.; de-Jesus-da-Silva, E.; Dezencourt, J.; Sousa-Passos, C.; Guimarães, J.; Mergler, D.; Amorin, M. The geochemistry of Hg in Central Amazonian soil developed on the Alter do Chão

formation of the lower Tapajós River Valley, Pará state, Brazil. *Science of the Total Environment* (1998) 223, 1-24.

[91] Zeidemann, V.K. A geoquímica do mercúrio em solos da bacia do Rio Negro e sua influência no ciclo regional do mercúrio. *Ecologia*, (1998) 75.

[92] Roulet, M.; Lucotte, M.; Farella, M.; Serique, G.; Coelho, H.; Souza-Passos, C.J.; de-Jesus-da-Silva, E.; Scavone de Andrade, P.; Mergler, D.; Guimarães, J.R.D.; Amorin, M.T. Effects of recent human colonization on the presence of Hg in Amazonian ecossytems. *Water, Air and Soil Pollution*, (1999) 112, 297-312.

[93] Dorea, J. G.; Barbosa, A. C.; Silva, G. S. Fish mercury bioaccumulation as a function of feeding behavior and hydrological cycles of the Rio Negro, Amazon. *Comparative Biochemistry and Physiology Part C: Toxicology Pharmacology*, (2006) 142, 275-283.

[94] Bergeron, C. M. *Individual and interactive effects of maternally- and trophically-derived mercury on early amphibian development.* PhD Thesis. Virginia State University Blacksburg, (2011), pp 180.

[95] Chumchal, M.M.; Rainwater, T.R.; Osborn, S.; Roberts, A.M.T.; Cobb, G.P.; Smith, P.N.; Baileyk, F.C. Mercury speciation and biomagnification in the food web of Caddo Lake, Texas and Lousiana, USA, a subtropical freshwater ecossytem. *Environmental Toxicology and Chemistry*, (2011) 30, 1153-1165.

[96] Goulding, M.; Carvalho, L.; Ferreira, E. Rio Negro: rich life in poor water. SPB Academic Publishing: The Hague, (1998).

[97] Meylan, A. Spongivory in hawksbill turtles: a diet of glass. *Science*, (1988) 239, 393-395.

[98] AMAP. Assessment report: Arctic pollution Issues. In: *Arctic monitoring and assessment program (AMAP)*: Oslo, Norway, (1998); pp. 859.

[99] Vogt, R.C.; Sever, G.R.; Moreira, G. Esophageal papillae in Pelomedusid turtles. *Journal of Herpetology*, (1998) 32, 279-281.

[100] Day, R.D.; Christopher, S.J.; Becker, P.R.; Whitaker, D.W. Monitoring mercury in loggerhead sea turtle, *Caretta caretta*. *Environmental Science and Technology*, (2005) 39, 437-346.

[101] Sakai, H.; Saeki, K.; Ichihashi, H.; Suganuma, H.; Tanabe, S.; Tatsukawa, R. Species-specific distribution of heavy metals in tissues and organs of loggerhead turtle (*Caretta caretta*) and green turtle (*Chelonia midas*) from Japanese coastal waters. *Marine Pollution Bulletin*, (2000) 40, 701-709.

[102] Jagoe, C. H.; Brant, H. A.; Kelsey-Wall, A.; Glenn, T. C.; Strickland, D. C.; Hauswaldt, S.; M.Sepulveda; Arnold, B.; Gross, T. Maternal transfer of mercury in fish and reptiles: concentrations in eggs versus somatic tissues. *24th Anual Meeting in North America of The Society of Environmental Toxicology and Chemistry*, SETAC (2003).

[103] Pfennig, D.W.; AMabry, A.; Orange, D. Environmental causes of correlations between age and size at metamorphosis in *Scaphiopus multiplicatus*. *Ecology*, (1991) 72, 2240-2248.

[104] Kiesecker, J.M; Blaustein, A.R. Effects of introduced bullfrogs and smallmouth bass on microhabitat use, growth and survival of native red-legged frogs (*Rana aurora*). *Conservation Biology*, (1998) 12, 776-787.

[105] Zangerl, R. The turtle shell. In: *Biology of Reptilia*. Bellairs, A.; Parsons, T. Eds.; New York: Academic Press: New York, (1969) 1, 311-339.

[106] Solomon, S.E.; Hendrickson, J.R.; Hendrickson, L.P.J. The structure of the carapace and plastron of juvenile turtles, *Chelonia mydas* (the green turtle) and *Caretta caretta* (loggerhead turtle). *Journal of Anatomy*, (1986) 145, 123-131.

[107] Burger, J.; Gochfeld, M. Effects of chemicals and pollution on seabirds. In; *Biology of marine birds.* Schreiber, E.; Burger, J. Eds.; CRC Press: Boca Raton, FL, (2001); pp. 485-526.

[108] Day, R. D.; Segars, A. L.; Arendt, M. D.; Lee, A. M.; Peden-Adams, M. M. Relationship of blood mercury levels to health parameters in the loggerhead sea turtle (*Caretta caretta*). *Environmental Health Perspectives*, (2007) 115, 1421-1428.

[109] Schneider, L.; Belger, L.; Burger, J.; Vogt, R. C.; Jeitner, C.; Peleja, J. R. P. Assessment of non-invasive techniques for monitoring mercury concentrations in species of Amazon turtle. *Toxicological and Environmental Chemistry*, (2011) 93, 238-250.

[110] Hall, B. D.; Bodaly, R. A.; Fudge, R. J. P.; Rudd, J. W. M.; Rosenberg, D. M. Food as the dominant pathway of methylmercury uptake by fish. *Water, Air, and Soil Pollution*, (1997) 100, 13-24.

[111] Tsui, M.T.K.; Wang, W. Uptake and elimination routes of inorganic mercury and methylmercury in *Paphnia magna*. *Environmental Science and Technology*, (2003) 38, 808-816.

[112] Barbosa, A. C.; Souza, J. d.; Dórea, J. G.; Jardim, W. F.; Fadini, P. S. Mercury biomagnification in a tropical black water, Rio Negro, Brazil. *Archives of Environmental Contamination and Toxicology*, (2003) 45, 235-246.

[113] Thomé-Souza, M.; Forsberg, B.R.; Vogt, R.C.; Sabbayrolles, M.G.P.; Peleja, R. The use of mercury and stable isotopes to investigate foodchain structure in aquatic ecossystems in the Amazon. VI International Congress on the Biology of Fish, Fish Communities and Fisheries: Manaus, AM, (2004); pp 235-239.

[114] Chumchal, M. M.; Hambright, K. D. Ecological factors regulating mercury contamination of fish from Caddo Lake, Texas, USA. *Environmental Toxicology and Chemistry*, (2009) 28, 962-972.

[115] Almonacid, J.; Carr, J.; Mittermeier, R.; Rodriquez Malecha, J.; Mast, R.; Vogt, R.; Rhodin, A.; Velasquez, J.; Rueda, J.; Mittermeier, C. Las tortugas y los cocodrilianos de los paises andinos del tropico. Conservacion Internacional: Bogata, DC, (2007).

[116] Vogt, R.; Platt, S.; Rainwater, T. *Rhinoclemmys areolata* (Duméril and Bibron 1851) - furrowed wood turtle, black-bellied turtle, mojina. Chelonian Research Monographs: (2009) 5, 022.1-022.7.

[117] McKay, G.; Cooke, F.; Hutchinson, S.; Vogt, R.C.; Dingle, H. The Encyclopedia of Animals: A Complete Visual Guide. University of California Press: Berkeley, CA, (2004).

[118] Vogt, R.; Villarreal-Benitez, J., Species abundance and biomass distributions in freshwater turtles, In: *Conservation restoration and management of tortoises and turtles-An international symposium.* New York Turtle and Tortoise Society: New York (1997).

[119] Vogt, R.; Villela, O. Aspectos de la ecología de la tortuga blanca *Dermatemys mawii* en La Reserva Montes Azules. In: *Reserva de la Biosfera Montes Azules: investigacion para su conservación.* Vázquez-Sánchez; M.A. Ramos. Eds.; San Cristoval de las casas, Chiapas, México, (1992), pp. 221-231.

[120] Bjorndal, K.A.; Bolten, A.B. Digestive processing in a herbivorous freshwater turtle: consequences of small intestine fermentation. *Physiological Zoology*, (1990) 63, 1232-1267.

[121] Legler, J. Morphology and physiology of the Chelonia. In: *Fauna of Australia*, Christopher, J.; Glasby, G.; Ross, J.; Beesley, P. Eds.; Australian Government Publishing Service: Canberra, (1993) Vol. 2A, Amphibia and Reptilia, pp. 108-119.

[122] Vogt, R. *Staurotypus triporcatus.* In: *Historia Natural de Los Tuxtlas*, Gonzalez, E.; Dirzo, R.; Vogt, R. Eds.; UNAM: Mexico City, (1997); pp. 494-495.

[123] Fachin-Terán, A.; Vogt, R. Food habits of an assembrage of five species of turtles in the rio Guaporé, Rondônia, Brazil. *Journal of Herpetology*, (1995) 29, 536-547.

[124] Vogt, R. *Claudius angustatus* (Chopontil, Taiman, Joloque). In: *Historia Natural de Los Tuxtlas.* Gonzàlez, E.; Dirzo, R.; Vogt, R.C. Eds.; UNAM: Mexico City, (1997), pp. 480.

[125] Redmond, A. Observations on the alligator snapping turtle. *Bulletin of the Georgia Herpetological Society*, (1979) 5, 5-7.

[126] Verdeja, S.; Vogt, R. *Kinosternon leucostomum* (Pochitoque). In: *Historia Natural de Los Tuxtlas.* Gonzàlez, E.; Dirzo, R.; Vogt, R.C. Eds.; UNAM: Mexico City, (1997), pp 488-490.

[127] Villarreal-Benitez, J.L.V.; Vogt, R. *Iguana iguana* (iguana verde, garrobo). In: *Historia Natural de Los Tuxtlas*. Gonzàlez, E.; Dirzo, R.; Vogt, R.C. Eds.; UNAM: Mexico City, (1997), pp 486-488.

[128] Troyer, K. Transfer of fermentation microbes between generations in a herbivorous lizard. *Science*, (1982) 216, 240-542.

[129] Soldin, O.; O'Mara, D.; Aschner, M. Thyroid hormones and methylmercury toxicity. *Biological Trace Element Research*, (2008) 126, 1-12.

[130] Wiener, J.G.; Spry, D.J. Toxicological significance of mercury in freshwater fish. In: *Environmental Contaminants in Wildlife: Interpreting Tissue Concentrations.* Beyer, W.; Heinz, G.; Redmon-Norwood, A. Eds.; Lewis Publisher: Boca Raton, FL, (1996), pp. 297-339.

[131] Oliveira Ribeiro, C. A.; Rouleau, C.; Pelletier, É.; Audet, C.; Tjälve, H. Distribution kinetics of dietary methylmercury in the Artic Charr (*Salvelinus alpinus*). *Environmental Science and Technology*, (1999) 33, 902-907.

[132] Pickhardt, P.C.; Fisher, N.S. Accumulation of inorganic and methylmercury by freshwater phytoplankton in two contrasting water bodies. *Environmental Science and Technology*, (2006) 41, 125-131.

[133] Wiener, J.G.; Krabbenhoft, D.P.; Heinz, G.H.; Scheuhammer, A.M. Ecotoxicology of Hg. In: *Handbook of Ecotoxicology.* Hoffman, D.; Rattner, B.; Burton, G. J.; Cairns, J. Eds.; Boca Raton, FL, (2003), pp. 409-464.

[134] Bridges, C. C.; Zalups, R. K. Molecular and ionic mimicry and the transport of toxic metals. *Toxicology and Applied Pharmacology*, (2005) 204, 274-308.

[135] Khan, M.A.K.; Wang, F. Mercury-selenium compounds and their toxicological significance: toward a molecular understanding of the mercury-selenium antagonism. *Environmental Toxicology and Chemistry*, (2009) 28, 1567-1577.

[136] Gonzalez, P.; Dominique, Y.; Massabuau, J. C.; Boudou, A.; Bourdineaud, J. P. Comparative effects of dietary methylmercury on gene expression in liver, skeletal

muscle and brain of zebrafish (*Danio rerio*). *Environmental Science and Technology*, (2005) 39, 3972-3980.

[137] Scheuhammer, A.M.; Basu, N.; Burgess, N.M.; Elliott, J.E.; Campbell, G.D.; Wayland, M.; Champoux, L.; Rodrigue, J. Relationship among Hg, selenium and neurochemical parameters in common loons (*Gavia immer*) and bald eagles (*Haliaeetus leucocephalus*). *Ecotoxicology*, (2008) 17, 93-101.

[138] Wiener, J.G.; Bodaly, R.A.; Brown, S.S.; Lucotte, M.; Newman, M.C.; Porcella, D.B.; Reash, R.J.; Swain, E.B. Monitoring and evaluating trends in methylHg accumulation in aquatic biota. In: *Ecosystem Responses to Hg Contamination.* Harris, R.; Krabbenhoft, D.; Mason, R.; Murray, M.; Reash, R.; Saltman, T. Eds.; SETAC: Pensacola, FL, (2007), pp. 87-122.

[139] Wolfe, M.; Atkenson, T.; Bowerman, W.; Burger, J.; Evers, D.; Murray, M.; Zillioux, E. Wildlife indicators. In: *Ecosystem Responses to Hg Contamination.* Harris, R.; Krabbenhoft, D.; Mason, R.; Murray, M.; Reash, R.; Saltman, T. Eds.; SETAC: Pensacola, FL, (2007); pp. 123-189.

[140] Drevnick, P.E.; Roberts, A.P.; Otter, R.R.; Hammerschmidt, C.R.; Klaper, R.; Oris, J.T. Hg toxicity in livers of northern pike (*Esox lucius*) from Isle Royale, USA. *Comparative Biochemistry and Physiology Part C: Toxicology and Pharmacology*, (2008) 147, 331-337.

[141] Grillitsch, B.; Schiesari, L. The ecotoxicology of metals in reptiles. In: *Ecotoxicology of Amphibians and Reptiles*, Sparling, D.; Linder, G.; Bishop, C.; Krest, S. Eds.; CRC Press: Boca Raton, FL, (2010).

[142] Jones, D.E.; Holladay, S.D. Excretion of three metals in the exuviae skin of exposed corn snakes (*Elaphe guttata*). *Ecotoxicology and Environmental Safety*, (2006) 64, 221-225.

[143] Wings, O. A review of gastrolith function with implications for fossil vertebrates and a revised classification. *Acta Palaeontologica Polonica*, (2007) 52, 1-16.

[144] Xu, Q.; Fang, S.; Wang, Z.; Wang, Z. Heavy metal distribution in tissues and eggs of Chinese alligator (*Alligator sinensis*). *Archives of Environmental Contamination and Toxicology*, (2006) 50, 580-586.

[145] Ultsch, G. Ecology and physiology of hibernation and overwintering among freshwater fishes, turtles and snakes. *Biological Reviews*, (1989) 64, 435-515.

[146] Nordberg, G.; Fowler, B.; Nordberg, M.; Friberg, L. Handbook on the toxicology of metals. Academic Press/ Elsevier: San Diego, CA, (2007).

[147] Sorensen, E. Metal Poisoning in Fish. CRC Press: Boca Raton, FL, (1991).

[148] Alibardi, L.; Mattia, T. Immunolocalization and characterization of beta-keratins in growing epidermis of chelonians. *Tissue Cell*, (2006) 38, 53-63.

[149] Vermeer, K.; Riseborough, R.W.; Spanns, A.L.; Reynolds, L.M. Pesticides effects on fishes and birds in rice fields of Surinam, South America. *Environmental Pollution*, (1974) 7, 217-236.

[150] Siegel, F.R. *Environmental Geochemistry of Pottentially Toxic Metals.* Springer: Berlin, (2002).

[151] McClellan-Green, P.; Celander, M.; Oberdoester, E. Hepatic, renal, and adrenal toxicology. In: *Toxicology of Reptiles: New Perspectives: toxicology and environment.* Gardner, S.; Oberdoester, E. Eds.; Taylor and Francis: Boca Raton, FL, (2006), pp. 123-148.

[152] Specker, J.; Sullivan, C. Vitellogenesis in fishes: status and perspectives, In: *Perspectives in Comparative Endocrinology*. Davey, K.; Peter, R.; Tobe, S. Eds.; NRC: Toronto, (1994).

[153] Bishop, C.A.; Gendron, A.D. Reptiles and amphibians: shy and sensitive vertebrates of the Great Lakes Basin and St. Lawrence River. *Environmental Monitoring and Assessment*, (1998) 53, 225-244.

[154] Burger, J.; Campbell, K. R.; Campbell, T. S.; Shukla, T.; Jeitner, C.; Gochfeld, M. Use of blood as nonlethal indicators of heavy metal contamination in northern water snakes (*Nerodia sipedon*). *Archives of Environmental Contamination and Toxicology*, (2005) 49, 232-238.

[155] Wolfe, M.F,; Schwarzbach, S.; Sulaiman, R.A. Effects of mercury on wildlife: a comprehensive review. *Environmental Toxicology and Chemistry*, (1998) 17, 146-160.

[156] Linder, G.; Grillitsch, B. Ecotoxicology of metals. In: *Ecotoxicology of Amphibians and Reptiles.* Sparling, D.; Linder, G.; Bishop, C. Eds.; Society of Environmental Toxicology and Chemistry Press: Pensacola, Florida, (2000), pp. 325-459.

[157] Heinz, G.H. Methylmercury: reproductive and behavioral effects on three generations of mallard ducks. *Journal of Wildlife Management*, (1979) 43, 394-401.

[158] Barr, J. Population dynamics of the Common Loon (*Gavia immer*) associated with mercury contaminated waters in northwestern Ontario. In: *Canadian Wildlife Service: Ottawa, Ontario, Canada*. (1986); Vol. Occasional Paper Number 56.

[159] Spalding, M.G.; Frederick, P.C.; McGill, H.; Bouton, S.N.; Richley, L.J.; Schumacher, I.M.; Blackmore, C.G.; Harrison, J. Histological, neurologic, and immunologic effects of methylmercury in captive great egrets. *Journal of Wildlife Diseases*, (2000) 36, 423-434.

[160] Friedmann, A.S.; Watzin, M.C.; Brinck-Johnsen, T.; Leiter, J.C. Low levels of dietary methylmercury inhibit growth and gonadal development in juvenile walleye (*Stizostedion vitreum*). *Aquatic Toxicology*, (1996) 35, 265-277.

[161] Hammerschmidt, C.R.; Sandheinrich, M.B.; Weiner, J.G.; Rada, R.G. Effects of dietary methylmercury on reproduction of fathead minnows. *Environmental Science and Technology*, (2002) 36, 877-923.

[162] Webber, H.M.; Haines, T.A. Mercury effects on predator avoidance behavior of a forage fish, golden shiner (*Notemigonus crysoleucas*). *Environmental Toxicology and Chemistry*, (2003) 22, 1556-1561.

[163] EPA, US. Special Report On Environmental Endocrine Disruption: An Effects Assessment and Analysis. In: *US Environmental Protection Agency* (1997); Vol. EPA/630/R-96/012.

[164] Kirubagaran, R.; Joy, K.P. Toxic effects of mercurials on thyroid function of the catfish, Clarias batracus. *Ecotoxicology and Environmental Safety*, (1989) 17, 265-271.

[165] Kirubagaran, R.; Joy, K. Changes in brain monoamine levels and monoamine oxidase activity in the catfish, *Clarias batrachus*. *Bulletin of Environmental Contamination and Toxicology*, (1990) 45, 88-90.

[166] Schneider, L.; Ferrara, C.; Vogt, R.C.; Guilhon, A.V. Nesting ecology and nesting predation of *Phrynops geoffroanus* (Testudines, Chelidae) in the Guapore River of the Brazilian and Bolivian Amazon. *Chelonian Conservation and Biology*, (2011) 10, 206-212.

[167] U.S. Environmental Protection Agency. Guidance for assessing chemical contaminant data for use in fish advisories. EPA 823-B-00-008. Volume 2: *Risk assessment and fish consumption limits* (2000).

[168] Ruckel, S. Mercury concentrations in alligator meat in georgia. *The Southeastern Association of Fish and Wildlife Agencies (SEAFWA) Annual Conference*, (1993), 47, pp. 287-292.

[169] Yanochko, G. M.; Jagoe, C. H.; Brisbin Jr, I. L. Tissue mercury concentrations in alligator (*Alligator mississippiensis*) from the Florida Everglades and the Savannah River Site, South Carolina. *Archives of Environmental Contamination and Toxicology*, (1997) 32, 323-328.

[170] Jagoe, C.H.; Arnold-Hill, U.B.; Yanochko, G.M.; Winger, P.V.; Brisbin, I.L.J. Mercury in alligators (*Alligator mississippiensis*) in the southeastern United States. *The Science of the Total Environment*, (1998) 213, 255-262.

[171] Rumbold, D. G.; Fink, L. E.; Laine, K. A.; Niemczyk, S. L.; Chandrasekhar, T.; Wankel, S. D.; Kendall, C. Levels of mercury in alligators (*Alligator mississippiensis*) collected along a transect through the Florida Everglades. *Science of the Total Environment*, (2002) 297, 239-252.

[172] Green, A.D.; Buhlmann, K.A.; Hagen, C.; Romanek, C.; Gibbons, J.W. Mercury contamination in turtles and implications for human health. *Journal of Environmental Health*, (2010) 72, 14-22.

[173] Jedrychowski, W.; Jankowski, J.; Flak, E.; Skarupa, A.; Mroz, E.; Sochacka–Tatara, E.; Lisowska–Miszczyk, I.; Szpanowska–Wohn, A.; Rauh, V.; Skolicki, Z.; Kaim, I.; Perera, F. Effects of prenatal exposure to mercury on cognitive and psychomotos function in one-year-old-infants: epidemiologic cohort study in Poland. *Annals of Epidemiology*, (2005) 16, 439-446.

[174] Kjellstrom, T.; Kennedy, P.; Wallis, S.; Mantell, C. Physical and mental development of children with prenatal exposure to mercury from fish: stage 1: preliminary tests at age 4. Solna, Sweden. *National Swedish Environmental Protection Board*: (1986).

[175] Grump, K.S.; Kjellström, T.; Shipp, A.M.; Silvers, A.; Stewart, A. Influence of prenatal mercury exposure upon scholastic and psychologia test performance: benchmark analysis of a New Zealand cohort. *Risk Analysis*, (1998) 18, 701-713.

[176] Grandjean, P.; Weihe, P.; White, R.; Debes, F.; Araki, S.; Yokoyama, K.; Murata, K.; Sorensen, N.; Dahl, R.; Jorgensen, P. Cognitive deficit in 7-year old children with prenatal exposure to methylmercury. *Neurotoxicology and Teratology*, (1997) 19, 417-428.

[177] Myers, G.J.; Davidson, P.W.; Cox, C.; Shamlaye, C.F.; Tanner, M.A.; Marsh, D.O.; Cernichiari, E.; Lapham, L.W.; Berlin, M.; Clarkson, T.W. Summary of seychelles child development study on the relationship of fetal methylmercury exposure to neurodevelopment. *Neurotoxicology and Teratology*, (1995) 16, 711-716.

[178] Davidson, P.; Myers, G.J.; Cox, C.; Axtells, C.; Shamlaye, C.; Sloane-Reeves, J.; Cernichiari, E.; Needham, L.; Choi, A.; Wang, Y.; Berlin, M.; Clarkson, T.W. Effects of prenatal and postnatal methylmercury exposure from fish consumption on neurodevelopment: outcomes at 66 months of age in the seychelles child development study. *Journal of the American Medical Association*, (1998) 280, 701-707.

[179] Daniels, J.L.; Longnecker, M.; Rowland, A.S.; Golding, J. Fish intake during pregnanxy and early cognitive development of offspring. *Epidemiology*, (2004) 15, 394-402.

[180] Lebel, J.; Mergler, D.; Branches, F.; Lucotte, M.; Amorim, M.; Larribe, F.; Dolbec, J. Neurotoxic effects of low-level methylmercury contamination in the Amazonian Basin. *Environmental Research*, (1998) Section A 79, 20-32.

[181] Dolbec, J.; Mergler, D.; Sousa Passos, C.J.; Sousa de Morais, S.; Lebel, J. Methylmercury exposure affects motor performance of a riverine population of the Tapajos River, Brazilian Amazon. *International Archives of Occupational and Environmental Health*, (2000) 173, 195-203.

[182] Yokoo, E.M.; Valente, J.G.; Grattan, L.; Schmidt, S.L.; Platt, I.; Silbergeld, E.K. Low level methylmercury exposure affects neuropsychological function in adults. *Environmental Health Perspectives*, (2003) 2, 8-13.

[183] Salonen, J.K.; Seppänen, K.; Nyyssönen, K.; Korpela, H.; Kauhanen, J.; Kantola, M.; Tuomilehto, J.; Esterbauer, H.; Tatzber, F.; Salonen, R. Intake of mercury from fish, lipid peroxidation, and the risk of myocardial infarctation and coronary, cardiovascular, and any death in Eastern Finnish Men. *Circulation*, (1995) 91, 645-655.

[184] Guallar, E.; Sanz-Gallardo, M.I.; Veer, P.V.; Bode, P.; Aro, A.; Gómez-Aracena, J.; Kark, J.D.; Riemersma, R.A.; Martín-Moreno, J.M.; Kok, F.J. Mercury, fish oils, and the risk of myocardial infarctation. *The New England Journal of Medicine*, (2002) 347, 1747-1755.

[185] Virtanen, J.K.; Voutilainen, S.; Rissanen, T.H.; Mursu, J.; Tuomainen, T.; Korhonen, M.J.; Valkonen, V.; Seppänen, K.; Laukkanen, J.A.; Salonen, J.T. Mercury, fish oils, and risk of acute coronary events and cardiovascular disease, coronary heart disease, and all-cause mortality in men in Eastern Finland. *Arteriosclerosis, Thrombosis, and Vascular Biology*, (2005) 25, 228-233.

[186] Sørensen, N.; Murata, K.; Budtz-Jørgensen, E.; Weihe, P.; Grandjean, P. Prenatal methylmercury exposure as a cardiovascular risk factor at seven years of age. *Epidemiology*, (1999) 10, 370-375.

[187] Blanvillain, G.; Schwenter, J. A.; Day, R. D.; Point, D.; Christopher, S. J.; Roumillat, W. A.; Owens, D. W. Diamondback terrapins, *Melaclemys terrapin*, as a sentinel species for monitoring mercury pollution of estuarine systems in South Carolina and Georgia, USA. *Environmental Toxicology and Chemistry*, (2007) 26, 1441-1450.

[188] Chaffin, K.; Norton, T. M.; Gilardi, K.; Poppenga, R.; Jensen, J. B.; Moler, P.; Cray, C.; Dierenfeld, E. S.; Chen, T.; Oliva, M.; Origgi, F. C.; Gibbs, S.; Mazzaro, L.; Mazet, J. Health assessment of free-ranging alligator snapping turtles (*Macrochelys temminckii*) in Georgia and Florida. *Journal of Wildlife Diseases*, (2008) 44, 670-686.

[189] Burger, J.; Gochfeld, M.; Rooney, A.; Orlando, E.; Woodward, A.; Guillette, L. J. Metals and metalloids in tissues of American alligators in three Florida lakes. *Archives of Environmental Contamination and Toxicology*, (2000) 38, 501-508.

[190] Turnquist, M. A.; Driscoll, C. T.; Schulz, K. L.; Schlaepfer, M. A. Mercury concentrations in snapping turtles (*Chelydra serpentina*) correlate with environmental and landscape characteristics. *Ecotoxicology,* (2011) 20, 1599-1608.

[191] Monteiro, L.R.; Furness, R.W. Accelerated increase in mercury contamination in north atlantic mesopelagic food chains as indicated by time series of seabird feathers. *Environmental Toxicology and Chemistry*, (1997) 16, 2489-2493.

[192] Bowles, K. C. *The cycling of mercury in Australasian aquatic systems.* Ph.D. thesis. University of Canberra Canberra, ACT, (1998), pp. 490.

[193] Schwartz, J.H. Heavy metal-induced alterations in ion transport by turtle urinary bladder. *American Journal of Physiology*, (1976) 230, 1582-1589.

[194] Bazar, M.; Holtzman, D.; Adair, B.; Gresens, S. Effectsof dietary methylmercury in juvenile corn snakes (*Elaphe guttata*). *SETAC 23rd Annual Meeting* (2002).

[195] Wobeser, G.; Nielsen, N.; Schiefer, B. Mercury and Mink. II. Experimental methyl mercury intoxication. *Canadian Journal of Comparative Medicine* (1976) 40, 34-45.

[196] de Solla, S. Organic contaminants in Reptiles. In: *Ecotoxicology of Amphibians and Reptiles.* Sparling, D.; Linder, G.; Bishop, C.; Krest, S. Eds.; CRC Press: Boca Raton, Fl, (2010), pp. 267-289.

[197] Schneider, L.; Belger, L.; Burger, J.; Vogt, R.C. Mercury bioaccumulation in four tissues of *Podocnemis erythrocephala* (Podocnemididae: Testudines) as a function of water parameters. *Science of the Total Environment*, (2009) 407, 1048-1054.

[198] Elsey, R.; Lance, V.; Campbell, L. Mercury levels in alligator meat in South Lousiana. *Bulletin of Environmental Contamination and Toxicology*, (1999) 63, 598-603.

[199] Delany, M.; Bell, J.; Sundlof, S. Concentrations of contaminants in muscle of the American alligator in Florida. *Journal of Wildlife Diseases*, (1988) 24, 62-66.

[200] [200] Vieira, L.M.; Nunes, V. d. S.; Amaral, M.C.A.; Oliveira, A.C.; Hauser-Davis, R. A.; Campos, R. C. Mercury and methylmercury rations in caimans (*Caiman crocodilus yacare*) from the Pantanal areas, Brazil. *Journal of Environmental Monitoring*, (2011) 13, 280-287.

[201] Jeffree, R. A.; Markich, S. J.; Twining, J. R. Element concentrations in the flesh and osteoderms of estuarine crocodiles (*Crocodylus porosus*) from the Alligator River region, Northern Australia. *Archives of Environmental Contamination and Toxicology*, (2001) 40, 236-245.

[202] Rainwater, T. R.; Wu, T. H.; Finger, A. G.; Cañas, J. E.; Yu, L.; Reynolds, K. D.; Coimbatore, G.; Barr, B.; Platt, S. G.; Cobb, G. P.; Anderson, T. A.; McMurry, S. T. Metals and organochlorine and pesticides in caudal scutes of crocodiles from Belize and Costa Rica. *Science of the Total Environment*, (2007) 373, 146-156.

[203] Almli, B.; Mwase, M.; Sivertsen, T.; M. Musonda, M.; Flåøyen, A. Hepatic and renal concentrations of 10 trace elements in crocodiles (*Crocodylus niloticus*) in the Kafue and Luangwa rivers of Zambia. *Science of the Total Environment*, (2005) 337, 75-82.

[204] Burger, J.; Campbell, K.; Campbell, T. Gender and spatial patterns in metal concentrations in brown anoles (*Anolis sagrei*) in Southern Florida, USA. *Environmental Toxicology and Chemistry*, (2004) 23, 712-718.

[205] Hsu, M. J.; Selvaraj, K.; Agoramoorthy, G. Taiwan´s industrial heavy metal pollution threatens terrestrial biota. *Environmental Pollution*, (2006) 143, 327-334.

[206] Marsili, L.; Casini, S.; Mori, G.; Ancora, S.; Bianchi, N.; D'Agostino, A.; Ferraro, M.; Fossi, M. C. The italian wall lizard (*Podarcis sicula*) as a bioindicator of oil field activity. *Science of the Total Environment*, (2009) 407, 3597-3604.

[207] [Marquez-Ferrando, R.; Santos, X.; Pleguezuelos, J. M.; Ontiveros, D. Bioaccumulation of heavy metals in the lizard *Psammodromus algirus* after a tailing-dam collapse in Aznalcollar (Southwest Spain). *Archives of Environmental Contamination and Toxicology,* (2009) 56, 276-285.

[208] Rainwater, T.; Adair, B.; Platt, S.; Anderson, T.; Cobb, G.; McMurry, S. Mercury in morelet´s crocodile eggs from Northern Belize. *Archives of Environmental Contamination and Toxicology,* (2002) 42, 319-324.

[209] Rainwater, T.; Reynolds, K.; Cañas, J.; Cobb, G.; Anderson, T.; McMurry, S.; Smith, P. Organochlorine pesticides and mercury in cotton mouths (*Agkistrodon piscivorus*) from Northeastern Texas, USA. *Environmental Toxicology and Chemistry*, (2005) 24, 665-672.

[210] Jones, D.; Holladay, S. Excretion of three heavy metals in the exuviae skin of exposed corn snakes (*Elaphe guttata*). *Ecotoxicology and Environmental Safety*, (2006) 64, 221-226.

[211] Burger, J.; Murray, S.; Gaines, K.; Novak, J.; Punshon, T.; Dixon, C.; Gochfeld, D. Element levels in snakes in South Carolina: differences between a control site and exposed site on the Savannah River site. *Environmental Monitoring and Assessment*, (2006) 112, 35-52.

[212] Campbell, K.; Campbell, T.; Burger, J. Heavy metal concentrations in northern water snakes (*Nerodia sipedon*) from East Fork Popular Creek and the Little River, East Tennessee, USA. *Archives of Environmental Contamination and Toxicology*, (2005) 49, 239-248.

[213] Heinz, G.; Haseltine, S.; Hall, R.; Krynitsky, A. Organochlorine and mercury residues in snakes from Pilot and Spider Islands, Lake Michigan -1978. *Bulletin of Environmental Contamination and Toxicology*, (1980) 25, 738-743.

[214] Burger, J.; Campbell, K. R.; Murray, S.; Campbell, T. S.; Gaines, K. F.; Jeitner, C.; Shukla, T.; Burke, S.; Gochfeld, M. Metal levels in blood, muscle and liver of water snakes (*Nerodia* spp.) from New Jersey, Tennessee and South Carolina. *Science of the Total Environment*, (2007) 373, 556-563.

[215] Burger, J.; Cooper, K.; Saliva, J.; Gochfeld, D.; Lipsky, D.; Gochfeld, M. Mercury bioaccumulation in organisms from three Puerto Rican estuaries. *Environmental Monitoring and Assessment*, (1992) 22, 181-197.

[216] Wylie, G.; Hothem, R.; Bergen, D.; Martin, L.; Taylor, R.; Brussee, B. Metals and trace elements in giante garter snakes (*Thamnophis gigas*) from the Sacramento Valley, California, USA. *Archives of Environmental Contamination and Toxicology*, (2009) 56, 577-587.

[217] Albrecht, J.; Abalos, M.; Rice, T. Heavy metals levels in ribbon snakes (*Thamnophis sauritus*) and anuran larvae from the Mobile-Tensaw River Delta, Alabama, USA. *Archives of Environmental Contamination and Toxicology*, (2007) 53, 647-654.

[218] Storelli, M.; Ceci, E.; Marcotrigiano, G. Comparison of total mercury, methylmercury, and selenium in muscle tissues and in the liver of *Stenella coeruleoalba* (Meyen) and *Caretta caretta* (Linnaeus). *Bulletin of Environmental Contamination and Toxicology and Applied Pharmacology*, (1998) 61, 541-547.

[219] Storelli, M.; Ceci, E.; Marcotrigiano, G. Distribution of heavy metals residues in some tissues of *Caretta caretta* (Linnaeus) specimen beached along the Adriatic Sea (Italy). *Bulletin of Environmental Contamination and Toxicology* (1998) 60, 546-552.

[220] Ley-Quinonez, C.; Zavala-Norzagaray, A.A.; Espinosa-Carreon, T.L.; Peckham, H.; Marquez-Herrera, C.; Campos-Villegas, L.; Aguirre, A.A. Baseline heavy metals and metalloid values in blood of loggerhead turtles (*Caretta caretta*) from Baja California Sur, Mexico. *Marine Pollution Bulletin*, (2011) 62, 1979-1983.

[221] Day, R.; Keller, J.; Harms, C.; Segars, S.; Cluse, W.; Godfrey, M.; Lee, A.; Peden-Adams, M.; Thorvalson, K.; Dodd, M.; Norton, T. Comparison of mercury burdens in chronically debilitated and healthy loggerhead sea turtles (*Caretta caretta*). *Journal of Wildlife Diseases*, (2010) 46, 111-117.

[222] Jerez, S.; Motas, M.; Cánovas, R. Á.; Talavera, J.; Almela, R. M.; del Río, A. B. Accumulation and tissue distribution of heavy metals and essential elements in loggerhead turtles (*Caretta caretta*) from the Spanish Mediterranean coastline of Murcia. *Chemosphere*, (2010) 78, 256-264.

[223] Deem, S.; Norton, T.; Mitchell, M.; Segars, A.; Alleman, A.; Cray, C.; Poppenga, R.; Dodd, M.; Karesh, W. Comparison of blood values in foranging, nesting and stranded loggerheads turtles (*Caretta caretta*) along the coast of Georgia, USA. *Journal Journal of Wildlife Diseases* (2009) 45, 41-56,

[224] Kampalath, R.; Gardner, S.; Mendez-Rodriguez, L.; Jay, J. Total and methylmercury species of sea turtles of Baja California Sur. *Marine Pollution Bulletin* (2006) 52, 1784–1832.

[225] Maffucci, F.; Caurant, F.; Bustamante, P.; Bentivegna, F. Trace elements (Cd, Cu, Hg, Se, Zn) accumulaion and tissue distribution in loggerhead turtles (*Caretta caretta*) from the Western Mediterranean Sea (Southern Italy). *Chemosphere*, (2005) 58, 535-542.

[226] Storelli, M.; Storelli, A.; D'Addabbo, R.; Marano, C.; Bruno, R.; Marcotrigiano, G. Trace elements in loggerhead turtles (*Caretta caretta*) from the eastern Mediterranean Sea: overview and evaluation. *Environmental Pollution* (2005) 135, 163-170.

[227] Godley, B.; Thompson, D.; Furness, R. Do heavy metal concentrations pose a threat to marine turtles from the Mediterranean Sea? *Marine Pollution Bulletin* (1999) 38, 497-502.

[228] Gordon, A.; Pople, A.; Ng, J. Trace metal concentration in livers and kidneys of sea turtles from south-eastern Queensland, Australia. *Marine and Freshwater Research* (1998) 49, 409-414.

[229] Sakai, H.; Ichihashi, H.; Suganuma, H.; Tatsukawa, R. Heavy metal monitoring in sea turtles using eggs. *Marine Pollution Bulletin* (1995) 30, 347-352.

[230] Komoroske, L.; Lewison, R.; Seminoff, J.; Deheyn, D.; Dutton, P. Pollutants and the health of green sea turtles resident to an urbanized estuary in San Diego, CA. *Chemosphere*, (2011) 84, 544-552.

[231] Anan, Y.; Kunito, T.; Watanabe, I.; Sakai, H.; Tanabe, S. Trace element accumulation in hawksbill turtles (*Eretmochelys imbricata*) and green turtles (*Chelonia mydas*) from Yaeyama Islands, Japan. *Environmental Toxicology and Chemistry*, (2001) 20, 2802-2814.

[232] Aguirre, A.; Balaz, J.; Zimmerman, B.; Galey, F. Organic contaminants and trace metals in the tissues of green turtles (*Chelonia mydas*) afflicted with fibropapillomas in the Hawaiian islands. *Marine Pollution Bulletin*, (1994) 28, 109-114.

[233] Turnquist, M. A.; Driscoll, C. T.; Schulz, K. L.; Schlaepfer, M. A. Mercury concentrations in snapping turtles (*Chelydra serpentina*) correlate with environmental and landscape characteristics. *Ecotoxicology* (2011) 20, 1599-1608.

[234] Meyers-Schone, L.; Shugart, L.; Beauchamp, J.; Walton, B. Comparison of two freshwater turtle species as monitors of a radionuclide and chemical contamination: DNA damage and residue analysis. *Environmental Toxicology and Chemistry* (1993) 12, 1487-1476.

[235] Helwig, D.; Hora, M. Polychlorinated biphenyl, mercury, and cadimium concentrations in Minnesota snapping turtles. *Bulletin of Environmental Contamination and Toxicology*, (1983) 30, 186-200.

[236] Golet, W.; Haines, T. Snapping turtles (*Chelydra serpentina*) as monitors for mercury contamination of aquatic environments. *Environmental Monitoring and Assessment*, (2001) 71, 211-220.

[237] Ashpole, S.; Bishop, C.; Brooks, R. Contaminant residues in snapping turtle (*Chelydra s. serpentina*) eggs from the Great Lakes-St. Lawrence River Basin (1999 to 2000). *Archives Environmental Contamination and Toxicology*, (2004) 47, 240-242.

[238] Bonin, J.; DesGranges, J. L.; Bishop, C. A.; Rodrigue, J.; Gendron, A.; Elliott, J. E. Comparative study of contaminants in the mudpuppy (Amphibia) and the common snapping turtle (Reptilia), St. Lawrence River, Canada. *Archives Environmental Contamination and Toxicology* (1995) 28, 184-194.

[239] Bergeron, C.; Husak, J.; Unrine, J.; Romanek, C.; Hopkins, W. Influences of feeding ecology on blood mercury concentrations in four species of turtles. *Environmental Toxicology and Chemistry*, (2007) 26, 1733-1740.

[240] Fu, J.; Wang, Y.; Zhou, Q.; Jiang, G. Trophic transfer of mercury and methylmercury in an aquatic ecosystem impacted by municipal sewage effluents in Beijing, China. *Journal of Environmental Sciences*, (2010) 22, 1189-1194.

[241] Henny, C.; Beal, K.; Bury, R.; Goggans, R. Organochlorine pesticides, PCBs, trace elements and metals in western pond turtle eggs from Oregon. *Northwest Science*, (2003) 77, 46-53.

[242] Guirlet, E.; Das, K.; Girondot, M. Maternal transfer of trace elements in leatherback turtles (*Dermochelys coriacea*) of French Guiana. *Aquatic Toxicology*, (2008) 88, 267-276.

[243] Davenport, J.; Wrench, J. Metal levels in a leatherback turtle. *Marine Pollution Bulletin* (1990) 21, 40-41.

[244] Perrault, J.; Wyneken, J.; Thompson, L.; Johnson, C.; Miller, D. Why are hatching and emergence succes low? Mercury and selenium concentrations in nesting leatherback sea turtles (*Dermochelys coriacea*) and their young in Florida. *Marine Pollution Bulletin* (2011) 62, 1671-1682.

[245] Harris, H.; Benson, S.; Gilardi, K.; Poppenga, R.; Work, T.; Dutton, P.; Mazet, J. Comparative health assessment of western Pacific leatherback turtles (*Dermochelys coriacea*) foraging off the coast of California, 2005-2007. *Journal of Wildlife Diseases*, (2011) 47, 321-337.

[246] Innis, C.; Tlusty, M.; Perkins, C.; Holladay, S.; Merigo, C.; Weber, S. Trace metal and organochorine pesticide concentrations in cold-stunned juvenile Kemp´s Ridley Turtles (*Lepidochelys kempii*) from Cape Cod, Massachusetts. *Chelonian Conservation and Biology*, (2008) 7 230-239.

[247] Páez-Osuna, F.; Calderón-Campuzano, M.; Soto-Jiménez, M.; Ruelas-Inzunza, J. Mercury in blood and eggs of the sea turtle (*Lepidochelys olivacea*) from a nesting colony in Oxaca, Mexico. *Marine Pollution Bulletin*, (2011) 3, 1320–1323.

[248] Chaffin, K.; Norton, T. M.; Gilardi, K.; Poppenga, R.; Jensen, J. B.; Moler, P.; Cray, C.; Dierenfeld, E. S.; Chen, T.; Oliva, M.; Origgi, F. C.; Gibbs, S.; Mazzaro, L.; Mazet, J. Health assessment of free-ranging alligator snapping turtles (*Macrochelys temminckii*) in Georgia and Florida. *Journal of Wildlife Diseases*, (2008) 44, 670-686.

[249] Blanvillain, G.; Schwenter, J. A.; Day, R. D.; Point, D.; Christopher, S. J.; Roumillat, W. A.; Owens, D. W. Diamondback terrapins, *Malaclemys terrapin*, as a sentinel species for monitoring mercury pollution of estuarine systems in South Carolina and Georgia, USA. *Environmental Toxicology and Chemistry*, (2007) 26, 1441-1450.

[250] Burger, J.; Gibbons, J. Trace elements in egg contents and eggshells of slider turtles (*Trachemys scripta*) from the Savannah River Site. *Archives of Environmental Contamination and Toxicology*, (1998) 34, 382-386.

[251] Schneider, L.; Peleja, R.P.; Kluczkovski, A.J.; Freire, G.M.; Marioni, B.; Vogt, R.C.; Da Silveira, R. Mercury concentration in the spetacled caiman and black caiman (Alligatoridae) of the Amazon: implications for human health. *Archives of Environmental Contamination and Toxicology,* (2012) 63, 270-279.

[252] Nagy, K. Field metabolic rate and body size. *The Journal of Experimental Biology,* (2005) 208, 1621-1625.

In: Mercury: Sources, Applications and Health Impacts
Editors: Ki-Hyun Kim and Richard J. C. Brown
ISBN: 978-1-62257-721-7

Chapter 10

ATMOSPHERIC DEPOSITION OF MERCURY FROM WAITES ISLAND, HORRY COUNTY, SOUTH CAROLINA, AS COMPARED TO REGIONAL MERCURY DEPOSITION AND SOUTH CAROLINA FISH CONSUMPTION ADVISORIES

Jane L. Guentzel*
Coastal Carolina University, Department of Marine Science, Conway, SC, US

ABSTRACT

Human health risks associated with the ingestion mercury (Hg) and the occurrence of high levels of mercury in fish from aquatic systems have resulted in the issuance of many fish consumption advisories throughout the state of South Carolina (SC). Such findings have created widespread interest in the harmful effects of Hg on wildlife and humans. Previous studies of Hg deposition and cycling in the environment have shown that wet and dry deposition from the atmosphere is the dominant source of Hg to many ecosystems. Therefore, the quantification of atmospheric Hg deposition to an ecosystem is necessary for studies designed to investigate the cycling and bioaccumulation of Hg within the ecosystem. This study quantified the atmospheric deposition of Hg in direct precipitation from an unobstructed salt marsh area and throughfall from an adjacent upland forested region at Waites Island, South Carolina from 2001-2003. There are no previous records of Hg deposition from this region of South Carolina. The concentration of total Hg in the marsh bulk deposition collectors ranged from 4.9 to-15.6 ng/L and the total Hg flux ranged from 0.195-2.45 μg/m^2/month for the time period ranging from July 2001-July 2003. The concentration of total Hg in the throughfall wet deposition samples ranged from 10.6-39.85 ng/L and the total Hg flux ranged from 0.23-4.0 μg/m^2/month for the time period ranging from July 2001-July 2003. The difference between the marsh and

* E-mail: jguentze@coastal.edu.

throughfall precipitation fluxes would represent the dry deposition to foliar surfaces that is "washed off" of the foliage or leached from the foliage during rainfall events. The estimated foliar deposition ranges from 0.035-0.84 μg/m^2/month. The significant correlation (R^2=0.7623; P=0.002) between total Hg and total organic carbon in event rainfall samples from Waites Island suggests that Hg in rainfall and throughfall may be associated with organic ligands. The Waites Island deposition data agrees well with several regional mercury deposition network (MDN) monitoring sites located in South Carolina (SC03, SC 19), North Carolina (NC08), Georgia (GA40), and Tennessee (TN11). The lack of an increasing west to east regional gradient in wet Hg deposition when compared the increasing west to east gradient in SC fish Hg concentrations suggests that while atmospheric deposition may be the source of Hg to SC ecosystems, other processes such as reservoir size, water chemistry, land use, hydrological modification, and proximity to wetlands contribute to the trend in fish Hg concentrations throughout South Carolina.

INTRODUCTION

Mercury is a metal that occurs naturally at low levels in the environment. The three most prevalent chemical fractions exist as gas phase elemental mercury (Hg^o), divalent inorganic mercury (Hg^{2+}), and methylmercury (CH_3Hg^+) which bioaccumulates in higher trophic levels and is most toxic to organisms near the top of the food chain. Sources of mercury to the environment are natural and anthropogenic. Natural sources of Hg include volcanic emissions, windblown dust from continental areas, and the emission of gaseous Hg from soils, vegetation, forest fires, and the ocean. Fossil fuel combustion, mining, smelting, and waste incineration are the major anthropogenic sources of mercury (Guentzel 2009; Guentzel et al., 2007; Friedli et al, 2003; NAS 2000; US EPA 1997). Atmospheric deposition of wet and dry species is the primary delivery mechanism of inorganic Hg to ecosystems. Mercury in wet and dry deposition originates from both local and global sources (Selin et al., 2008; Guentzel et al., 2001). It is estimated that North American anthropogenic Hg emissions contribute on average, 20% of the total Hg deposition in United States (US), with the other 80% originating from emissions outside North America (Selin et al., 2008). In highly industrialized regions within the US local sources may contribute up to 50% of the inorganic Hg deposition (Selin et al., 2008) while Hg from the global pool contributes more significantly to the deposition in remote non-industrialized regions (Selin et al., 2008; Guentzel et al., 2001). In addition to anthropogenic Hg emissions, mobilization of mercury from soils as a result of deforestation and storm events can be a source of inorganic and organic Hg to ecosystems (Friedli et al., 2003; Porvari et al., 2003; Munthe et al., 2007).

There are several processes that can affect inorganic Hg subsequent to its deposition to an ecosystem. These processes include reduction to gas phase Hg and release back to the atmosphere via volatilization, complexation with terrestrial soils and vegetation, complexation with aquatic organic/particulate matter and sedimentation, or transformation to methylmercury via anerobically mediated microbial processes in soils and sediments (Gilmour et al., 1992; Mason et al., 1995; Guentzel et al., 1996; Guentzel et al., 1998). Mercury methylation is mediated primarily by iron and sulfate reducing bacteria present in organic rich anaerobic soils and sediments (Gilmour et al., 1992; Kerin et al., 2006). The net production of methyl mercury in an ecosystem can result in its biomagnification in higher trophic level organisms. Methylmercury comprises 95-100% of the total mercury in predatory fish (Bloom, 1992). Processes that can affect the net production of methyl Hg in the environment include: microbial demethylation, photochemical destruction of methyl Hg, and atmospheric loadings of inorganic Hg (Compeau et al., 1984; Guentzel et al., 1996; Sellers et al., 1996; Guentzel et al., 1998; Orithel et al., 2006; Guentzel et al., 2007).

Landscape characteristics, proximity to wetlands, and hydrological flow patterns can also influence the cycling of methylmercury in an ecosystem. Fish mercury concentrations in a hydrologically unregulated Alabama coastal plain river that drains marshes and wetlands were significantly higher than concentrations in an adjacent regulated reservoir with little to no floodplain connectivity (Rypel et al., 2008). Concentrations of total and methyl Hg in rivers and streams located in the southern coastal plain Mobile Alabama River Basin were highest in catchments impacted by agriculture and wetlands (Warner et al., 2005). Watersheds containing low percentages of wetlands have been shown to be sinks for total and methyl Hg. A preliminary study of two North Carolina blue ridge watersheds containing little to no wetlands reported low concentrations of total and methyl Hg and retention of >90% total Hg and >80% methyl Hg within the catchment area (Allan et al., 1998). Studies conducted in Lake Superior have demonstrated that proximity to wetlands increases export of methyl Hg and organic matter to tributaries and results in increased concentrations of methyl Hg in aquatic biota (Weiner et al., 2008; Back et al., 2002). A 2009 study in South Carolina demonstrated that increasing percentages of wetland area within a watershed contribute to increased concentrations of methylmercury in largemouth bass (Guentzel, 2009). Other factors within South Carolina, such as reservoir size, water chemistry, land use, and hydrological modification were significant predictors of fish mercury concentrations rather than atmospheric source strength (Glover et al., 2010).

Methylmercury is a neurotoxin that can cross the placental barrier and can thus cause pre- and/or post-natal damage in children and negative health affects in adults (Yoshizawa et. al., 2002; Clarkson et al., 2003; Oken et al., 2005, Virtanen et al., 2007, Murata, et al., 2004; Murata et al., 1999; Grandjean et al., 1998). Humans are exposed most commonly to methylmercury through the consumption of fish, shellfish, and sea mammals (NAS, 2000; Clarkson et al., 2003). The health risks associated with methylmercury poisoning and the occurrence of elevated (>0.3 ppm) levels of mercury in marine and freshwater fish have resulted in consumption advisories for 48 states, 92% of the Atlantic coast, and 100% of the Gulf coast in the U.S. The state of South Carolina currently has issued more than 50 fish consumption advisories, with approximately 80% of these advisories located in the coastal plain region where mercury levels in largemouth bass exceed 1 ppm (Glover, et al., 2010). The issuance of consumption advisories for fish with high mercury concentrations results from the consideration of fish as a healthy alternative to other protein sources because they are low in saturated fats and contain omega-3polyunsaturated fatty acids which can reduce the risk of developing cardiovascular disease (Domingo et al. 2007; Sidhu 2003; American Heart Association 2009; Smith and Guentzel, 2010). However, negative health effects can occur when humans consume fish with high concentrations of contaminants. The consumption of fish containing elevated levels of mercury can increase the risk of developing cardiovascular disease (Choi et al., 2009). Adults may experience blurred vision, tremors, ataxia, myocardial infarction, and coronary heart disease as a result of mercury poisoning (Yoshizawa et. al., 2002; Clarkson et al., 2003; Virtanen et al., 2007). Exposure to methylmercury in children, both pre and/or post-natal can be associated with decreased visual memory recognition in infants (Oken et al., 2005), delayed auditory response, and deficits in motor function in children through age 14 (Murata, et al., 2004; Murata et al., 1999; Grandjean et al., 1998).

The objective of this paper is to report the atmospheric deposition of Hg in direct precipitation and throughfall at Waites Island, South Carolina from 2001-2003. Waites Island is an undeveloped tract of land located off the northeastern coast of South Carolina. There are no previous records of Hg deposition from this region of South Carolina. Rainfall mercury concentrations and deposition from this region of South Carolina will be compared to other atmospheric monitoring locations in South Carolina, North Carolina, Georgia, and Tennessee. The regional trend in atmospheric mercury deposition will be compared to mercury concentrations in largemouth bass and the resulting fish consumption advisories issued throughout South Carolina.

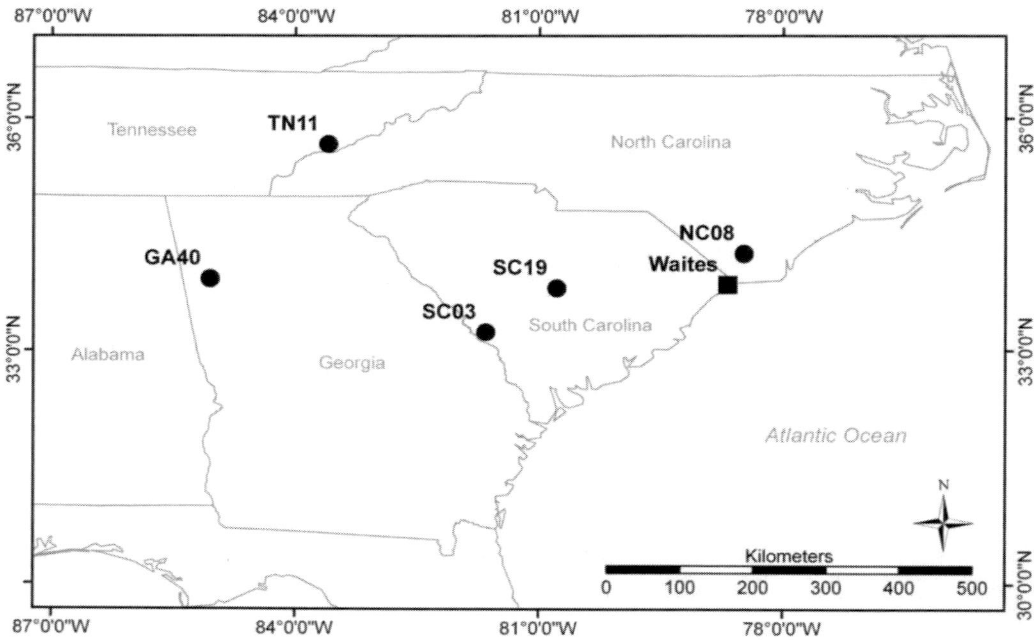

Figure 1. The salt marsh and throughfall sites on Waites Island, SC are denoted by the square. The surrounding mercury deposition network (MDN) sites are denoted by the circles (http://nadp.sws.uiuc.edu/mdn/).

Experimental Methods

Site Location

The study site was located on Waites Island within the Anne Tilghman Boyce (ATB) Coastal Reserve located above Cherry Grove Beach, South Carolina (Figure 1). Waites Island is an undeveloped tract of land located off the northeastern coast of South Carolina. It is comprised of ocean frontage, tidal creeks, salt marshes, hardwood/pine forests, and freshwater streams. The marsh rain samples were collected in an open area of the salt marsh and the throughfall samples were collected under a hardwood/pine forest.

The Waites Island data will be compared to regional mercury deposition network monitoring sites in South Carolina (SC03, SC 19), North Carolina (NC08), Georgia (GA40), and Tennessee (TN11). The latitude and longitude for the MDN sites denoted on Figure 1 were obtained from the mercury deposition network (http://nadp.sws.uiuc.edu/mdn/).

Sample Collection

Precipitation samples were collected using rigorous ultra clean sample collection protocols coupled with processing and analysis under "clean lab" conditions. In the clean lab, extensive detergent, solvent and acid washing procedures were used to clean the sampling equipment. Thorough discussions of the QA/QC procedures, cleaning procedures, equipment design, and sampling methods have been published by (Guentzel et al., 1995; Guentzel et al., 1998; Guentzel et al., 2001 and Landing et al., 1998).

Direct, or unobstructed precipitation samples were collected for monthly integrated wet-only and bulk deposition in the marsh proper. Wet-only precipitation samples were collected from July 2001-July 2002 and bulk precipitation samples were collected from July 2001- July 2003. Wet only deposition samples were collected only during rainfall events and bulk deposition samples were collected using a collection funnel that was continuously open to the atmosphere.

Thus the bulk deposition samples contain 100% of the wet deposition and whatever dry deposition collects in the funnels and is rinsed into the receiving bottle during rainfall events. Monthly integrated wet deposition throughfall samples were collected from July 2001 to July 2003.

The throughfall samples were collected from an upland-forested region of the island and represent rainfall and the contribution of dry deposition on foliage or foliar leaching that was rinsed into the collection bottle during rainfall events. There were 12 event rain samples collected during rainfall events that occurred in the summer of 2002.

The monthly integrated bulk deposition samples were fabricated using Teflon components, FEP Teflon receiving bottles, and polycarbonate funnels. Duplicate bulk deposition samplers were deployed at the marsh site and the upward-facing polycarbonate funnels were left uncovered for the entire month-long deployment period. Immediately before the samples were recovered, the funnels were each rinsed with two 70 mL aliquots of ultrapure acidified water (0.045 M 3xQ HCl). Thus, the bulk deposition samples represent wet deposition plus any dry deposition that may have accumulated on the inner surface of the funnel.

The wet deposition samples at the marsh and throughfall sites were collected using a modified Aerochem Metrics 301 wet/dry deposition sampler equipped with a polycarbonate roof, splash guard, Teflon-covered seal, and Teflon-coated arms two 1L FEP Teflon receiving bottles were nested inside the wet bucket and were attached to polycarbonate funnels using PTFE Teflon couplings (Guentzel et al., 1995; Guentzel et al., 1998; Guentzel et al., 2001 and Landing et al., 1998).

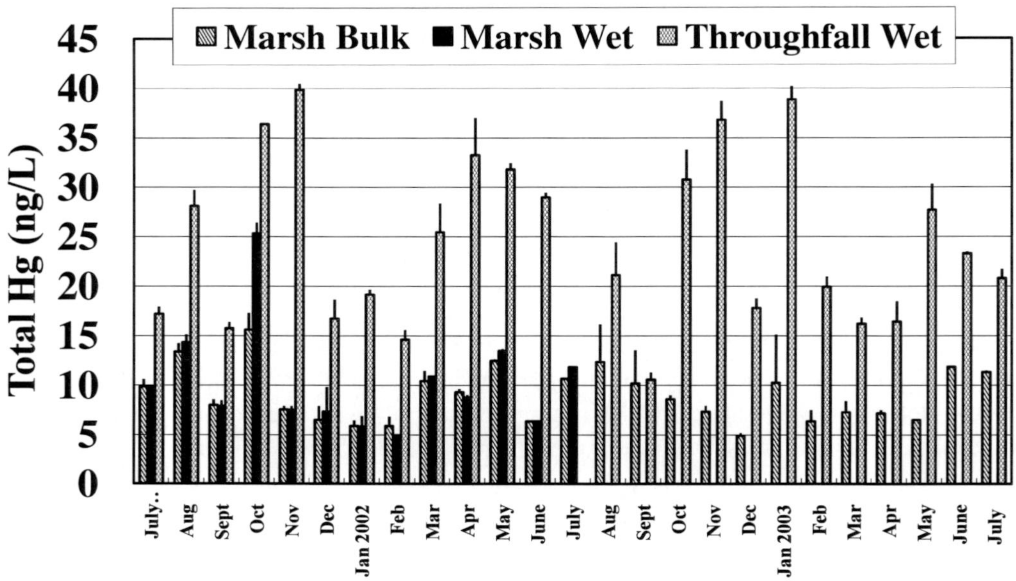

Figure 2. The concentration of total Hg (ng/L) in rainfall from the marsh and throughfall sites on Waites Island. The samples with error bars represent the mean and standard deviation of replicate field samples. There was no statistically significant difference between the concentrations ($P=0.580$) and fluxes ($P=0.592$) of total Hg in the marsh bulk and marsh wet deposition samples so the collection of marsh wet samples was discontinued in July of 2002.

Sample Analysis

The rain samples for mercury analysis were acidified with 7.5 mL 6M 3xQ HCl and 3 mL 7.5M 1xQ HNO_3 per liter of sample, hermetically sealed in the FEP Teflon bottles and digested in a low wattage UV flux (730 $\mu W/cm^2$; 254 nm) for 48 hours. Sub-samples (100-300 mL) were reduced with 25 mL 1 M $NaBH_4$ followed by gas phase stripping and Hg amalgamation onto gold coated quartz grains. Total Hg was quantified using cold vapor atomic fluorescence.

The analytical detection limit, based on three times the standard deviation of the purge blank, was 0.05 ng/L. Aqueous phase standards gave recoveries of 95±5% relative to injections of gas phase elemental Hg. (Guentzel et al., 1995; Guentzel et al., 1998; Guentzel et al., 2001 and Landing et al., 1998). The rain samples for total organic carbon (TOC) were analyzed using a Shimadzu 5000A total organic carbon analyzer. The instrument measures total carbon (TC) and total inorganic carbon (TIC). Total organic carbon (TOC) was determined by subtracting the TIC concentration from the TC concentration. Statistical calculations were performed using SigmaStat 3.5 (SysStat Software, Inc., Richmond, CA). The calculations included t-test, mean, standard deviation, linear regression, and normality test to determine statistical significance. Statistical results were considered significant at $P \leq 0.005$.

RESULTS AND DISCUSSION

Rainfall in South Carolina

The monthly integrated Hg concentrations and fluxes for the Waites Island salt marsh and throughfall sites are shown in Figures 2 and 3. The monthly integrated rainfall amounts for Waites Island are shown in Figure 4. The annual volume weighted mean rainfall Hg concentrations and fluxes are summarized in Tables 1 and 2. The concentration of total Hg in the continuously open marsh bulk deposition collectors ranged from 4.9 to 15.6 ng/L and the Hg flux ranged from 0.195-2.45 $\mu g/m^2$/month for the time period ranging from July 2001-July 2003. The concentration of total Hg in the marsh wet collectors ranged from 4.9-25.3 ng/L and the Hg flux ranged from 0.2-2.1 $\mu g/m^2$ for the time period ranging from July 2001-July 2002. There was no statistically significant difference between the concentrations ($P=0.580$) and fluxes ($P=0.592$) of total Hg in the marsh bulk and marsh wet deposition samples so the collection of marsh wet samples was discontinued in July of 2002. There are seasonal trends in the Hg concentration (Figure 2), Hg deposition (Figure 3), and rainfall amounts (Figure 4), at Waites Island. The rainfall amounts are 2 to 3 times higher and the Hg concentrations in the marsh bulk and wet samples are 2 times higher during the spring and summer months (April-October). This results in values of Hg deposition that are 2 to 3 times higher during the spring and summer months. The 2002 (January-December) Hg marsh bulk deposition was 10.72 $\mu g/m^2$, with 67% of this deposition occurring from April-October. Plots of the monthly Hg concentrations vs. monthly rainfall amounts for the winter and summer marsh rainfall samples do not show evidence of washout behavior, which is indicative of particulate Hg as the source term for the Hg in the rain (Figure 5).

The large amounts of rainfall that coincide with increased Hg rainfall concentrations and increased Hg deposition during the summer months suggest that the source term for the Hg is convective thunderstorm entrainment of RGM from the boundary layer and the free troposphere. Convective activity and convective thunderstorm entrainment of RGM, rather than particulate Hg scavenging, has been demonstrated to represent the dominant source term for Hg in rainfall from other regions within the United States (Guentzel et al., 1995; Guentzel et al., 2001).

The Waites Island concentrations and fluxes compare well with rainfall samples collected in other areas of South Carolina, North Carolina, and Georgia for the same time period. The concentration of total Hg ranged from 0.36-28 ng/L and the fluxes ranged from 0.01-1.1 $\mu g/m^2$/month for the MDN site SC 19 located in Richland County.

The concentration of total Hg ranged from 3.5-37 ng/L and the fluxes ranged from 0.03-1.2 $\mu g/m^2$/month for the MDN site SC 03 located in Barnwell County. The concentration of total Hg ranged from 2.0-46 ng/L and the fluxes ranged from 0.2-1.8 $\mu g/m^2$/month for the North Carolina MDN site NC 08 located in Columbus County. The concentration of total Hg ranged from 2.9-45 ng/L and the fluxes ranged from 0.1-1.7 $\mu g/m^2$/month for the Georgia MDN site GA 40 located in Paulding County. During the January 2002-July 2003 time period, the concentration of total Hg ranged from 2-42 ng/L and the fluxes ranged from 0.1-1.4 $\mu g/m^2$/month for the Tennessee MDN site TN 11 located in Sevier County. (http://nadp.sws.uiuc.edu/mdn/).

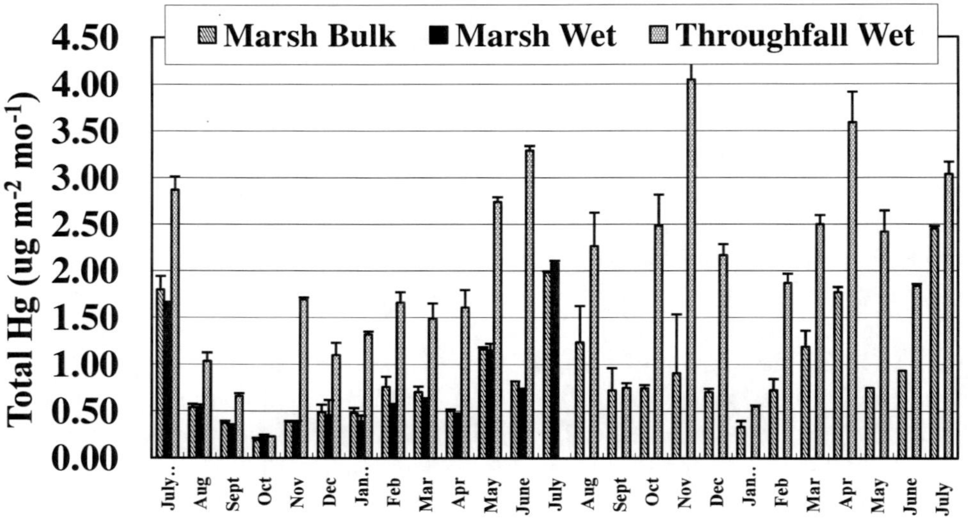

Figure 3. The flux of total Hg (μg/m^2/mo) in rainfall from the marsh and throughfall sites on Waites Island. The samples with error bars represent the mean and standard deviation of replicate field samples. There was no statistically significant difference between the fluxes (P=0.592) of total Hg in the marsh bulk and marsh wet deposition samples so the collection of marsh wet samples was discontinued in July of 2002.

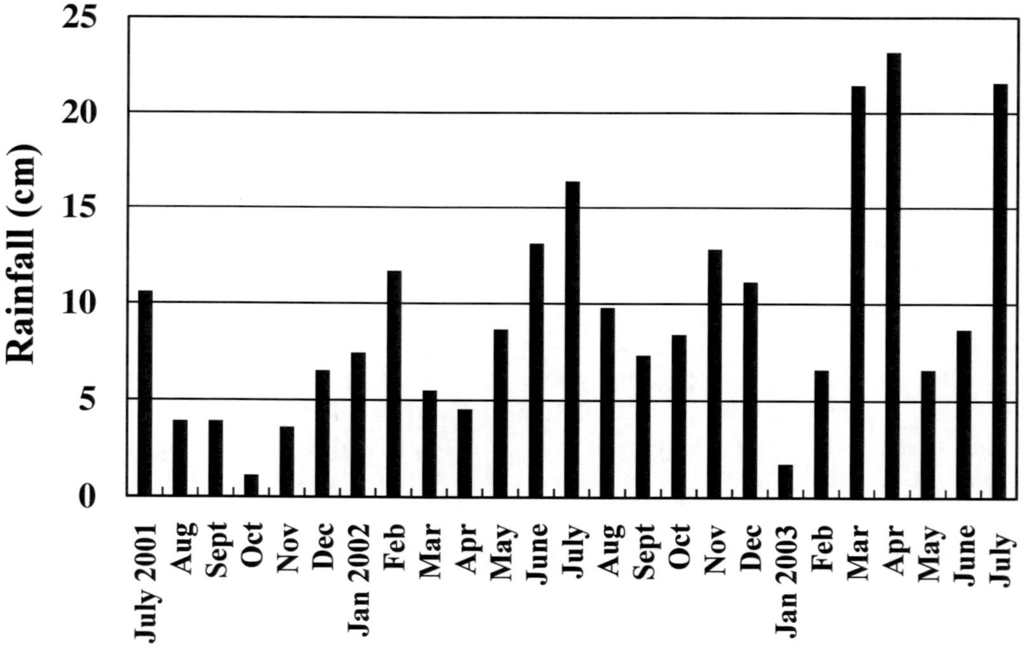

Figure 4. Monthly rainfall amounts at Waites Island, South Carolina.

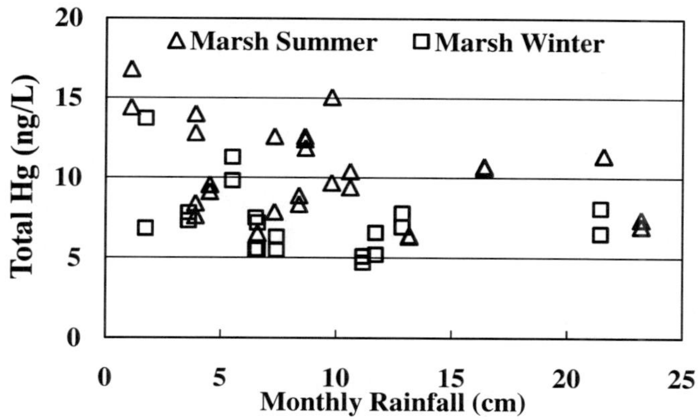

Figure 5. The monthly Hg concentration vs. monthly rainfall amount for summer (April-October) and winter (November-March) marsh bulk deposition samples.

The concentration of total Hg in the throughfall wet deposition samples ranged from 10.6-39.85 ng/L and the total Hg flux ranged from 0.23-4.0 μg/m^2/month for the time period ranging from July 2001-July 2003 at the Waites Island site. These values are similar to throughfall concentrations and fluxes reported recently for the Great Smoky Mountains National Park in North Carolina. Throughfall total Hg concentrations ranged from 13.6-18.3 ng/L and fluxes ranged from 2-4 μg/m^2 for the 2008-2009 time period (Fisher and Wolfe, 2012). At the Waites Island site, the concentrations ($P \leq 0.001$) and fluxes ($P \leq 0.001$) of total Hg in the throughfall samples are significantly greater than the concentrations and fluxes of total Hg in the marsh samples. This is similar to precipitation and throughfall samples collected in New York for the 2005-2006 time period. Total Hg concentrations in rainfall ranged from 1-11 ng/L and throughfall concentrations ranged from 2-14 ng/L (Choi et al., 2008). Monthly fluxes of total Hg in precipitation ranged from 0.1-1.2 μg/m^2/month and throughfall fluxes ranged from 0.1-1.4 μg/m^2/month (Choi et al., 2008).

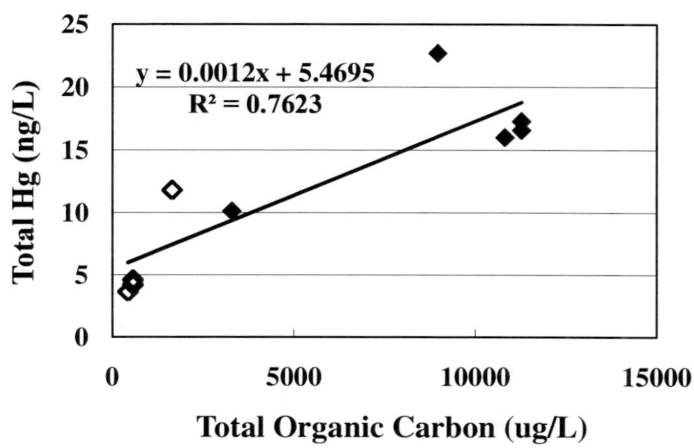

Figure 6. The relationship between total Hg and total organic carbon in event rainfall samples collected at the marsh and throughfall sites on Waties Island. The marsh samples are open triangles and the throughfall samples are filled triangles. The correlation is significant ($P=0.002$) and the data are normally distributed (Kolmogorov-Smirnov normality test).

The volume weighted Hg concentrations in Waites Island salt marsh samples ranged from 9.08-12.77 ng/L (Table 1). The volume weighted Hg fluxes in Waites Island salt marsh samples ranged from 3.78-10.72 μg/m^2 (Table 2). The volume weighted Hg concentrations in Waites Island throughfall samples ranged from 20.02-25.64 ng/L (Table 1). The volume weighted Hg fluxes in Waites Island throughfall ranged from 7.59-26.69 μg/m^2 (Table 2). The difference between the volume weighted marsh and throughfall precipitation fluxes would represent the Hg dry deposition to foliar surfaces that is "washed off" of the foliage or leached from the foliage during rainfall events. The estimated foliar deposition fluxes ranged from 3.8-15.97 μg/m^2 and are almost equivalent to or greater than the salt marsh Hg fluxes (Table 2). Thus, the volume weighted throughfall fluxes that range from 7.59-26.69 μg/m^2 (Table 2) represent a more comprehensive estimate of the total Hg deposition (wet+dry+foliar leaching) to the Island ecosystem.

Table 1. Volume Weighted Hg Concentration (ng/L)

	NADP/MDN Wet Deposition	Waites Island Salt Marsh Bulk Depostion	Waites Island Throughfall Wet Deposition	Precipitation (cm)
2001; July-December				
Waites Island, SC (n=11)		12.77	25.64	29.6
NC 08 (n=14)	8.86			57.7
SC 19 (n=15)	11.23			33.5
SC 03 (n=15)	17.9			27.8
TN 11	-----			-----
GA 40 (n=28)	9.36			34.9
2002; January-December				
Waites Island, SC (n=24)		9.18	22.62	118
NC 08 (n=46)	10.25			108
SC 19 (n=39)	11.41			79.1
SC 03 (n=41)	12.47			92.2
TN 11 (n=42)	13.77			119
GA 40 (n=63)	9.72			145.1
2003; January-July				
Waites Island, SC (n=14)		9.08	20.02	78.94
NC 08 (n=47)	9.72			87.3
SC 19 (n=45)	10.47			83.7
SC 03	-----			-----
TN 11 (n=31)	10.58			94.6
GA 40 (n=41)	13.76			95.7

"----" indicates that the MDN site was not operational at the time. (http://nadp.sws.uiuc.edu/mdn/).

The concentration of total Hg in event based marsh and throughfall samples collected in 2002 ranged from 3.6-11.7 and 10.0-22.7 ng/L, respectively. The concentration of total organic carbon (TOC) in event based marsh and throughfall samples collected in 2002 ranged from 430-1650 μg/L and 3,292-11,273 μg/L, respectively (Table 3). These TOC concentrations are similar to rainfall dissolved organic carbon (DOC) samples collected in

North Carolina. The North Carolina samples ranged from 144—5,500 μg/L for event based precipitation collected in 2009 (Southwell et al., 2010). The total Hg and TOC in the event rainfall samples from Waites Island are significantly correlated (P=0.002) (Figure 6). This correlation explains 76% of the variance in the data and suggests that Hg in rainfall and throughfall is associated with organic ligands.

Table 2. Hg Deposition (ug/m^2)

	NADP/MDN Wet Deposition	Waites Island Salt Marsh Bulk Depostion	Waites Island Throughfall Wet Deposition	Waites Island Estimated Foliar Depositon
2001; July-December				
Waites Island, SC (n=11)		3.78	7.59	3.81
NC 08 (n=14)	5.12			
SC 19 (n=15)	3.77			
SC 03 (n=15)	4.97			
TN 11	------			
GA 40 (n=28)	3.27			
2002; January-December				
Waites Island, SC (n=24)		10.72	26.69	15.97
NC 08 (n=46)	9.46			
SC 19 (n=39)	9.04			
SC 03 (n=41)	11.5			
TN 11 (n=42)	11.56			
GA 40 (n=63)	14.1			
2003; January-July				
Waites Island, SC (n=14)		8.15	15.85	7.7
NC 08 (n=47)	8.49			
SC 19 (n=45)	8.78			
SC 03	------			
TN 11 (n=31)	10.01			
GA 40 (n=41)	13.16			

"----" indicates that the MDN site was not operational at the time.

The difference between the marsh and throughfall precipitation fluxes would represent the dry deposition to foliar surfaces that is "washed off " of the foliage or leached from the foliage during rainfall events.

(http://nadp.sws.uiuc.edu/mdn/).

Comparison to Regional Rainfall and SC Fish Consumption Advisories

Atmospheric deposition of wet and dry species is the primary mechanism for the delivery of inorganic mercury to ecosystems. Once deposited to the ecosystems, inorganic mercury can be microbialy transformed to methyl mercury which subsequently biomagnifies in piscivorous fish and poses a risk to humans who consume these fish. Therefore, the quantification of atmospheric mercury deposition to an ecosystem is necessary for studies designed to investigate the aquatic and terrestrial cycling of Hg within the ecosystem.

Table 3. Hg and TOC concentration in event based rain collection at Waites Island, SC

x	Waites Island Salt Marsh Hg (ng/L)	Waites Island Salt Marsh TOC (µg/L)	Waites Island Throughfall Hg (ng/L)	Waites Island Throughfall TOC (µg/L)	Precipitation (cm)
2002					
June 18	4.39±0.32	567	16.98±0.56	11273	5.54
June 20	11.76	1650	22.79	8966	1.02
June 21	3.66	430	16.1	10819	1.55
June 22	5.79	------	10.12	3292	3.71
July 5	19.6	------			1.65
July 11	6.42	------			0.76

"------" indicates that there was not enough sample to analyze for TOC.
For the Hg samples n=2 for June 18 samples; for all other samples n=1.
For the DOC samples n=1.

The Waites Island deposition data agrees well with several regional mercury deposition network (MDN) monitoring sites located in South Carolina (SC03, SC 19), North Carolina (NC08), Georgia (GA40), and Tennessee (TN11). The 2002 annual deposition at TN11, GA40, SC03, SC19, NC08, and Waites Island was 11.56, 14.1, 11.5, 9.04, 9.46, and 10.72 $\mu g/m^2$, respectively (Table 2). The 6 month 2003 deposition (January-July) at TN11, GA40, SC19, NC08, and Waites Island was 10.01, 13.16, 8.78, 8.49, and 8.15 $\mu g/m^2$, respectively (Table 2). The 6 month 2001 deposition (July-December) at GA40, SC03, SC19, NC08, and Waites Island was 3.27, 4.97, 3.77, 5.12, and 3.78 $\mu g/m^2$, respectively (Table 2). For the 2001-2003 time period, there is no strong gradient in mercury wet deposition as the sites move geographically from west to east or north to south across the region (Table 2). The 2001-2003 Waites Island and MDN values agree well with current 2009-2010 MDN values. The 2009 annual wet Hg deposition for MDN sites NC08, SC19, SC03, TN11, and GA40 was 8.83, 9.71, 8.49, 10.1, and 12.98 $\mu g/m^2$, respectively (Table 4). The 2010 annual wet Hg deposition for MDN sites NC08, SC19, TN11, and GA40 was 10.08, 10.15, 11.7, and 10.17 $\mu g/m^2$, respectively (Table 4). In addition to the 2001-2003 deposition, the 2009 and 2010 deposition from the MDN sites does not display a strong east to west or north to south gradient in mercury wet deposition.

A previous study has linked the concentrations of methylmercury in largemouth bass to wet deposition of inorganic Hg to ecosystems. Statewide averages of Hg concentrations in largemouth bass from 22 states (excluding SC) have been significantly correlated to wet deposition of Hg with fish Hg concentrations increasing from ~ 0.2 to 0.6 ug/g as wet deposition of Hg increased from ~ 4 to 17 $\mu g/m^2 yr$ (Hammerschmidt et al., 2006). If atmospheric deposition was the sole factor in determining SC fish mercury concentrations then one would expect to find similar fish mercury concentrations throughout the state because there is no strong gradient in atmospheric Hg wet deposition across the state. While there is no strong gradient in Hg wet deposition, there is an observable west to east gradient in fish Hg concentrations throughout SC. Currently, the western SC Ppiedmont and Bblue ridge ecoregions have the lowest fish Hg concentrations (0.11-0.62 ppm Hg) and the Ssoutheastern Pplains, Mmiddle Aatlantic Ccoastal Pplain, and Ssouthern Ccoastal Pplain ecoregions located in the central and eastern region of the state have the highest fish Hg concentrations (0.28-1.05) (Glover et al., 2010).

Table 4. Hg Deposition (ug/m²)

	NADP/MDN Wet Deposition
2009; January-December	
NC 08	8.83
SC 19	9.71
SC 03	8.49
TN 11	10.1
GA 40	12.98
2010; January-December	
NC 08	10.08
SC 19	10.15
SC 03	***
TN 11	11.7
GA 40	10.17

"***" indicates that the site was operational but there was too much missing data to calculate an annual flux (http://nadp.sws.uiuc.edu/mdn/).

South Carolina has three distinct physiographic provinces with the blue ridge and piedmont regions in the north west and central portion of the state and the coastal plain region in the central to south eastern region of the state (Murphy, 1995; Stewart and Roberson, 2007). The percentage of wetland area in these provinces increases from <1% in the western upstate blue ridge/piedmont region to 11-25% in the central, southern, and eastern coastal flood plain systems (http://www.scdhec.net/environment/water/shed). The mean value of percent wetland area (16.3±5%) in the coastal plain ecoregions is significantly higher ($P<0.001$) than the blue ridge/piedmont region (1.14±1.6%) (Guentzel et al., 2009). Correspondingly, ~80% of the South Carolina fish consumption advisories are located within these coastal plain regions (http://www.scdhec.net/environment/water/fish) that contain higher percentages of wetlands. The median concentration of Hg in largemouth bass (0.940 μg/g; n=309) from the SC coastal plain ecoregions that contain higher percentages of wetlands is significantly higher ($P<0.001$) than the median value (0.405 μg/g; n=126) from the SC blueridge piedmont ecoregions. (Guentzel et al., 2009; http://www.epa.gov/ waterscience/fish/advisories/ index.html).

A 2009 study in South Carolina demonstrated that increasing percentages of wetland area within a watershed contribute to increased concentrations of mercury in largemouth bass (Guentzel, 2009). The mean fish Hg concentrations in largemouth bass from each watershed were significantly correlated with percent wetland in each watershed ($P=0.003$). This correlation explained 66% of the variance in the data and suggested that increasing percentages of wetland area contribute to increased fish Hg concentrations in SC (Guentzel et al., 2009). In addition to wetlands, other factors within South Carolina, such as reservoir size, water chemistry, land use, and hydrological modification were also significant predictors of fish mercury concentrations rather than atmospheric source strength (Glover et al., 2010). This suggests that while the atmospheric source signal of mercury in wet deposition to ecosystems in this region is similar, other aquatic and or terrestrial Hg cycling processes, in addition to possible differences in Hg dry deposition may contribute to the trend in SC fish Hg concentrations.

CONCLUSION

The concentration of total Hg in the marsh bulk deposition collectors ranged from 4.9 to 15.6 ng/L and the total Hg flux ranged from 0.195-2.45 $\mu g/m^2$/month for the time period ranging from July 2001-July 2003. These concentrations and fluxes compare well with rainfall samples collected in other areas of South Carolina for the same time period. The concentration of total Hg in the throughfall wet deposition samples ranged from 10.6-39.85 ng/L and the total Hg flux ranged from 0.23-4.0 $\mu g/m^2$/month for the time period ranging from July 2001-July 2003. The difference between the marsh and throughfall precipitation fluxes would represent the dry deposition to foliar surfaces that is "washed off" of the foliage or leached from the foliage during rainfall events. The estimated foliar deposition ranges from 0.035-0.84 $\mu g/m^2$/month.

The volume weighted throughfall fluxes are a measure of the Hg wet deposition, any particulate Hg that is "washed off" of the foliage, and any Hg that is leached from the foliage during rainfall events, and thus represent a more comprehensive estimate of the total Hg deposition (wet+dry+foliar leaching) to the Island ecosystem. The concentration of total Hg in event based marsh and throughfall samples collected in 2002 ranged from 3.6-11.7 and 10.0-22.7 ng/L, respectively.

The concentration of total organic carbon (TOC) samples in event based marsh and throughfall samples collected in 2002 range from 430-1650 μg/L and 3,292-11,273 μg/L, respectively. The total Hg and TOC in the event rainfall samples from Waites Island are significantly correlated ($P=0.002$) and this suggests that Hg in rainfall and throughfall is associated with organic ligands.

The Waites Island deposition data agrees will with several regional mercury deposition network (MDN) monitoring sites located in South Carolina (SC03, SC 19), North Carolina (NC08), Georgia (GA40), and Tennessee (TN11).

The lack of an increasing west to east regional gradient in wet Hg deposition when compared the increasing west to east gradient in SC fish Hg concentrations suggests that while atmospheric deposition may be the source of Hg to SC ecosystems, other processes such as reservoir size, water chemistry, land use, hydrological modification, and proximity to wetlands contribute to the trend in fish Hg concentrations throughout South Carolina.

ACKNOWLEDGMENTS

The assistance of individuals who contributed to this effort are gratefully acknowledged. Field sampling and laboratory assistance was provided by Coastal Carolina University undergraduate students Griffin Meincke, Aaron Beck, Eric Mazzone, Christopher Weeks, Yoko Tsukamoto, Devra Blandon, James Patanio, Heather Young, and Lacie Armstrong. Dr. Eric Wright of Coastal Carolina University constructed the map of the sampling locations (Figure 1).

REFERENCES

Allan, C.J. and Heyes, A., A preliminary assessment of wet deposition and episodic transport of total and methyl mercury from low order blue ridge watersheds, S.E. USA. *Water Air Soil Poll 1998*; 105: 573-592.

American Heart Association (AHA). *Fish and Omega*-3 Fatty Acids [Internet]. [updated 2009]. [cited 2009 Feb 25]. Available at: http://www. americanheart. org/presenter. jhtml?identifier=4632

Back, R.C., Hurley, J.P., Rolfhus, K.R., Watershed influences on the transport, fate, and bioavailability of mercury in Lake Superior: field measurements and modeling approaches. *Lakes and Res. Resear and Manag.* 2002: 7: 201-206.

Bloom, N. S. On the chemical speciation of mercury in edible fish muscle issue. *Canadian Journal of Fisheries and Aquatic Sciences* 1992; 49; 1010.

Choi, H-D., Sharac, T.J., Holsen. T.M. Mercury deposition in the Adirondacks: A comparison between precipitation and throughfall. *Atmos Environ* 2008; 42; 1818-1827.

Clarkson T, Magos L, Myers G. 2003. The Toxicology of Mercury—Current Exposures and Clinical Manifestations. *New England Journal of Medicine* 349 (18): 1731-1737.

Compeau, G., Bartha, R. Methylation and demethylation of mercury under controlled redox, pH, and salinity conditions. *Appl Environ Micro.* 1984; 48: 1203-1207

Domingo J, Bocio A, Falco G and Llobet J. 2007a. Benefits and risks of fish consumption Part I. A quantitative analysis of the intake of omega-3 fatty acids and chemical contaminants. T*oxicology* ; 230: 219-226.

Fisher, L.S., Wolfe, M.H. Examination of mercury inputs by throughfall and litterfall in the Great Smoky Mountains National Park. *Atmos Environ* 2012; 47; 554-559.

Friedli, H., Radke, L., Lu, J., Banic, C., Leaitch, W., MacPherson, J. Mercury emissions from burning of biomass from temperate North American forests: laboratory and airborne measurements. *Atmospheric Environment* 2003; 37: 253-267.

Gilmour, C. C., Henry, E. A. and Mitchell, R. Sulfate stimulation of mercury methylation in freshwater sediments. *Environ Sci Technol* 1992; 26**:** 281-2287.

Glover, J.B., Domino, M.E., Altman, K.C., Dillman, J.W., Castleberry, W.S., Eidson, J.P., Mattocks, M. *Mercury in South Carolina Fishes,* USA. Ecotoxicology 2010; 9: 781-795.

Grandjean, P., Weihe, P., White, R., Debes, F. Cognitive performance of children prenatally exposed to "safe" levels of methylmercury. *Environmental Research Section A* 1998; 77:165-172.

Guentzel, J., Landing, W., Gill, G., Pollman, C. Atmospheric deposition of mercury in Florida: the FAMS project (1992-1994). *Water Air Soil Poll* 1995; 80: 393-402.

Guentzel, J., Powell, R., Landing, W., Mason, R. Mercury associated with colloidal material in an estuarine and open ocean environment. *Mar Chem* 1996; 55: 177-188.

Guentzel, J., Landing, W., Gill, G., Pollman, C., Mercury and major ions in rainfall, throughfall, and foliage from the Florida Everglades. *Sci Total Environ* 1998: 213; 43-51.

Guentzel, J., Landing, W., Gill, G., Pollman, C. Processes influencing rainfall deposition of mercury in Florida. *Environ Sci Technol* 2001; 35: 863- 873.

Guentzel, J., Portilla, E., Keith, K., Keith, E. Mercury Transport and bioaccumulation in riverbank communities of the Alvarado Lagoon System, Veracruz State, Mexico. *Sci Total Environ* 2007: 388; 316-324.

Guentzel, J. Wetland influences on mercury transport and bioaccumulation in South Carolina. *Sci Total Environ* 2009: 407; 1344-1353.

Hammerschmidt, C.R., Fitzgerald, W.F., Methyl mercury in freshwater fish linked to atmospheric mercury deposition. *Environ Sci Technol.* 2006; 40: 7764-7770.

Kerin, E.J., Gilmour, C.C., Roden, E., Suziki, M.T., Coates, J.D., Mason, R.P. Mercury methylation by dissimilatory iron-reducing bacteria. *Appl Environ Micro.* 2006; 72: 7919-7921.

Landing, W., Guentzel, J., Gill, G., Pollman, C. Methods for Measuring Mercury in Rainfall and Aerosols in Florida. *Atmos Environ* 1998; 32: 909-918.

Mason, R. P., Morel, F. M. M., and Egmond, H. F. The role of microorganisms in elemental mercury formation in natural waters. *Water Air Soil Poll* 1995; 80: 915-921.

Munthe, J., Hellsten, S., Zetterberg, T., Mobilization of mercury and methymercury from forest soils after a severe storm-fell event. *2007 Ambio;* 36: 111-113.

Murata, K., Weihe, P., Rensoni, A., Debes, F., Vasconcelos, R., Zino, F., Araki, S., Jorgensen, P., White, R., Grandjean, P. Delayed evoked potentials in children exposed to methylmercury from seafood. *Neurotoxicology and Teratology* 1999; 21: 343-348.

Murata, K., Weihe, P., Budtz-Jorgensen, E., Jorgensen, P., Grandjean, P. Delayed brainstem auditory evoked potential latencies in 14-year old children exposed to methylmercury. *The Journal of Pediatrics* 2004; 144: 177-173.

Murphy, C.H. *Carolina Rocks.* Sandlapper Publishing Co., Inc., Orangeburg, SC., 1995, 261pp.

National Academy of Sciences (NAS), 2000. *Toxicological effects of methylmercury.* National Academy Press, Washington, D.C.

National Atmospheric Deposition Program. *Mercury Deposition Network.* (http://nadp.sws.uiuc.edu/mdn/).

Oken E, Wright R, Kleinman K, Bellinger D, Amarasiriwardena C, Hu H, Rich-Edwards J and Gillman M. 2005. Maternal Fish Consumption, Hair Mercury, and Infant Cognition in a U.S. Cohort. *Environmental Health Perspectives;* 113: 1376-1380.

Orithel., D.M., Paterson, M.J., Gilmour, C.C., Bodaly, R.A., Blanchfield, P.J., Hintelmann, H., Harris., R.C., Rudd, J.M. Effect of mercury loading rate on the fate of mercury in littoral mesocosms. *Environ Sci Technol.* 2006; 40: 5992-6000.

Porvari, P., Verta, M., Munthe, J., Haapanen, M. Forestry practices increase mercury and methylmercury output from boreal forest catchments. *Environmental Science and Technology* 2003: 37: 2389-2393.

Rypel, A.L., Arrington, D.A., Findaly, R.H., Mercury in southeastern U.S. riverine fish populations linked to water body type. *Environ Sci Technol* 2008; 42: 5118-5124.

Selin, N.E., Jacob, D.J. Seasonal and spatial patterns of mercury wet deposition in the Unitesd States: Constraints on the contribution from North American anthropogenic sources. *Atmos. Environ;* 42, 5193-5204.

Sellers, P., Kelly, C.A., Rudd, J.W., MacHutchon, A.R. Photodegredation of methyl mercury in lakes. *Nature.* 1996; 380: 694-697.

Sidhu, K. Health benefits and potential risks related to consumption of fish or fish oil. 2003. *Regulatory Toxicology and Pharmacology;* 38: 336-344.

Smith, K.L., Guentzel, J.L. Mercury concentrations and omega-3 fatty acids in fish and Shrimp: Preferential consumption for maximum health benefits. *Marine Pollution Bulletin.* 2010; 60: 1615-1618.

Southwell, M.W., Smith, J.D., Avery, G.B., Kieber, R.J., Willey, J.D. Seasonal variability of formaldehyde production from photolysis of rainwater dissolved organic carbon. *Atmos Environ* 2010; 44; 3638-3643.

South Carolina Department of Health and Environmental Control. http:// www.scdhec.net/ environment/water/shed.

South Carolina Department of Health and Environmental Control. http:// www.scdhec.net/ environment/water/fish.

Stewart, K.G., Roberson, M.R. Exploring the Geology of the Carolinas. The University of North Carolina Press, Chapel Hill, NC, 2007, 298 pp.

U.S. EPA Mercury Study Report to Congress 1997. http://www.epa.gov/ mercury/report.htm.

US EPA National Listing of Fish Advisories (NLFA) http://www.epa.gov/ waterscience/ fish/advisories/index.html.

Wiener, J.G., Rolfhus, K.R., Haro, R.J., Wetland proximity affects methyl mercury in the aquatic food web of Chequamegon Bay (Lake Superior). *ASLO/AGU Ocean Sciences Meeting,* March 2-7, 2008. Orlando, FL.

Virtanen J, Rissanen T, Voutilainen S, Tuomainen T. 2007. Mercury as a risk factor for cardiovascular diseases. *Journal of Nutritional Biochemistry;* 18: 75-85.

Yoshizawa K, Rimm EB, Morris JS, Spate VL, Hsieh C, Spiegelman D, Stampfer MJ, Willet WC. 2002. Mercury and the risk of coronary heart disease in men. *New England Journal of Medicine* 347 (22): 1755-1760.

INDEX

#

20th century, 157

A

abiotic parameter, 220
absorption spectra, 38, 48
absorption spectroscopy, 157
access, 40, 103
accessibility, 26, 42, 44, 46, 52, 116
acetaminophen, 88, 95, 96, 97
acetonitrile, 96
acid, 1, 14, 34, 39, 40, 45, 56, 63, 101, 102, 110, 121, 123, 125, 137, 138, 140, 141, 143, 144, 145, 146, 147, 149, 152, 153, 154, 159, 169, 190, 211, 237
acidic, 42, 95, 110, 120
acidity, 156, 215
activated carbon, 29, 30, 31, 55, 56, 118, 124
active oxygen, 153
acute renal failure, 101, 102, 105
ADA, 132
adaptations, 154
additives, 121, 123, 126, 130
adenine, 76
adenosine, 102
adipose tissue, 175, 211
adrenal gland, 208
adsorption, 25, 29, 30, 31, 33, 44, 46, 51, 52, 53, 55, 56, 115, 117, 118, 124, 190
adulthood, 193
adults, 113, 189, 193, 197, 205, 209, 211, 212, 227, 235
adverse effects, 28, 109, 110, 112, 113, 128, 176, 214
aerosols, 169
agar, 137, 139
age, 191, 193, 194, 206, 210, 211, 214, 221, 226, 227, 236
agencies, 28
aggregation, 40
agriculture, 154, 235
air pollutants, 114
air quality, 127
Alaska, 217
albumin, 40, 101, 142
alcohol consumption, 211
alfalfa, 154
algae, 29, 57, 193, 199
aliphatic amines, 58
alters, 138, 207
aluminium, 119, 151
amalgam, 99, 100, 111, 113, 114, 119, 120, 130, 133
American Heart Association, 235, 247
amine(s), 12, 58, 101
amine group, 101
amino, 101, 204
amino acid(s), 101, 204
ammonium, 44, 45
amphibia, 221
amphibians, 174, 177, 199, 215, 225
ANC, 190
aniline, 63
ANOVA, 142
anoxia, 191
antagonism, 223
anti-inflammatory drugs, 102, 104
antioxidant, 137, 138, 139, 146, 148, 149, 151, 152, 154
apoptosis, 103
aquatic systems, 110, 192, 219, 228, 233
aqueous solutions, 29, 36, 39, 56, 57, 59, 63, 90
Arabidopsis thaliana, 36, 63
Argentina, 99, 105
arsenic, 217

ascorbic acid (ASA), 121, 137, 143, 144, 146, 152, 153
Asia, 189, 199, 216, 218
assessment, 26, 52, 107, 131, 177, 221, 226, 227, 231, 247
assimilation, 176, 192, 194, 201, 203
ataxia, 207, 235
atmosphere, 26, 42, 122, 125, 131, 171, 175, 176, 213, 216, 233, 234, 237
atmospheric deposition, 233, 236, 244, 246
atomic emission spectrometry, 28, 55
atoms, 1, 2, 3, 4, 5, 6, 7, 8, 9, 10, 11, 12, 13, 14, 15, 16, 17, 18, 19, 20, 21, 22, 23, 34, 42, 67, 71, 74, 76, 77, 78, 79, 81, 82, 88
atrophy, 207
Au nanoparticles, 36
avoidance behavior, 207, 225
Azerbaijan, 177, 186, 217

B

bacteria, 1, 26, 27, 29, 168, 190, 191, 194, 234, 248
ban, 114
barriers, 189, 202
base, 8, 9, 10, 21, 34, 36, 53, 54, 63, 77, 83, 137, 194
batteries, 100, 111, 113, 114, 122, 130
beams, 117
beef, 178
Beijing, 218, 231
beneficial effect, 212
benefits, 212, 248
bias, 214
bile, 203
biliary obstruction, 106, 107
bioaccumulation, 27, 29, 173, 174, 176, 177, 190, 191, 192, 194, 197, 213, 214, 216, 217, 221, 228, 229, 233, 247, 248
bioavailability, 247
biodegradation, 125
bioindicators, 154, 173, 174, 215
biological activity, 115
biological processes, 111
biological samples, 55
biological systems, 18, 69, 97, 220
biomarkers, 104, 107
biomass, 29, 57, 58, 157, 178, 222, 247
biomolecules, 148
bioremediation, 53, 129
biosensors, 54
biosphere, 1
biosynthesis, 148
biotic, 194
birds, 26, 28, 54, 199, 200, 204, 206, 208, 222, 224
blends, 124
blindness, 28, 207
blood, 31, 100, 101, 102, 103, 106, 174, 181, 192, 194, 201, 202, 204, 207, 210, 211, 212, 213, 222, 225, 229, 230, 231
blood flow, 106
blood pressure, 212
blood urea nitrogen, 103
blood-brain barrier, 100, 101, 204
bloodstream, 202
boas, 199
body size, 201, 232
body weight, 103, 104, 194, 209
bonding, 6, 7, 9, 13, 14, 15, 17, 23, 44, 88, 119, 124
bonds, 2, 3, 15, 101, 120
bone, 204
boreal forest, 248
brain, 1, 99, 100, 113, 152, 202, 204, 211, 224, 225
brainstem, 248
Brazil, 137, 139, 140, 141, 153, 173, 182, 185, 187, 190, 197, 210, 211, 217, 219, 220, 221, 222, 223, 228
breakdown, 125
breast milk, 113
breeding, 189
Britain, 210
bromine, 5, 7, 14, 15
by-products, 56, 124

C

Ca^{2+}, 124
cadmium, 1, 61, 151, 152, 153, 154, 217
calcification, 206
calcium, 59, 121, 123, 126
calcium carbonate, 123
calibration, 140
candidates, 208
carapace, 181, 182, 183, 184, 185, 186, 187, 188, 192, 194, 195, 203, 212, 222
carbon, 8, 9, 10, 16, 29, 30, 31, 55, 56, 57, 88, 101, 111, 116, 117, 118, 123, 124, 125, 127, 128, 134, 149, 152, 159, 175, 176, 177, 190, 192, 219, 234, 238, 241, 242, 246, 249
carbon atoms, 10, 88
carbon dioxide, 123
carbon monoxide, 149, 152
carbonization, 123
carboxyl, 101
carboxylic acid, 39
cardiovascular disease(s), 211, 212, 227, 235, 249
cardiovascular disorders, 211
cardiovascular risk, 227

carnivores, 193, 197
carotenoids, 137, 139, 148, 149, 152
casting, 123
catalysis, 18, 28, 32, 69, 147
catalyst, 39, 177
catchments, 235, 248
catfish, 206, 208, 225
cation, 13, 30, 63, 68, 76, 100, 110
cationic surfactants, 33, 44
C-C, 9, 63
cell culture, 152
cell division, 28
cellulose, 199
central nervous system (CNS), 1, 28, 68, 100, 112, 113
ceramic, 124
cerebral palsy, 28
challenges, 42, 53, 197
chemical(s), 26, 28, 29, 30, 31, 32, 33, 42, 45, 52, 54, 58, 61, 71, 74, 75, 78, 88, 89, 99, 100, 111, 120, 122, 123, 124, 125, 138, 156, 171, 175, 176, 177, 206, 208, 216, 220, 222, 226, 230, 234, 247
chemical bonds, 120
chemical characteristics, 122
chemical properties, 206
chicken, 178
child development, 226
children, 105, 111, 113, 173, 209, 210, 211, 226, 235, 247, 248
Chile, 87
China, 129, 182, 189, 210, 218, 231
chitin, 29
chitosan, 29, 56
chlorine, 5, 7, 8, 10, 12, 13, 20, 21, 22, 23, 77, 79, 100, 124, 156, 177
chlorophyll, 138, 139, 148, 152, 154
chloroplast, 151
cholestasis, 102, 107
chromium, 107, 148, 217
chromosome, 28
chronic renal failure, 102, 106
circulation, 26, 202, 203
citizens, 128
cladding, 54
classes, 6, 7, 8, 9, 11, 12, 14, 17, 18, 20, 23, 82
classification, 123, 224
Clean Air Act, 114
cleaning, 237
climate, 174, 215, 216
climate change, 174, 215, 216
clothing, 2
clusters, 2, 12, 13, 15, 16, 17
C-N, 9, 13
CNS, 28, 112, 113
CO_2, 125, 134
coal, 29, 57, 111, 114
cobalt, 13, 17
Code of Federal Regulations, 134
cognition, 210
cognitive development, 227
colonization, 221
combustion, 25, 26, 111, 118, 128, 131, 234
commercial, 53, 115, 178, 189
community(s), 151, 214, 219, 247
compensatory effect, 137
competition, 194
complexity, 118
complications, 113, 155
composites, 33
composition, 32, 33, 43, 123, 169
compounds, 2, 3, 5, 6, 7, 17, 25, 26, 28, 30, 31, 40, 47, 54, 57, 68, 69, 88, 91, 96, 100, 101, 102, 103, 104, 109, 110, 111, 112, 113, 115, 118, 120, 121, 124, 125, 128, 129, 130, 146, 149, 157, 164, 223
comprehension, 156, 168
Concise, 129
condensation, 54, 138
conductivity, 110, 140
conference, 172
configuration, 3, 13, 15
Congress, 114, 170, 171, 219, 222, 249
connectivity, 33, 235
consciousness, 28
consent, 114
conservation, 179, 215, 217, 218
constituents, 120, 123, 203
construction, 45, 115
consumers, 174, 189, 194, 198
consumption, vi, 28, 96, 100, 111, 112, 173, 174, 178, 179, 189, 193, 208, 209, 210, 211, 212, 214, 226, 233, 235, 236, 245, 247, 248
contact time, 49
contaminant, 100, 116, 121, 173, 174, 177, 213, 226
contaminated sites, 177
contaminated soil(s), 114, 115, 116, 126, 127, 131, 132
contaminated water, 225
contamination, 26, 28, 115, 123, 129, 132, 133, 138, 140, 173, 174, 176, 177, 189, 197, 204, 205, 213, 214, 216, 222, 225, 226, 227, 230, 231
Continental, 215
contrast sensitivity, 211
control group, 211
cooling, 116, 122, 125, 140

coordination, 1, 2, 3, 5, 6, 7, 8, 9, 10, 11, 13, 17, 18, 19, 20, 21, 22, 23, 36, 42, 67, 69, 71, 74, 76, 77, 78, 79, 81, 82, 83, 88, 89, 211
copolymer(s), 43, 58, 63
copper, 97, 119, 120, 125
coronary heart disease, 211, 212, 227, 235, 249
correlation(s), 5, 49, 67, 145, 148, 167, 168, 177, 190, 194, 221, 234, 241, 243, 245
correlation coefficient, 167, 168
corrosion, 120
cortex, 101
cortisol, 208
cosmetic, 114
cost, 29, 33, 39, 52, 115, 116
Costa Rica, 182, 228
cotton, 229
cotyledon, 141
Council of Europe, 113, 130
covalent bond, 2, 3, 5, 101
covering, 67
creatinine, 104
crocodile, 229
crocodilians, 174, 199, 206, 208
crystal structure, 10
crystalline, 43, 59, 60, 110, 122, 128, 133
culture, 153, 189
CVD, 212
cycles, 26, 27, 51, 52, 221
cycling, 26, 27, 129, 175, 214, 215, 216, 228, 233, 235, 243, 245
cyclopentadiene, 125
Cyprus, 184, 185
cysteine, 28, 87, 88, 89, 91, 93, 96, 101, 102, 106, 141, 204
cystine, 88, 89, 96
cytochrome, 87, 93
cytoplasm, 104
cytotoxicity, 89

D

data processing, 33
database, 150
DDT, 179, 216
decomposition, 90, 97, 123
decontamination, 131, 132
defects, 1, 28
defence, 213
defense mechanisms, 148
deficiencies, 210
deficit, 226
deforestation, 234
deformation, 120
degradation, 102, 174
dehydration, 56
delirium, 28
Delta, 141, 147, 153, 229
delta-aminolevulic acid dehydratase (ALA-D), 137
Denmark, 210
dental care, 113
Department of Energy, 131, 132, 134, 135
Department of Health and Human Services, 54
deposition, vi, 26, 28, 111, 152, 176, 202, 219, 233, 234, 236, 237, 238, 239, 240, 241, 242, 243, 244, 245, 246, 247, 248
deposits, 176
depth, 127, 159, 161, 167, 168
derivatives, 2, 3, 4, 5, 6, 8, 10, 11, 12, 13, 16, 17, 18, 19, 20, 21, 22, 23, 34, 36, 42, 69, 71, 74, 76, 78, 79, 82, 83
desorption, 115, 116, 117, 118, 119, 131, 132, 169
destruction, 122, 124, 133, 159, 235
detectable, 214
detection, 28, 32, 34, 36, 39, 40, 42, 44, 46, 47, 50, 51, 52, 53, 54, 55, 58, 61, 62, 63, 64, 96, 104, 154, 157, 159, 188, 238
detection system, 47
detoxification, 57, 132, 138, 148, 152, 178, 203, 204, 206
developed countries, 174, 179, 189
developing countries, 173, 189
deviation, 3, 18, 71, 74, 75, 78, 81, 238, 240
diastolic blood pressure, 212
diet, 177, 178, 192, 193, 199, 201, 211, 214, 221
dietary intake, 28
diffusion, 33, 44, 46, 59
diffusivity, 44
digestion, 54
dimethylsulfoxide, 141
dipeptides, 97
direct observation, 34, 39
discharges, 68, 176
discs, 44, 49
diseases, 197
dispersion, 44, 111, 169
dissociation, 31
dissolved oxygen, 220
distillation, 120
distilled water, 140
distribution, 28, 36, 44, 47, 105, 145, 151, 154, 155, 160, 161, 163, 164, 165, 168, 169, 172, 204, 206, 212, 218, 221, 224, 230
diuretic, 100
divergence, 215
diversity, 1, 23, 192, 215
DNA, 36, 39, 62, 63, 151, 152, 177, 230

DNA damage, 152, 177, 230
DOC, 177, 190, 192, 214, 242, 244
docosahexaenoic acid, 211
DOI, 53, 54, 55, 56, 57, 59, 60, 61, 62, 63, 64, 65, 153
donors, 2, 8
doping, 45
down-regulation, 102, 103, 104
drainage, 164, 190, 219
drawing, 116
drinking water, 57
drought, 197
drugs, 102, 104, 106
drying, 179
durability, 135
dyes, 34, 44
dyspnea, 28

E

earthworms, 199
ecology, 225, 231
economic status, 189
ecosystem, 26, 155, 157, 171, 231, 233, 234, 235, 242, 243, 246
ecotoxicology, 224
ectothermic, 201
edema, 103
EEA, 130
effluent, 29, 56
effluents, 231
egg, 181, 205, 206, 232
Egypt, 25
electric current, 110, 127
electrical conductivity, 110, 140
electricity, 127
electrodes, 127
electrolysis, 124
electrolyte, 137, 142, 145
electrolyte leakage percentage (ELP), 137, 145
electron(s), 3, 42, 47, 148
emission, 28, 35, 38, 55, 64, 95, 126, 138, 168, 170, 234
encapsulation, 109, 110, 114, 119, 122, 123, 124, 125, 133
encoding, 103
endangered, 174, 179
endangered species, 174
endocrine, 176, 207, 208, 214, 217
endocrine system, 207, 208
endogenous thiols, 87, 88, 89, 93, 96
endoskeleton, 194
endothermic, 178, 201
energy, 31, 40, 47, 49, 116, 117, 120, 122, 126, 127, 128, 133, 155, 178, 194, 206
England, 171, 227
environmental characteristics, 191, 192
environmental conditions, 139, 190, 192
environmental contamination, 213
environmental control, 176
environmental factors, 176, 177, 192, 194
environmental organizations, 180
Environmental Protection Agency (EPA), 114, 126, 129, 131, 134, 173, 174, 181, 209, 219, 225, 226, 249
environmental variables, 192
enzymatic activity, 149
enzyme(s), 88, 90, 101, 102, 138, 139, 140, 141, 142, 143, 146, 147, 149, 152, 154, 208
epidemiologic, 226
epidermis, 194, 204, 224
epinephrine, 140
epithelial cells, 89, 101, 102
epithelial transport, 106
EPR, 87, 90, 91, 93
equilibrium, 56
equipment, 32, 33, 117, 120, 130, 237
erosion, 169
erythrocytes, 100, 112
estuarine systems, 227, 232
ethanol, 48
Europe, 113, 130
European Commission, 133
European Parliament, 110, 130
European Union (EU), 113
evaporation, 156
Everglades, 181, 189, 195, 226, 247
evidence, 26, 46, 52, 87, 89, 90, 102, 104, 138, 239
evoked potential, 248
examinations, 210
exchange rate, 202
excitation, 95
exclusion, 31
excretion, 99, 103, 104, 105, 106, 107, 203, 204
experimental condition, 91, 119
exploitation, 165, 189, 197, 218
exports, 110, 113, 130
extinction, 34, 140, 179, 189
extraction, 28, 56, 58, 111, 119, 133, 152

F

fabrication, 31, 32, 33, 43, 44
factories, 177
fatty acids, 235, 247, 248
fermentation, 199, 223

fertilizers, 138
fetus, 111, 113
fiber, 54, 133
fillers, 123
films, 58
filters, 57, 159
filtration, 28, 30, 31, 101, 159
financial, 83, 170
financial support, 83, 170
Finland, 211, 219, 227
fires, 26, 111, 234
fish, 26, 28, 54, 100, 111, 112, 113, 114, 170, 174, 177, 179, 190, 192, 193, 198, 199, 202, 206, 208, 209, 210, 211, 212, 214, 216, 217, 219, 221, 222, 223, 225, 226, 227, 233, 235, 236, 243, 244, 245, 246, 247, 248, 249
fish oil, 227, 248
fishing, 211
fitness, 205
fixation, 124
flame, 128, 171
flexibility, 33, 42
flooding, 191
flora, 197, 199
flotation, 30
flue gas, 115
fluid, 203, 204
fluorescence, 28, 34, 36, 40, 54, 55, 63, 64, 91, 92, 95, 238
fluorophores, 34, 42
food, 25, 27, 99, 100, 112, 130, 173, 174, 177, 178, 179, 185, 187, 189, 191, 192, 193, 194, 197, 200, 201, 208, 211, 214, 215, 219, 221, 227, 234, 249
food additive(s), 130
Food and Drug Administration (FDA), 114, 130, 173, 174, 209
food chain, 25, 27, 191, 200, 227, 234
food intake, 211
food products, 28
food web, 194, 197, 200, 201, 214, 219, 221, 249
forest fire, 234
formaldehyde, 249
formation, 27, 44, 47, 50, 87, 88, 89, 90, 91, 95, 96, 97, 119, 121, 122, 123, 124, 125, 126, 138, 139, 140, 141, 156, 168, 221, 248
formula, 110
fossil fuel combustion, 25, 26
fouling, 31
fragments, 12, 16
free energy, 120
free radicals, 148, 153
freshwater, 28, 178, 179, 191, 197, 213, 215, 218, 219, 220, 221, 222, 223, 224, 230, 235, 237, 247, 248
fructose, 151
fruits, 197, 199, 208
functionalization, 28, 34
fungi, 29

G

gastrointestinal tract, 28, 100, 101, 111, 112, 113, 202
gel, 32
gene expression, 223
genes, 103, 168
genetic defect, 28
genus, 199
geology, 111
geometry, 3, 4, 8, 9, 12, 13, 14, 15, 16, 33, 44, 88
Georgia, 129, 181, 184, 187, 223, 227, 230, 231, 232, 234, 236, 237, 239, 244, 246
germanium, 13
germination, 146
gland, 207, 208
globalization, 189, 219
glomerulus, 101
glutathione, 87, 88, 89, 96, 97, 98, 101, 138, 139, 151, 153
glycine, 87, 88, 89, 91, 93, 96, 140
gold nanoparticles, 40, 63, 64
grants, 105
graph, 163
graphite, 127
grasses, 197
gravity, 30
grazers, 174
growth, 137, 138, 139, 146, 149, 151, 153, 194, 207, 221, 225
growth rate, 207
guidance, 130
guidelines, 179, 214
Guinea, 184

H

habitat(s), 173, 174, 179, 192, 197, 213
habitat quality, 213
hair, 113, 210, 211, 212
half-life, 100, 101, 200, 203
hallucinations, 28
halogen, 3
hardness, 123, 127

harmful effects, 233
harvesting, 189
hazardous air pollutants, 114
hazardous substances, 130
hazardous waste(s), 117, 119, 122, 123, 133, 134
hazards, 53, 99, 100, 130
H-bonding, 44
HE, 93
health, 26, 28, 52, 54, 99, 100, 107, 109, 110, 112, 113, 125, 128, 129, 130, 138, 173, 174, 176, 179, 206, 207, 208, 209, 211, 212, 214, 215, 217, 222, 226, 230, 231, 232, 233, 235, 248
Health and Human Services, 54
health effects, 26, 176, 206, 207, 208, 235
health problems, 26, 27, 174, 208
health risks, 125, 173, 217, 233, 235
heart disease, 211, 212, 227, 235
heart rate, 212
heat loss, 117
heat transfer, 115
heavy metals, 1, 26, 27, 28, 29, 30, 31, 54, 57, 102, 130, 132, 138, 150, 152, 155, 221, 228, 229, 230
heterogeneous catalysis, 32
hexane, 16
high strength, 33, 124
histology, 104
history, 174, 178, 211, 213, 217
HM, 155, 156
homeostasis, 151, 208, 212
homocysteine, 40, 87, 88, 89, 96, 101
hormone(s), 205, 207, 208, 223
human activity, 26
human exposure, 28, 100, 113, 114
human health, 26, 27, 52, 100, 109, 110, 113, 128, 138, 173, 214, 215, 226, 232
humus, 156, 219
hunting, 179, 189, 217, 218
hybrid, 57, 58, 61
hydrocarbons, 42, 179
hydrogen, 88, 90, 95, 96, 97, 120, 121, 128, 137, 138, 139, 142, 151, 153
hydrogen chloride, 128
hydrogen gas, 120
hydrogen peroxide, 88, 90, 95, 96, 97, 137, 138, 139, 142, 151, 153
hydrogen sulfide, 121
hydroxide, 100, 119, 124, 126
hydroxyl, 90, 97, 138, 139
hypersensitivity, 207
hypothesis, 200
hysteresis, 31

I

ideal, 4, 18, 29, 34, 42, 213
identification, 215
identity, 71, 74, 75, 78
image(s), 35, 37, 41
immobilization, 42, 44, 45, 46, 60, 109, 110, 119, 120, 123, 124, 125, 126
immobilize chromophore probes, 25, 42, 52
impurities, 121
in utero, 210
in vitro, 64, 101, 103, 150, 177
in vivo, 64, 102, 103, 177
income, 189
indirect effect, 169, 208
individuals, 177, 193, 195, 212, 213, 246
Indonesia, 187
induction, 204
industrial revolution, 26
industrial wastes, 176
industry(s), 25, 56, 100, 111, 128, 130, 131
infants, 113, 210, 212, 226, 236
infarction, 211, 235
ingest, 178, 202
ingestion, 99, 100, 112, 174, 189, 202, 233
inhibition, 102, 147, 148, 150, 208, 211
inhibitor, 149
injections, 238
injury, 103, 104, 105, 107, 139, 142, 148
inorganic fillers, 123
insects, 193, 198, 199, 208
insertion, 34
insomnia, 113
integration, 203, 217
integrity, 138, 145, 148
integument, 202, 203
interface, 27, 151
interference, 40, 42
intermolecular interactions, 3
international trade, 208, 214
interrelations, 155
intestine, 99, 204, 223
intoxication, 103, 145, 228
inversion, 16
invertebrates, 177, 193, 199
iodine, 5, 7, 8, 13
ion transport, 228
ionization, 18, 89
ions, 25, 26, 29, 30, 31, 32, 33, 34, 36, 38, 39, 40, 41, 42, 44, 45, 46, 47, 48, 49, 50, 51, 52, 56, 57, 61, 62, 63, 64, 65, 87, 88, 89, 91, 97, 101, 103, 106, 110, 111, 124, 149, 204, 220, 247
iris, 130

iron, 22, 29, 31, 81, 114, 124, 133, 234, 248
irrigation, 138
ischemia, 102, 106
islands, 230
isolation, 49, 122
isomers, 18, 19, 20, 21, 67, 69, 78, 79, 80, 81, 82, 83
Israel, 211
Italy, 172, 182, 183, 184, 229, 230

J

Japan, 25, 105, 115, 172, 184, 186, 230

K

keratin, 174, 194, 203
kidney(s), 1, 28, 99, 100, 101, 102, 103, 104, 105, 106, 107, 112, 113, 181, 195, 202, 203, 204, 207, 212, 213, 230
kidney failure, 113
kinetic studies, 57
kinetics, 51, 56, 223

L

laboratory studies, 191
lakes, 27, 111, 190, 215, 219, 227, 248
land disposal, 119, 122, 126
landscape, 227, 230
lanthanide, 11, 12
large snakes, 199
larvae, 199, 229
Latin America, 179, 180, 189
laws, 189
leaching, 45, 46, 120, 125, 176, 191, 237, 242, 246
lead, 28, 33, 42, 54, 55, 58, 69, 97, 113, 119, 151, 152, 168, 217
leakage, 137, 142, 145, 147
legislation, 179
legs, 15
liberation, 123
libido, 179
life cycle, 192, 193
lifetime, 202
ligand, 1, 3, 4, 5, 7, 8, 9, 11, 14, 15, 16, 17, 18, 19, 20, 23, 58, 67, 69, 71, 74, 75, 76, 77, 78, 79, 81, 82, 83, 89, 123
light, 22, 81, 82, 100, 111, 122, 139, 169
lignin, 124
lipid peroxidation, 138, 152, 154, 227
lipids, 110, 113, 138, 139, 147, 178, 208
lipoproteins, 205
liquids, 1
liver, 1, 99, 112, 113, 152, 181, 182, 183, 184, 185, 186, 187, 188, 192, 194, 195, 202, 203, 204, 207, 208, 210, 212, 213, 216, 223, 229
longevity, 194
Louisiana, 181, 182, 188, 197
lumen, 101
luminescence, 32

M

macrophages, 204
magnesium, 31, 123, 126
magnitude, 190, 191
Malaysia, 185, 186, 187
mammals, 26, 28, 54, 111, 112, 192, 199, 200, 202, 204, 206, 235
management, 110, 114, 133, 218, 222
manganese, 217
manufacturing, 157, 176, 177
marsh, 151, 154, 233, 236, 237, 238, 239, 240, 241, 242, 243, 246
mass, 28, 33, 59, 89, 116, 120, 121, 122, 159, 164, 165, 168, 170, 171, 178, 188, 189, 195, 196, 200, 201, 204, 205, 206, 212
mass spectrometry, 28, 89
materials, 28, 29, 32, 33, 42, 46, 53, 54, 58, 61, 120, 123, 124, 125, 126, 128, 133, 135, 138, 194
materials science, 32, 53
matrix, 120, 123, 124, 126, 127, 134, 194
measurement(s), 159, 169, 171, 177, 194, 203, 210, 247
meat, 179, 189, 210, 214, 226, 228
mechanical properties, 126
media, 40, 53, 64, 97, 177
median, 210, 245
medical, 153, 176
medicine, 2, 208, 214
Mediterranean, 171, 184, 185, 230
medulla, 101
melanin, 204
melt, 127, 128
melting, 39, 122, 127, 156
melting temperature, 39, 127
melts, 125, 127
membrane permeability, 140
membranes, 31, 32, 49, 56, 102, 103, 104, 145, 152
memory, 55, 113, 210, 235
memory loss, 113
mental development, 210, 226
mental retardation, 28
mesoporous materials, 32
mesoporous silica, 26, 42, 52, 55, 60

metabisulfite, 121
metabolism, 29, 101, 138, 139, 146, 148, 151, 152, 153, 178, 207, 208
metabolized, 202
metabolizing, 178
metal ion(s), 25, 26, 29, 31, 33, 34, 38, 39, 41, 42, 44, 46, 49, 50, 51, 59, 97, 204
metals, 1, 14, 17, 26, 27, 28, 29, 30, 31, 50, 54, 57, 102, 107, 111, 119, 124, 127, 128, 130, 132, 138, 139, 148, 150, 152, 154, 155, 202, 205, 221, 223, 224, 225, 228, 229, 230, 231
metamorphosis, 221
meter, 114, 140, 161
methodology, 95, 157, 159
methyl group(s), 8, 9, 10, 13, 20, 76, 77
methylation, 28, 68, 110, 168, 175, 176, 177, 190, 191, 192, 204, 214, 216, 219, 220, 234, 247, 248
metropolitan areas, 179
Mexico, 179, 183, 184, 186, 199, 223, 229, 231, 247
mice, 103, 152
microfabrication, 33
microhabitats, 194
micronucleus, 177
microorganisms, 29, 111, 248
migration, 155, 157, 159, 167, 169, 172
milligrams, 141
mimicry, 107, 223
mine soil, 118
mineralization, 156
Ministry of Education, 83
Minneapolis, 215
missions, 111, 129, 216
Mississippi River, 172
Mississippi River Delta, 172
mixing, 89, 90, 91, 119, 120, 121, 125, 126, 127, 133, 141, 161, 163, 165, 169
model system, 220
models, 105, 131
modifications, 102, 103, 140
moisture content, 115
mold, 126
molecular mass, 204
molecular oxygen, 87, 90, 95, 97
molecular structure, 6, 17
molecular weight, 103
molecules, 11, 20, 21, 22, 32, 33, 40, 44, 45, 46, 47, 69, 70, 71, 74, 76, 77, 79, 82, 89, 97, 101, 138, 146
mollusks, 193
molybdenum, 11, 21, 79
monolayer, 55
monomers, 2, 3, 10, 19, 71
morphology, 33, 203, 219
mortality, 227
Moscow, 170, 171
motor control, 28
motor skills, 211
MR, 103, 106, 107
mRNA, 102, 104
mucosa, 203
muscles, 157
museums, 213
myocardial infarction, 211, 235

N

Na^+, 101, 207
NaCl, 152
nanocomposites, 59, 60
nanomaterials, 32, 33
nanometer scale, 33
nanoparticles, 36, 39, 40, 63, 64, 133
nanoreactors, 28
nanoscale materials, 33
nanoscale structures, 28
nanotechnology, 53
naphthalene, 57
National Academy of Sciences (NAS), 234, 235, 248
NATO, 58
necrosis, 103
negative consequences, 157
negative effects, 137, 149
nephropathy, 107
nervous system, 1, 25, 28, 68, 100, 112, 113, 176, 190, 207
neurotoxicity, 204, 210
neutral, 33, 39
New England, 227, 247, 249
New Zealand, 210, 226
nickel, 16, 57, 119
nitrogen, 8, 17, 42, 103
NMR, 89
non-protein thiols (NPSH), 137
non-steroidal anti-inflammatory drugs, 102
North America, 193, 199, 221, 234, 247, 248
Norway, 221
NRC, 225
nucleic acid, 138, 139
nursing, 209
nutrients, 174, 205
nutrition, 138

O

obstruction, 102, 106, 107

oceans, 27
OH, 9, 90, 123, 138, 139, 168
oil, 114, 228
oligomers, 125
omega-3, 211, 235, 247, 248
operations, 115, 134
optical chemical sensors, 32, 52, 54
optical density, 93
optical fiber, 54
optical parameters, 32
optimization, 32, 53
ores, 119
organ(s), 1, 2, 7, 8, 18, 20, 22, 68, 69, 99, 105, 110, 174, 176, 190, 202, 203, 213, 221
organic compounds, 40, 88, 109, 111
organic matter, 29, 56, 115, 155, 156, 172, 191, 216, 235
organism, 204
OSHA, 114, 130
osmium, 15, 17
osmosis, 28, 31
oxidability, 156
oxidation, 1, 3, 6, 8, 9, 11, 12, 19, 23, 26, 28, 53, 88, 89, 91, 92, 93, 95, 96, 97, 100, 109, 110, 121, 138, 140, 146, 147, 148, 156, 169, 204
oxidative damage, 138, 139, 147
oxidative stress, 138, 147, 148, 149, 151, 153, 154, 204
oxygen, 5, 7, 8, 13, 21, 42, 79, 87, 88, 89, 90, 95, 96, 97, 115, 138, 139, 143, 151, 152, 153, 155, 156, 159, 160, 161, 168, 204, 220
ozone, 152

P

Pacific, 231
pain, 28
paints, 2
palladium, 11, 56
Parliament, 110, 130
patents, 122
pathways, 111, 175
PCA, 219
PCBs, 179, 231
peat, 29
penicillin, 104
peptide(s), 138, 204
perchlorate, 18
permafrost, 156
permeability, 125, 126, 140
permit, 123
peroxidation, 138, 152, 154, 227
peroxide, 88, 90, 95, 96, 97, 137, 138, 139, 142, 147, 151, 152, 153
personality, 28
pesticide, 231
petroleum, 111
pH, 29, 30, 39, 47, 48, 49, 50, 51, 56, 57, 58, 110, 124, 139, 140, 141, 155, 156, 159, 160, 166, 168, 176, 177, 190, 191, 192, 214, 219, 220, 247
phenol, 47
phenotypes, 103
phosphate, 123, 124, 133, 134, 140
photolysis, 249
photosynthesis, 138
physical characteristics, 192
physical properties, 44, 127
physiological, 151, 207, 223
physiology, 106, 151, 206, 223, 224
phytoplankton, 145, 154, 157, 194, 223
pigs, 199
pith, 56
pituitary gland, 207, 208
placenta, 99, 100, 101, 111, 113
placental barrier, 235
plankton, 219
plants, 30, 114, 137, 138, 148, 149, 150, 151, 152, 153, 154, 176, 197, 199
plasma levels, 104
plasma membrane, 102, 106
plasma proteins, 202, 204
plastics, 2
platform, 25
platinum, 11, 13, 15, 22, 60, 79
Poland, 210, 226
polar, 1, 155
polarity, 44, 45
policy, 130, 131, 217
pollutants, 26, 28, 32, 114, 155, 156, 157
pollution, 54, 151, 154, 156, 168, 170, 174, 215, 217, 221, 222, 227, 228, 232
polycarbonate, 237
polychlorinated biphenyls (PCBs), 179
polymer(s), 10, 19, 21, 23, 29, 32, 38, 39, 77, 124, 125, 126, 134, 135
polymerization, 1, 18, 23, 69, 83
polypeptides, 150
polysulfides, 121
polyunsaturated fat, 235
polyunsaturated fatty acids, 235
ponds, 170
pools, 178
population, 100, 111, 112, 113, 165, 174, 176, 199, 211, 214, 227
porosity, 29, 124, 125, 126

porphyrins, 138
portal vein, 202, 203
positive correlation, 148, 167, 190
potassium, 140
potato, 152
power plants, 114, 176
precipitation, 28, 29, 30, 31, 52, 58, 156, 233, 236, 237, 241, 242, 243, 246, 247
predation, 194, 208, 225
predators, 174, 178, 193, 194, 197, 199, 208
pregnancy, 210
preparation, 32, 39, 132, 141, 159, 169
preservative, 99, 100
prevention, 110, 114, 121
principal component analysis, 219
principles, 31
probability, 142, 165, 169
probe, 32, 34, 36, 37, 39, 40, 42, 44, 45, 46, 47, 48, 49, 61, 63, 64, 91, 92, 93
producers, 198
project, 247
proliferation, 138
proline, 148
protection, 103, 113, 128, 152, 179, 189, 204, 218
protective mechanisms, 151
proteins, 28, 33, 54, 59, 88, 138, 139, 141, 153, 178, 202, 204, 214
proximal tubules, 101, 102, 107
psychological development, 210
PTFE, 237
publishing, 214
pumps, 103
purification, 124, 134
purity, 122

Q

quality control, 127
quantification, 49, 157, 233, 243
quantum yields, 34
quartz, 56, 238
Queensland, 184, 186, 230

R

race, 231
radiation, 117, 156, 170, 190, 215
radicals, 90, 91, 97, 138, 139, 147, 148, 153, 204
radicle, 139
radius, 3, 4, 5, 6, 8, 10, 11, 12, 13, 14, 19, 20, 75
rainfall, 234, 237, 238, 239, 240, 241, 242, 243, 246, 247
rape, 152
reactants, 120
reaction time, 48
reactions, 28, 109, 111, 123, 125, 154
reactive oxygen, 89, 96, 138, 139, 143, 152
reading, 140
reagents, 55, 119, 159
reality, 179
receptors, 33, 111
recognition, 33, 42, 45, 52, 130, 235
recovery, 28, 110, 114, 115, 117, 131, 132
recrystallization, 123
Red List, 215
redistribution, 104, 203
reflectance spectra, 48, 49, 50
reflectivity, 32
reflexes, 197
refractive index, 32
regeneration, 51, 52
Registry, 54, 129
regression, 238
regulations, 114, 131
relevance, 130
remediation, 115, 116, 131, 132, 134
renal dysfunction, 104
renal failure, 101, 102, 104, 106
renin, 101
reproduction, 28, 83, 193, 194, 203, 206, 208, 217, 225
reptile, 173, 174, 179, 189, 190, 192, 194, 195, 196, 197, 200, 203, 204, 206, 209, 212, 214, 215, 217
reptile species, 173, 174, 192, 197, 200, 204, 214
researchers, 122, 145, 157
reserves, 178
residues, 61, 204, 205, 217, 229, 231
resins, 30
resistance, 53, 57
resorcinol, 47
resources, 177
respiration, 176, 219
response, 31, 32, 36, 40, 42, 45, 48, 49, 104, 106, 107, 137, 147, 148, 149, 151, 152, 206, 207, 208, 211, 214, 218, 236
response time, 31, 32, 45
restaurants, 179, 210
restoration, 134, 222
restrictions, 113, 126
retardation, 28
reusability, 26, 52
reverse osmosis, 28, 31
Rhizopus, 54
rice field, 224
rice husk, 29

Richland, 239
rings, 4, 9, 13, 15, 19, 22
risk(s), 105, 109, 110, 122, 125, 131, 173, 174, 179, 189, 206, 209, 210, 211, 212, 214, 215, 217, 227, 233, 235, 243, 247, 248, 249
risk factors, 211
rodents, 103
room temperature, 110, 121, 122, 140
root, 36, 37, 145, 154, 220
root cap, 36, 37
root system, 145
Rouleau, 223
routes, 112, 173, 202, 203, 205, 222
rules, 11, 114
runoff, 156, 169
Russia, v, 155, 156, 157, 159, 164, 170
ruthenium, 11, 13, 16, 17, 42

S

safety, 130
salinity, 155, 156, 159, 160, 161, 167, 168, 169, 220, 247
saliva, 111
salts, 10, 31, 64, 100, 202
saturated fat, 235
scaling, 127
scarcity, 206
scattering, 62
scavengers, 153
school, 210
science, 32, 53, 54, 59
scope, 174
Scots pine, 153
seafood, 208, 211, 248
secretion, 208
sediment(s), 27, 29, 30, 114, 122, 127, 133, 134, 156, 157, 159, 167, 168, 169, 170, 176, 177, 190, 191, 202, 219, 220, 234, 247
sedimentation, 30, 234
seed, 174
seedlings, 137, 138, 139, 140, 141, 142, 143, 144, 145, 146, 147, 148, 149, 151
selectivity, 31, 32, 33, 34, 36, 39, 42, 45, 49, 50, 51, 52, 63, 64
selenium, 7, 121, 122, 133, 152, 216, 217, 223, 224, 229, 231
seller, 189
semiconductors, 59
sensing, 28, 32, 33, 34, 35, 42, 44, 45, 46, 47, 48, 50, 51, 52, 53, 58, 59, 60, 62, 63
sensitivity, 31, 32, 33, 39, 40, 42, 45, 49, 51, 52, 139, 211
sensors, 25, 31, 32, 33, 36, 42, 46, 47, 50, 51, 52, 53, 54, 58
serum, 31, 40, 142, 212
sewage, 29, 138, 231
sex, 194
Seychelles, 210, 218
shelf life, 31
shellfish, 28, 114, 235
shoot, 145
signals, 33, 42, 91
silane, 44
silanol groups, 44
silica, 26, 29, 31, 42, 44, 45, 50, 52, 55, 58, 59, 60, 124
silicon, 59, 124
silver, 119, 177, 217
simulations, 91, 220
SiO_2, 124
skeletal muscle, 202, 224
skeleton, 15
skin, 100, 112, 113, 174, 181, 182, 183, 184, 185, 186, 187, 188, 195, 202, 203, 212, 224, 229
slag, 124, 125, 128, 133, 134
sludge, 29, 68, 115, 118, 120, 122, 124, 125, 126, 134, 138
small intestine, 223
smoking, 211
snakes, 174, 178, 188, 195, 197, 198, 199, 206, 212, 224, 225, 228, 229
sodium, 55, 57, 100, 121, 123, 126
sodium hydroxide, 100, 126
solid matrix, 121
solid solutions, 119
solid waste, 26, 121, 124, 125, 133
solidification, 109, 110, 114, 115, 119, 120, 124, 126, 127, 133, 134, 135
solubility, 42, 113, 119, 124, 128
solution, 29, 31, 34, 36, 37, 39, 40, 41, 48, 51, 53, 55, 56, 57, 61, 62, 89, 90, 91, 93, 95, 96, 124, 140, 141, 159, 170, 189
solvents, 39, 120
sorption, 56, 57, 169
South America, 177, 197, 199, 224
Southeast Asia, 189
Soviet Union, 177
Spain, 109, 116, 127, 128, 154, 182, 184, 228
specialists, 197
speciation, 131, 215, 220, 221, 247
spectroscopic techniques, 32
spectroscopy, 28, 36, 49, 89, 157
spin, 87, 90
St. Petersburg, 171
stability, 26, 44, 52, 119, 125, 148

stabilization, 109, 110, 114, 119, 120, 121, 122, 124, 125, 126, 127, 132, 133, 134, 135
stable complexes, 97
standard deviation, 238, 240
starvation, 178
state(s), 1, 3, 6, 8, 9, 11, 12, 19, 23, 32, 88, 109, 110, 111, 113, 125, 168, 177, 204, 221, 233, 235, 244, 245
steel, 114, 116, 117
stereospecificity, 69
stimulation, 247
storage, 32, 109, 110, 113, 114, 120, 126, 128, 130, 133, 194, 202, 203, 204
stress, 105, 137, 138, 139, 147, 148, 149, 150, 151, 152, 153, 154, 189, 204
structural protein, 204
structure, 2, 5, 6, 10, 12, 13, 14, 15, 16, 17, 21, 22, 31, 34, 36, 60, 79, 106, 204, 206, 222
subgroups, 15, 31, 82
substrate(s), 148, 205, 206
sulfate, 27, 31, 102, 107, 123, 168, 234
sulfur, 29, 31, 36, 120, 121, 122, 124, 125, 126, 128, 133, 134, 135, 204
sulfuric acid, 56, 110
sulphur, 5, 7, 8, 11, 101, 133, 148
superoxide dismutase (SOD), 90, 137, 143
supplier, 189
surface area, 29, 30, 32
surface layer, 156, 160, 161, 163, 164, 165, 166, 167, 168, 169
surface modification, 45
surface properties, 28
surface tension, 110
surfactant(s), 33, 42, 44, 51
survival, 28, 103, 138, 197, 218, 221
survival rate, 28
susceptibility, 101
Sustainable Development, 189, 217, 218
Sweden, 115, 226
swelling, 57
Switzerland, 131
symmetry, 13, 14, 16
symptoms, 113, 206
synthesis, 26, 33, 42, 44, 52, 60, 104, 138, 148, 149, 152, 208
systolic blood pressure, 212

T

Taiwan, 182, 183, 228
target, 39, 89, 99, 101, 105, 148, 150, 157, 207
target organs, 99
taxa, 204
techniques, 25, 28, 31, 32, 42, 52, 115, 134, 157, 174, 212, 213, 218, 222
technology(s), 28, 33, 42, 52, 109, 110, 114, 115, 116, 117, 119, 120, 122, 125, 127, 129, 131, 132, 133, 135, 214
temperature, 39, 69, 78, 82, 95, 109, 110, 115, 116, 117, 118, 119, 120, 121, 122, 125, 127, 128, 139, 140, 155, 156, 159, 160, 166, 167, 168, 170, 176, 194
tension, 110
terrestrial ecosystems, 174, 213
territory, 156, 159
testing, 210
Thalassiosira, 145
theoxidation state, 1
therapeutics, 99
thermal treatment, 109, 118, 119, 131
thimerosal, 100
thin films, 58
threats, 197
thymine, 36
thyroid, 207, 208, 225
time series, 227
tin, 30, 57, 119
tissue, 1, 141, 147, 152, 153, 175, 177, 190, 193, 194, 202, 203, 204, 206, 209, 210, 212, 216, 217, 230
tobacco, 152
toluene, 57
tooth, 2, 111
tortoises, 174, 178, 197, 218, 222
toxic effect, 25, 103, 137, 151, 197
toxic metals, 107, 139, 223
toxic substances, 129
toxicity, 26, 28, 89, 100, 102, 105, 106, 109, 129, 137, 138, 139, 145, 149, 151, 153, 154, 168, 176, 177, 178, 204, 214, 223, 224
toxicology, 54, 105, 139, 152, 224
trace elements, 129, 228, 229, 231
trade, 173, 174, 186, 187, 189, 208, 214, 215, 218, 219
traits, 178, 192, 206
trajectory, 205
transduction, 31, 32, 42
transformation(s), 26, 110, 122, 155, 156, 157, 167, 190, 234
transition metal, 12, 17, 38, 39, 50, 88
transition metal ions, 38, 39
transmission, 35
transpiration, 138
transport, 26, 28, 30, 33, 46, 53, 101, 102, 103, 104, 106, 107, 111, 115, 131, 148, 151, 155, 176, 204, 205, 207, 208, 216, 223, 228, 247, 248

transport processes, 103
treatment, 28, 29, 30, 31, 51, 52, 99, 101, 102, 103, 104, 114, 115, 116, 117, 119, 120, 121, 122, 125, 127, 128, 131, 133, 134, 139, 142, 144, 145, 146, 149, 153, 167, 168
turbulent mixing, 165
turtle, 174, 179, 180, 192, 193, 194, 197, 198, 200, 207, 208, 210, 212, 213, 215, 218, 219, 221, 222, 223, 228, 230, 231

U

umbilical cord, 210
unintentional occupational exposure, 99, 100
unique features, 32
United Nations, 114, 115, 110, 114, 131, 209
United States (US), 129, 134, 140, 179, 181, 182, 183, 184, 185, 186, 187, 188, 189, 195, 197, 200, 210, 221, 222, 224, 227, 228, 229, 230, 232, 247
uranium, 132
urban, 159, 176
urbanization, 155
urea, 34, 36, 103, 104
urinary bladder, 228
urine, 100, 101, 104, 107, 202, 211
UV, 28, 34, 37, 38, 39, 48, 49, 50, 96, 140, 238

V

vacuole, 138
vacuum, 115, 116, 118, 121
valence, 3
validation, 99
vapor, 26, 28, 55, 58, 109, 111, 112, 114, 115, 116, 118, 119, 120, 126, 130, 131, 157, 159, 171, 238
variations, 14, 161, 169
vasoconstriction, 101
vegetation, 234
vein, 202, 203, 213
Vermeer, 224
versatility, 33
vertebrates, 174, 178, 192, 193, 200, 201, 202, 203, 206, 207, 214, 224, 225
vessels, 121
vision, 207, 235
visual field, 211
visualization, 33
vitrification temperature, 127
volatility, 117, 128
volatilization, 115, 176, 234
vulnerability, 205

W

wages, 179
washing procedures, 237
Washington, 129, 248
waste, 26, 29, 30, 56, 57, 109, 110, 114, 115, 117, 119, 121, 122, 123, 124, 125, 127, 128, 129, 130, 131, 132, 133, 134, 153, 157, 159, 169, 177, 234
waste disposal, 177
waste incineration, 26, 234
waste treatment, 153
wastewater, 29, 30, 31, 55, 56, 115, 129, 138, 157, 159, 169
water chemistry, 191, 219, 234, 235, 245, 246
water policy, 130
water supplies, 57
water vapor, 26
watershed, 219, 235, 245
wavelengths, 95
wealth, 178
web, 177, 194, 197, 200, 201, 221
weight gain, 207
weight ratio, 104
wetlands, 190, 191, 234, 235, 245, 246
wild animals, 189
wildlife, 28, 173, 176, 177, 178, 179, 189, 191, 192, 194, 208, 214, 216, 217, 225, 233
wood, 29, 222
workers, 111
World Health Organization (WHO), 54, 129, 209
worldwide, 26, 173, 174, 189, 213

X

X-ray analysis, 78

Y

yang, 151
yeast, 29, 54, 56
yield, 101, 115, 120, 141

Z

zeolites, 125
zinc, 1, 30, 57, 119, 134, 149, 152
zooplankton, 157